Springer Series in Microbiology

Editor: Mortimer P. Starr

Ascomycete Systematics
The Luttrellian Concept

Edited by

Don R. Reynolds

With Contributions by

A. Beckett, A. Bellemère, A. Henssen, M. C. Janex-Favre, G. Keuck, J. W. Kimbrough, M. A. Letrouit-Galinou, E. S. Luttrell, D. Malloch, E. Müller, A. Parguey-Leduc, B. Renner, J. van Brummelen, G. Vobis

With 122 Illustrations

Springer-Verlag
New York Heidelberg Berlin

Don R. Reynolds
Natural History Museum
900 Exposition Boulevard
Los Angeles, California 90007/USA

Series Editor:
Mortimer P. Starr
University of California
Department of Bacteriology
Davis, California 95616/USA

Library of Congress Cataloging in Publication Data
Main entry under title:
Ascomycete systematics.
(Springer series in microbiology)
Bibliography: p.
Includes indexes.
1. Ascomycetes. 2. Botany—Classification.
I. Reynolds, Don R. II. Title: Luttrellian concept.
QK623.A1A73 589.2′3′012 80-27675

Printed in the United States of America

9 8 7 6 5 4 3 2 1

ISBN 0-387-90488-3 Springer-Verlag New York Heidelberg Berlin

ISBN 3-540-90488-3 Springer-Verlag Berlin Heidelberg New York

Contents

Contributors

A. Beckett, Department of Botany, Bristol University, Bristol, England, *Chapter 2*

A. Bellemère, Laboratoire de Mycologie, Ecole Normale Supérieure de Saint-Cloud, Saint-Cloud, France, *Chapter 5*

A. Henssen, Fachbereich Biologie, Botanik, Marburg-Lahn, F.R.G., *Chapter 10*

M. C. Janex-Favre, Laboratoire de Cryptogamie, Université Pierre et Marie Curie, Paris, France, *Chapter 8*

G. Keuck, Institut für Landwirtschaftliche Botanik der Universität, Bonn, F.R.G., *Chapter 10*

J. W. Kimbrough, Department of Botany, University of Florida, Gainesville, Florida, USA, *Chapter 7*

M. A. Letrouit-Galinou, Laboratoire de Cryptogamie, Université Pierre et Marie Curie, Paris, France, *Chapter 5*

E. S. Luttrell, Department of Plant Pathology, University of Georgia, Athens, Georgia, USA, *Chapter 9*

D. Malloch, Department of Botany, University of Toronto, Toronto, Ontario, Canada, *Chapter 6*

E. Müller, Institut für Spezielle Botanik, Technische Hochschule, Zurich, Switzerland, *Chapter 4*

A. Parguey-Leduc, Laboratoire de Cryptogamie, Université Pierre et Marie Curie, Paris, France, *Chapter 8*

B. Renner, Fachbereich Biologie, Botanik, Marburg-Lahn, F.R.G., *Chapter 10*

J. van Brummelen, Department of Mycology, Rijksherbarium, Leiden, Netherlands, *Chapter 3*

G. Vobis, Fachbereich Biologie, Botanik, Marburg-Lahn, F.R.G., *Chapter 10*

Acknowledgment

Many thanks to Sterling Keeley and David M. Minor for their assistance in the preparation of this volume.

Chapter 1

The Luttrellian Concept: Introductory Remarks

D. R. REYNOLDS

The Concept

In his 1951 treatise *Taxonomy of the Pyrenomycetes* Luttrell wrote that the classification of the ascomycetes is a difficult taxonomic problem. Nevertheless, he undertook the analysis of ascomycete literature to that time with a focus on pyrenomycete systematics and with the intent of taxonomic revision. In doing so, he noted a shift in the emphasis from artificial criteria—such as the habitat or fruit body wall color and consistency—to more fundamental criteria—such as those derived from the ascus. In a contemporary mode, Luttrell (1951) revised the systematics of pyrenomycetous fungi, emphasizing the relationship between the ascus and other components of the ascocarp occupying the perithecial cavity, or locule, within which the asci develop. The resultant Luttrellian concept in fungal systematics can be described as *an emphasis on the relationship between the ascus and the totality of ascocarp components.*

Luttrell utilized three major ascocarp characters in determining the basic relationships among pyrenomycete species. One was the nature of the ascocarp wall—whether false and derived of somatic tissue and called "stromatic," or true and developed as a specialized component of the sexually committed hyphal system. A second character was the structure of the ascus: one in which the wall is organized into an inner, usually extendable layer and an outer nonextendable layer at the time of ascospore ejaculation, versus a comparably undifferentiated wall. A third aspect was the development of the centrum and its mature components.

Influential contributions to ascomycete systematics, aside from those of Luttrell, have continued during the last three decades. Many have been studies of ascocarp development. Others have dealt with ascus and ascospore development and structure. Taxonomic revisions of taxa at generic, familial, and ordinal levels have continued in most groups. However, the overall impact of the Luttrellian concept on ascomycete systematics has not received a general assessment. Several issues raised by Luttrell's work have yet to be resolved. Controversy exists concerning the term "loculoascomycetes," which was created (Luttrell, 1955) to contain "*Ascis bitunicatis, in ascostromate evolutis.*" The level of this taxon is variously accepted as the original subclass or elevated to a higher position separate from the euascomycetes (Barr, 1976, 1979). Luttrell's contention that the taxon Loculoascomycetes is the equivalent of Nannfeldt's (1932) Ascoloculares is controversial. The shift in emphasis from the stromatic ascocarp containing a particular ascus to the bitunicate ascus in a stromatic fruit body could be a reason. The acceptance of the reality of the ascomata is a related concern.

This Volume

The topics addressed in this volume have "haustoria" in my service as Program Committee Member for Ascomycetes for the 1977 Second International Mycological Congress. In order to determine relevant symposium topics for that event, I canvassed ascomycetologists throughout the world. The initial responses and subsequent refined suggestions led to themes involving the basic elements of the Luttrellian concept. One theme concerned the ascus; second was the interpretation of various ascomycete groups from the perspective of the Luttrellian concept. This volume is the result of symposium contributions edited as chapters on the ascus and the centrum.

In the ascus chapters, the form and function of the ascus are emphasized. The ascus that produces a specialized pore in the apex through which the ascospores are discharged is analyzed (A. Beckett); the discomycetous ascus with a lid or operculum is discussed (J. van Brummelen); the basic concept of the bitunicate ascus is evaluated (E. Müller); and the ascus in a group of fungi similar to the discomycetous ascomycetes, the lichenized species in the order Lecanorales, is reviewed (A. Bellemére and M. A. Letrouit-Galinou).

The centrum chapters are presented somewhat differently than are those on the ascus. Here the authors take the attitude that the Luttrellian concept can be applied to ascomycete groups where the centrum character is not normally utilized. The plectomycete centrum is defined and elaborated upon (D. Malloch); a centrum is considered as a possibility in discomycetous fungi (J. Kimbrough); the centrum approach is taken in the lichenized ascomycetes placed in the order Lecanorales (A. Henssen); and the fungi with a pyrenomycete centrum are reviewed according to whether the component ascus is unitunicate within a nonlichenized or a lichenized fruit body (A. Parguey-Leduc and M. C. Janex-Favre). Finally, the classic centrum containing the bitunicate ascus is considered (E. S. Luttrell).

The general consensus from the chapters presented here is that the basic concepts of asci and the centrum are viable in ascomycete taxonomy. Concept refinement to date is through application of contemporary techniques and subsequent comparative reanalysis of ideas formulated in the last several decades. In the ascus chapters, electron microscopy is stressed as a useful tool for generation of data, and when compared to data derived from the light microscope, such data are useful in interpretive and practical taxonomy. In the centrum chapters, the call is for a broader data base, to include some emphasis on the development of the ascocarp walls as well. The centrum approach seems to be an effective analytical device for evaluation of heterogeneous groups, such as the cleistothecial pyrenomycetes. In other groups, such as the discomycetes, the centrum approach can show evolutionary relationships but is difficult to apply in a stricter systematic manner because of the traditional emphasis on other criteria. Nevertheless, the basic concept that Luttrell detected and utilized is operative in ascomycete systematics. The overall trend in systematic analysis now seems to be an emphasis on the relationships between ascocarp components rather than on the individual components themselves.

References

Barr, M. E., 1976. Perspectives in the ascomycotina. Mem. New York Bot. Garden 28: 1–8.

Barr, M. E., 1979. A classification of the loculoascomycetes. Mycologia 71: 935–957.

Luttrell, E. S., 1951. Taxonomy of the pyrenomycetes. Univ. Missouri Studies 24(3): 1–120.

Luttrell, E. S., 1955. The ascostromatic ascomycetes. Mycologia 47: 511–532.

Nannfeldt, J. A., 1932. Studien über die Morphologie und Systematik der nicht-lichenisierten inoperculaten Discomyceten. Nova Acta Regiae Soc. Sci. Upsaliensis (4)8: 1–368.

The Ascus

Chapter 2

The Ascus with an Apical Pore: Development, Composition, and Function

A. BECKETT

Introduction

In those ascomycetes distinguished by the possession of a functionally unitunicate ascus with some kind of apical annulus, the morphology and structure of the ascus shows considerable variation. The presence or absence of certain apical structures, variations in their morphology, and the structural details of the ascus wall are now regarded as important diagnostic features in modern schemes of classification.

Considerable work has been done on the structure and morphology of the wall and apex of the unitunicate ascus (Beckett and Crawford, 1973; Chadefaud, 1973; Griffiths, 1973). These studies and personal observations on *Neurospora* Shear and Dodge, *Sordaria* Cesati and de Notaris, *Xylaria* Hill ex Greville, and *Rosellinia* de Notaris provide the basis for this evaluation of the current status of the development, structure, and composition of the ascus wall and apex. The possible functions of the ascus apical apparatus in relation to ascus dehiscence and spore discharge are also worthy of consideration.

The Ascus Wall

Information on the mode of development of both the walls and apices of asci is generally sparse, and because the structures involved are largely beyond the resolution of the light microscope, the available information comes from the relatively few ultrastructural studies that have been made. Schrantz (1970) showed that in the young ascus of *Xylaria polymorpha* (Persoon ex Mirat) Dumortier, the wall was composed of a thin layer of fibrillar material that was electron opaque when fixed with osmium tetroxide. During maturation, the ascus wall differentiated to form an outer and an inner layer. Formation of the inner layer apparently involved the fusion of vesicles with the ascus plasma membrane. Lomasomes were observed, and it may be that these play a role in the formation of the inner layer (Schrantz, 1970). A similar method of wall formation occurs in the asci of *Xylaria longipes* Nitschke (Fig. 2.1), *Xylaria hypoxylon* (Linneus) Greville, *Rosellinia aquila* (Fries) de Notaris, and *Sordaria humana* (Fuckel) Winter

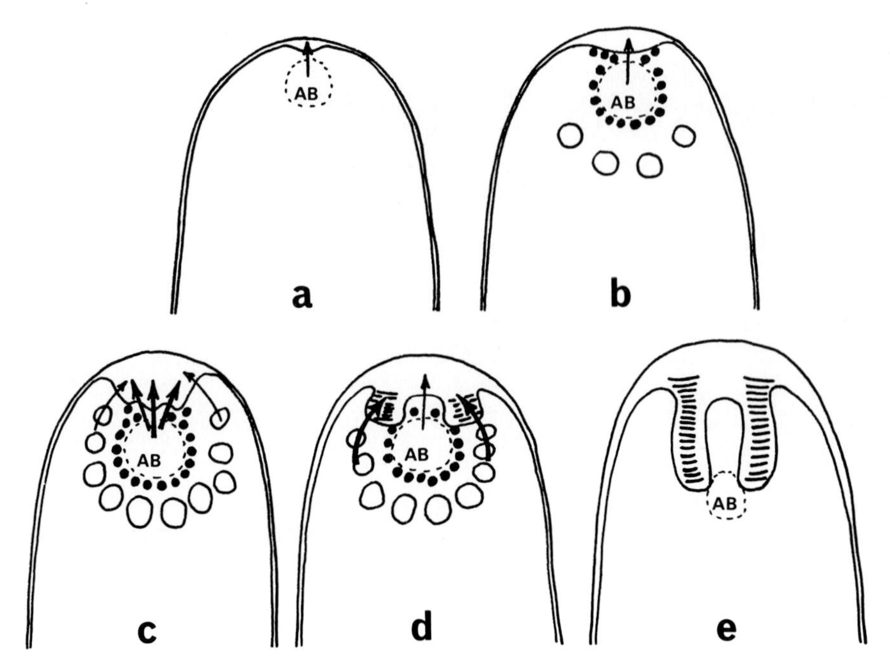

Fig. 2.1. Diagrammatic interpretation of the possible role of apical vesicles in the development of the apical apparatus in *Xylaria longipes.* The thickness of the arrows indicates a measure of activity of the particular vesicle type, i.e., small versus large. **a** and **b** Early stages, apical wall thickens to form "apical cushion" or "plug." Apical body (AB) and small vesicles involved. **c** Major activity by apical body and small vesicles. Large vesicles play minor role. **d** Apical body and small vesicles less active; major deposition by large vesicles to form pendulous apical ring. **e** Apical apparatus fully formed; apical vesicles no longer evident. Apical body remains.

(Pl. I, Figs. 2.2, 2.3, 2.5, and 2.8). In all these fungi, vesicles and paramural bodies (Marchant and Robards, 1968) are evident during the stages of ascus elongation up until the time at which the apical apparatus is complete. In *Sordaria humana* (Read, 1977), this is when the ascus measures 140–162.5 μm in length. In *X. longipes,* this occurs when the ascus reaches 130–160 μm in length (Beckett and Crawford, 1973).

The layering and ultrastructural appearance of the ascus wall in unitunicate ascomycetes varies considerably. In *X. longipes* (Beckett and Crawford, 1973) and *R. aquila* (Pl. I, Fig. 2.4), it is a single, granulofibrillar layer. In *S. humana* and in *Neurospora crassa* Shear and Dodge, the wall is also one homogeneous layer, but it is granular throughout and, after aldehyde fixation, no fibrils are visible. In *X. hypoxylon* and *X. polymorpha* (See Pl. III, Figs. 2.17 and 2.18) and *Hypoxylon fragiforme* (Persoon ex Fries) Kickx (Greenhalgh and Evans, 1967, 1968), the ascus wall is distinctly two-layered. In *Sclerotinia sclerotiorum* (Libert) de Bary (Codron, 1974), the wall is apparently multilayered. Little is known of the details of wall formation in these multilayered types or of the development of the fine fibrils found at the outer edge of the apical wall in asci of *Bulgaria inquinans* Fries (Bellemére, 1969), *Geoglossum* sp. (Bellemére, 1975), and *S. sclerotiorum* (Codron, 1974). However, in *S. sclerotiorum,* lomasomes are present in the young wall at the time of nuclear fusion in the ascus, and numerous small vesicles have been seen in the ascus cytoplasm (Codron, 1974).

Information is lacking on the rates at which asci grow during the period from the ascus initial to the stage at which postmeiotic mitosis occurs, but it can be expected that until the ascus initial has reached a given length and width for a particular species, material is added to the wall over a large part of the cell surface. The numerous paramural bodies may be a morphological manifestation of such an overall accretion of wall material and/or plasma membrane synthesis.

As yet, very few studies have been made on the composition of the ascus wall in unitunicate, inoperculate ascomycetes. In some cases, the procedures used have been nonspecific, and control tests have not been made (Chadefaud, 1969). In other instances, more specific tests have been carried out, but controls apparently have not been run (Schrantz, 1970). It has been suggested that chitin is absent from the ascus wall in *X. polymorpha* (Schrantz, 1970), although the inner wall layer (endoascus) (Schrantz, 1970) was shown and described as being fibrillar. However, the fibrillar or granulofibrillar structure seen in the ascus wall of *Xylaria* and *Rosellinia* after chemical fixation (Pl. I, Figs. 2.4 and 2.5; see also Pl. IV, Figs. 2.21, 2.23, and 2.25) has not been confirmed either by shadowed preparations of chemically cleaned wall material or by enzymatic digestion. The presence or absence of true fibrils, and possibly therefore of chitin, remains in doubt. Mannan could be present in the ascus wall in a granulofibrillar form. It has been demonstrated to exist as such in certain algal cell walls (Frei and Preston, 1961). There is as yet, however, no evidence for it in asci. Schrantz (1970) proposed that callose was a major component of the "endoascus" in *X. polymorpha* and that the outer layer of the wall (exoascus) was composed of pectin. The latter substance, however, gave only a weak reaction with alcian blue, and controls were not given for either the callose or the pectin tests. In a recent preliminary study on the developing ascus of *S. humana,* Read (1977) has shown that the ascus wall reacts positively for neutral polysaccharides and for B-hexopyranose polymers but is negative for lipid. These were respectively tested for by the periodic acid Schiff's (PAS) reaction (Jensen, 1962), the calcofluor fluorescence method (Maeda and Ishida, 1967), and the benzapyrene–caffeine method (Jensen, 1962). At the ultrastructural level, use of the periodic acid–thiosemicarbazide–silver proteinate technique (Thiéry, 1967) has shown that the ascus wall in *X. hypoxylon, R. aquila* (Pl. I, Figs. 2.4 and 2.8; Pl. III, Fig. 2.15; Pl. IV, Fig. 2.25), and *S. humana* (Pl. I, Fig. 2.3) are polysaccharide positive. The results also show that the deposition of silver grains is greater toward the outer edge of the ascus wall, particularly in maturing asci of *X. hypoxylon.* This enhances the impression of a two-layered wall (Pl. III, Fig. 2.15). When 10% acetic acid was used in place of the thiosemicarbazide, no staining of the wall occurred. The observations presented here, together with those of Read (1977), suggest that in *S. humana,* at least, the ascus wall is composed in part of R-glucan. The ninhydrin–Schiff's reaction for total protein (Jensen, 1962) also gave a positive result for the ascus wall of *S. humana* (Read, 1977).

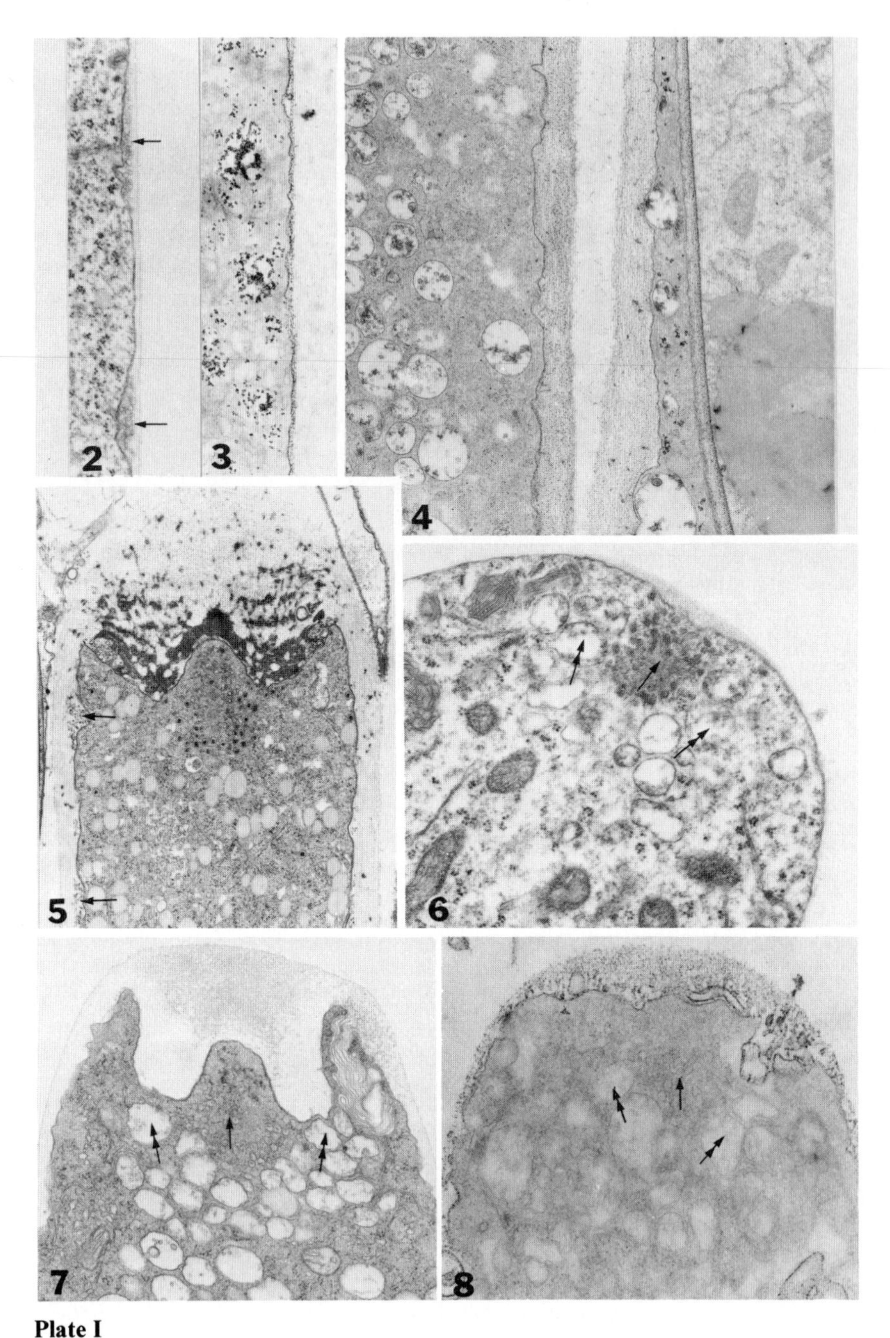

Plate I

Fig. 2.2. *Xylaria longipes.* Longitudinal section through part of the ascus wall showing paramural bodies (arrows). Glutaraldehyde/acrolein fixation. ×2100.

Fig. 2.3. *Sordaria humana.* Longitudinal section through part of the ascus wall showing a very early stage in cell wall deposition. The cell wall material is polysaccharide positive. Glutaraldehyde/formaldehyde fixation. Thiosemicarbazide 1 hour ×20 800.

The Apical Apparatus

The organization of the ascus protoplast in terms of wall vesicles and Spitzenkorper's characteristic of tip-growing, vegetative hyphae (Grove and Bracker, 1970) is not found in young, elongating asci. However, organized vesicles are found in the apical region of the protoplast in those asci characterized at maturity with an apical ring. A distinct apical body (Pl. I, Figs. 2.5–2.8; Pl. II, Figs. 2.9–2.11) is found at a later stage of ascus development. This event is associated with localized wall growth leading to the elaboration of the wall in the form of an apical ring.

The precise arrangement and structural characteristics of these wall vesicles vary from species to species (compare Pls. I and II, Figs. 2.5–2.11). Generally, they may be grouped under two main types: (1) small, spherical to ellipsoidal vesicles (0.07–0.17 μm in diameter) with dense contents; (2) larger vesicles, irregular in shape (0.12–0.45 μm in diameter) with either diffuse, granular contents or electron-transparent contents. These larger vesicles are particularly evident in *X. longipes* (Pl. I, Fig. 2.6) and *X. hypoxylon* (Pl. I, Fig. 2.7). In *R. aquila,* only one morphological vesicle type is found associated with the developing ascus apex. Here the vesicles are more uniform in both size and shape, and after glutaraldehyde/acrolein fixation, the center of each vesicle is occupied by a spherical, electron-opaque deposit similar in appearance to the material that comprises the apical ring (Pl. I, Fig. 2.5).

The essentially spherical nature of the apical body (Pl. I, Figs. 2.5–2.7, single arrows) in xylariaceous genera may be a functional feature linked with the extensive, pendulous apical ring of, for example, *Xylaria* and *Rosellinia.* It may alternatively be the result of physical constraints imposed upon the apical body by the

---

Fig. 2.4. *Rosellinia aquila.* Longitudinal section through parts of two asci. The walls are polysaccharide positive and more or less homogeneous throughout. There is slight indication of denser staining at the outer edges. Glutaraldehyde/acrolein fixation. Thiosemicarbazide 24 hours. ×13 000.

Fig. 2.5. *Rosellinia aquila.* Longitudinal section through the ascus apex at a "midstage" in its development. Note the compact apical body and uniformly small vesicles. Paramural bodies are also present (arrows). Glutaraldehyde/acrolein fixation. ×13 000.

Fig. 2.6. *Xylaria longipes.* Median longitudinal section through a very young ascus apex showing the dense apical body (single arrow) surrounded by small, dense vesicles and large vesicles (double arrows). Compare with Fig. 2.lb. Glutaraldehyde/acrolein fixation. ×24 000.

Fig. 2.7. *Xylaria hypoxylon.* Median longitudinal section through a "midstage" ascus apex showing small and large (double arrows) apical vesicles and the dense apical body (single arrow). Glutaraldehyde fixation. ×22 400.

Fig. 2.8. *Xylaria hypoxylon.* Median longitudinal section through a young ascus apex showing small apical vesicles with dense grains in them (single arrow) and large vesicles with unstained contents. The ascus wall is positively stained. Glutaraldehyde/formaldehyde fixation. Thiosemicarbazide 24 hours. ×28 000. All magnifications are approximate.

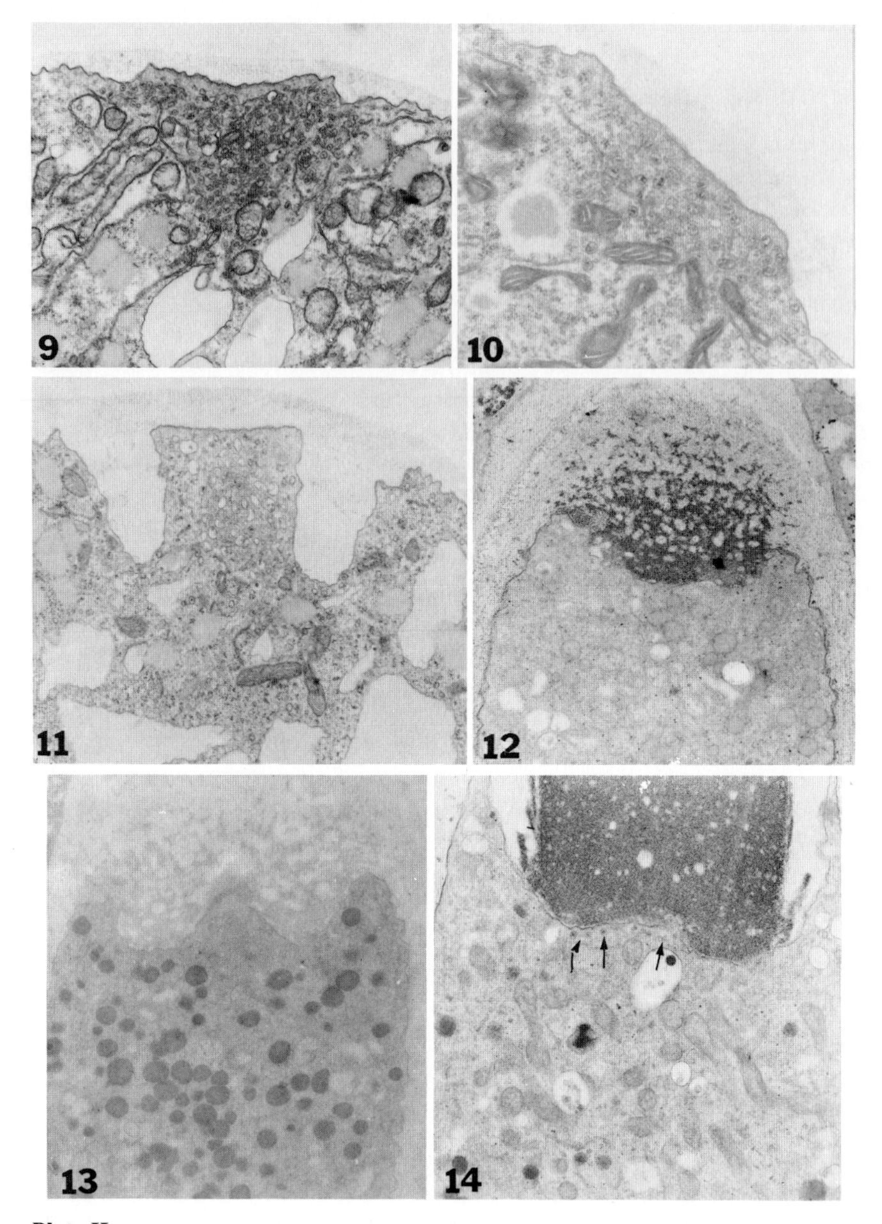

Plate II

Fig. 2.9. *Neurospora crassa.* Median longitudinal section through a young ascus apex showing the mass of small apical vesicles with a dense granule in each of them. Glutaraldehyde fixation. ×17 600.

Fig. 2.10. *Sordaria humana.* Same as Fig. 2.9. Note multiple granules in the vesicles. Glutaraldehyde/formaldehyde fixation. ×22 000.

Fig. 2.11. *Sordaria humana.* Median longitudinal section through a fully formed ascus apex showing a pendulous ring, penetrated by a cytoplasmic channel within which is a compact mass of vesicles. Glutaraldehyde fixation. ×14 300.

ring as it extends down into the ascus lumen. Whether there are functional differences between the two vesicle types seen in *Xylaria* (Pl. I, Figs. 2.6 and 2.7) is unknown, but it is possible that the vesicles beneath the spherical apical body (the larger ones) may fuse with the ascus plasma membrane and deposit their contents in a radial pattern around the tip of the ascus (Beckett and Crawford, 1973). Such a pattern of deposition could result in a localized annular thickening of the ascus wall and ultimately could produce the characteristically pendulous apical ring, open-ended at its lower end and attached to the thickened, fibrogranular "apical cushion" at its upper end (Fig. 2.1). A similar mechanism has been postulated for the development of the "pore cylinder" in the ascus apical apparatus of *Ciboria acerina* by Corlett and Elliott (1974). Chadefaud (1973) figured a proposed method of apical ring formation based on the electron micrographs of Bellèmere (1969) of *Bulgaria inquinans*. This method involves the development of a unique layer of the ascus wall termed by Chadefaud the "strate annellogene." However, the relationships between this scheme (Chadefaud, 1973) and the micrographs of Bellemére (1969) must be considered somewhat tenuous since the micrographs illustrate only mature apical rings and not developmental stages.

The smaller vesicles with electron-opaque contents (Pl. I, Fig. 2.6) may be concerned with the deposition of dense matrix material which forms the bulk of the substance within the ring. Alternatively, these smaller vesicles may deposit their contents into the morphologically distinct "apical cushion" or "plug" region (Beckett and Crawford, 1973, and Pl. I, Fig. 2.1). The apparent absence of large-type vesicles from the apical body of *Rosellinia* (Pl. I, Fig. 2.5) implies that here, at least, they are not essential for the successful development of a complex, pendulous apical ring. However, at the present time, any property ascribed to the apical vesicles can at best be speculative, and a full understanding of their functional significance awaits more detailed cytological and cytochemical studies.

A difference in the polysaccharide content of the apical ring in asci of *X. hypoxylon* and *R. aquila,* on the one hand, and *S. humana,* on the other, is already indicated (Table 2.1) (Pl. III, Figs. 2.15 and 2.19; Pl. IV, Fig. 2.23). Members of the family Xylariaceae have long been characterized by the positive amyloid reaction at the ascus apex when stained in Melzer's reagent (Dennis, 1968), although some may need to be pretreated with KOH solution to provoke a positive reaction (Kohn and Korf, 1975; Nannfeldt, 1976). In contrast to this, genera

---

Fig. 2.12. *Rosellinia aquila.* Longitudinal section through a young ascus apex showing polysaccharide-positive staining. Glutaraldehyde/acrolein fixation. Thiosemicarbazide 24 hours. ×13 000.

Fig. 2.13. *Rosellinia aquila.* Serial section to that shown in Fig. 2.12. Control. No staining is evident in the apical ring. Glutaraldehyde/acrolein fixation. Minus Thiosemicarbazide. ×13 000.

Fig. 2.14. *Rosellinia aquila.* Longitudinal section through more mature ascus apex showing positive polysaccharide reaction both in the apical ring material and in the apical vesicles (arrows). Glutaraldehyde/acrolein fixation. Thiosemicarbazide 1 hour. ×16 000. All magnifications are approximate.

Table 2.1. Comparison of the Ascus Wall and Apical Apparatus in Six Unitunicate, Inoperculate Ascomycetes

	Xylaria polymorpha[a]	*Xylaria longipes*[a]	*Xylaria hypoxylon*[a]	*Rosellinia aquila*[a]	*Sordaria humana*[a]	*Neurospora crassa*[a]
Ascus wall	Two layers, granular and fibrillar, G–Ep	One layer, granulofibrillar, G/A–Ep	Two layers, granulofibrillar, G/A–Ep, G/F–Sp, G–Sp	One layer, granulofibrillar, G/A–Ep	One layer, granular, G/F–Sp, G/F–Ep	One layer, granular, G–Sp
Apical ring	Electron opaque, alveolate, G–Ep	Electron opaque, alveolate, G/A–Ep	Electron opaque, alveolate, polysaccharide +[b], C/A–Ep Electron transparent, G/F–Sp Electron transparent with fine fibrils, G–Sp	Electron opaque, alveolate, polysaccharide + G/A–Ep	Electron transparent, polysaccharide –[c], Lipid ?[d] G/F–Sp G/F–Ep	Electron transparent, G–Sp G/F–Sp
Apical vesicles	Small: 0.03–0.05 μm[e], dense contents Large: 0.25–0.7 μm[f], dense contents	Small: 0.03–0.07 μm, dense throughout Large: 0.12–0.45 μm, diffuse granular contents	Small: 0.04–0.06 μm, containing single, dense granule Large: 0.15–0.37 μm, diffuse granular contents	Small: 0.08–0.12 μm, with single, dense globule	Small: 0.06–0.1 μm, containing several dense granules Large: 0.13–0.23 μm, electron-transparent contents	Small: 0.05–0.08 μm, containing single, dense granule

Notes [a]Fixation and embedding: G, glutaraldehyde; G/A, glutaraldehyde/acrolein mixture; G/F, glutaraldehyde/formaldehyde mixture; Ep, Epon; Sp, Spurrs resin.
[b]Using periodic acid–thiosemicarbazide–silver proteinate procedure (Thiéry, 1967). Control: 10% acetic acid in place of thiosemicarbazide.[c]Lower region of apical ring negative with Thiéry technique and with PAS but positive with calcofluor. (Calcofluor and PAS data from Read, 1977.)
[d]Data from Read (1977). Benzapyrene–caffeine fluorescence.[e]Calculated from Schrantz (1970), Pl. X, Fig. 1, Os04 fixation.
[f]Sizes estimated on the assumption that the dumbbell-shaped cisternae (Schrantz, 1970) are in fact the large vesicles, which have collapsed.

belonging to the Sordariaceae show a negative reaction. The positive or bluing reaction as seen in the ascus apices of *X. hypoxylon, X. polymorpha, X. longipes,* and *R. aquila* has never been specifically correlated with a particular structure, although Griffiths (1973) has suggested that it is, "almost certainly a staining reaction of the electron-dense component of these rings."

An exception to this would be the interesting case of *Sordaria fimicola* (Reeves, 1971), in which the apical ring is composed of electron-opaque material similar to that shown here for members of the Xylariaceae. *Sordaria fimicola,* however, gives a nonamyloid reaction with Melzer's reagent. A further point of interest is related to the apparent difference in electron opacity of the apical ring between *S. fimicola* (Reeves, 1971) and *S. humana* described here (Pl. II, Fig. 2.11, and Pl. III, Fig. 2.19). Whether this reflects a real (compositional?) difference at the species level or whether it reflects differences in preparatory techniques is not clear at present.

Nannfeldt (1976) has suggested that the amyloid reaction is localized to an apical "plug" in the more or less complex apical apparatuses such as those found in members of the Xylariaceae. Nannfeldt has speculated that it is linked in some way with the gun function of this type of ascus and suggests that the direct positive reaction as found in *Xylaria* may be a "primitive" feature as opposed to the reaction that is only positive after pretreatment with KOH or NaOH.

At the ultrastructural level, the most obvious difference between the apical apparatuses of members of the Xylariaceae and those of the Sordariaceae is the presence in the former of an extensive electron-opaque, usually alveolate, component of the apical ring. Periodic acid–thiosemicarbazide–silver proteinate staining shows this component to be positive for polysaccharide (Pl. II, Figs. 2.12 and 2.14; Pl. III, Fig. 2.15; Pl. IV, Fig. 2.23), whereas in *S. humana* it is absent and gives a correspondingly negative reaction (Pl. III, Fig. 2.19, double arrows). In both cases, however, this technique shows that the small wall vesicles described previously are involved in the synthesis of at least part of the apical apparatus and small granules of silver deposit can readily be seen within them (Pl. II, Fig. 2.14, and Pl. III, Fig. 2.19).

When viewed with ultraviolet light, the apical ring of *S. humana* and that part of the ascus wall immediately proximal to it autofluoresce strongly. The significance of this is not fully understood at present, but it indicates some property of the wall that is highly localized.

Griffiths (1973) reported that attempts to stain the apical rings of various members of the Xylariaceae and Diatrypaceae with Sudan III were inconclusive. She suggested, however, that at the ultrastructural level, because the rings were electron opaque following osmium tetroxide fixation, they may contain lipid. Read (1977) used the highly sensitive benzapyrene–caffeine method (Table 2.1), but because the apical ring of *S. humana* exhibited autofluorescence, a positive reaction could not be demonstrated unequivocally. The Sudan black B method (Jensen, 1962) for total lipids, phospholipids, and free fatty acids gave a negative reaction. It will be informative, however, to use the benzapyrene–caffeine technique with *Xylaria* and *Rosellinia* to determine whether or not the osmiophilic apical rings of these genera show positive fluorescence.

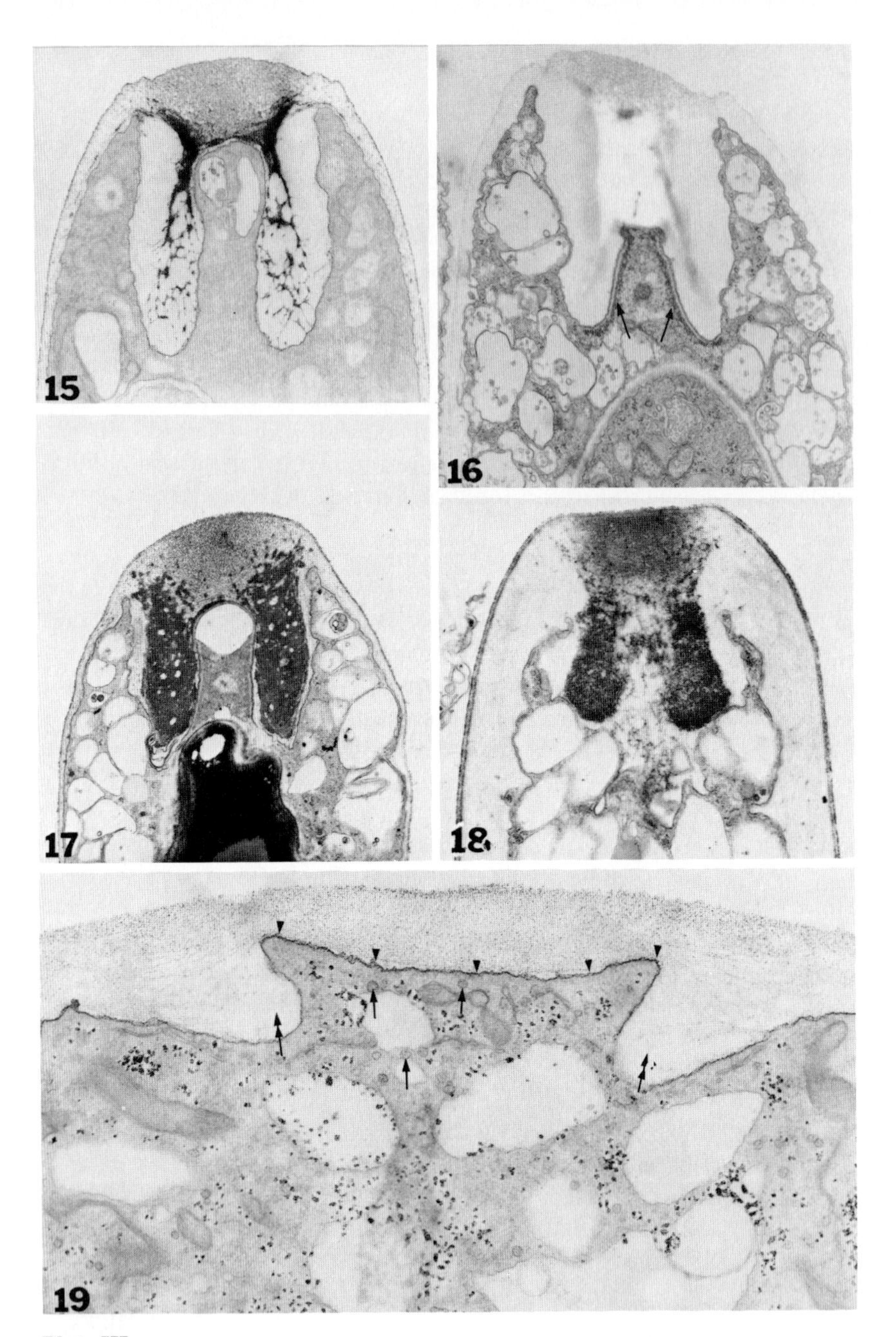

Plate III

Fig. 2.15. *Xylaria hypoxylon.* Median longitudinal section through a fully formed ascus apex showing the localization of polysaccharide material within the apical ring region. Glutaraldehyde/formaldehyde fixation. Thiosemicarbazide 24 hours. ×16 000.

Fig. 2.16. *Xylaria hypoxylon.* Slightly oblique longitudinal section through ascus apex showing the electron-opaque, granular material lining the lumen within the apical ring (arrows). Glutaraldehyde/formaldehyde fixation. ×16 000.

Pectin and callose have been reported to be major components of the apical apparatuses of unitunicate, inoperculate asci (Chadefaud, 1969; Schrantz, 1970). There is a need, however, for more specific cytochemical tests to be done, particularly in conjunction with fluorescein-conjugated lectins and with enzymatic digestion techniques at the ultrastructural level.

Ascus Function

Of the inoperculate, unitunicate fungi for which information on ascus dehiscence is available, the major group is the pyrenomycetes. Ingold (1953) has described the *Sordaria* type of ascus dehiscence, and because this is probably representative of the majority of perithecial ascomycetes that actively discharge their spores, a brief description of this type will help clarify any subsequent discussion on the mechanics of ascus dehiscence.

In *Podospora minuta* (Fuckel) Niessl and *Podospora curvula* (de Bary) Niessl, which are examples of the *Sordaria* type, asci remain attached to the pseudoparenchymatous cells at the base of the perithecium and mature sequentially. The most advanced ascus in this sequence elongates into the upper region of the perithecium, along the neck, and eventually it protrudes just beyond the ostiole. The spores are discharged and the dehisced ascus retracts into the lower region of the perithecium. Under normal conditions, this process is repeated, presumably until all the asci have discharged their spores.

The unitunicate ascus is bounded by what is now generally regarded as a structurally and functionally single wall (Luttrell, 1951; Griffiths, 1971; Beckett and Crawford, 1973). The ability of the ascus to elongate, and later to retract, has been taken as evidence that this single wall is elastic and therefore highly extensible (Ingold, 1933; Burnett, 1968). Ingold (1939) suggested that the final elongation of the ascus prior to dehiscence was "probably in part due to actual growth and in part to the pressure of the surrounding turgid cells." What is meant by "actual growth" is not clear, but an extensive synthesis of wall material would seem unlikely because it would tend to counteract the ability to retract after dehiscence. However, that the wall is under strain because it is being stretched by internally generated pressure seems to be a reasonable deduction. The possible

---

Fig. 2.17. *Xylaria hypoxylon.* Median longitudinal section through ascus apex showing dense material within lumen of apical ring. Glutaraldehyde/acrolein fixation. ×16 000.

Fig. 2.18. *Xylaria polymorpha.* Oblique longitudinal section through ascus apex. Note marked layering of ascus wall. Glutaraldehyde fixation. ×15 000.

Fig. 2.19. *Sordaria humana.* Median longitudinal section through an almost fully formed ascus apex. Note granular staining deposits in the "plug" region of the wall above the densely stained plasma membrane (arrowheads). Small apical vesicles with minute, stained granules in them (single arrows) can be seen beneath the apical ring. The lower region of the ring (double arrows) is unstained. Glutaraldehyde/formaldehyde fixation. Thiosemicarbazide 1 hour. ×24 700. All magnifications are approximate.

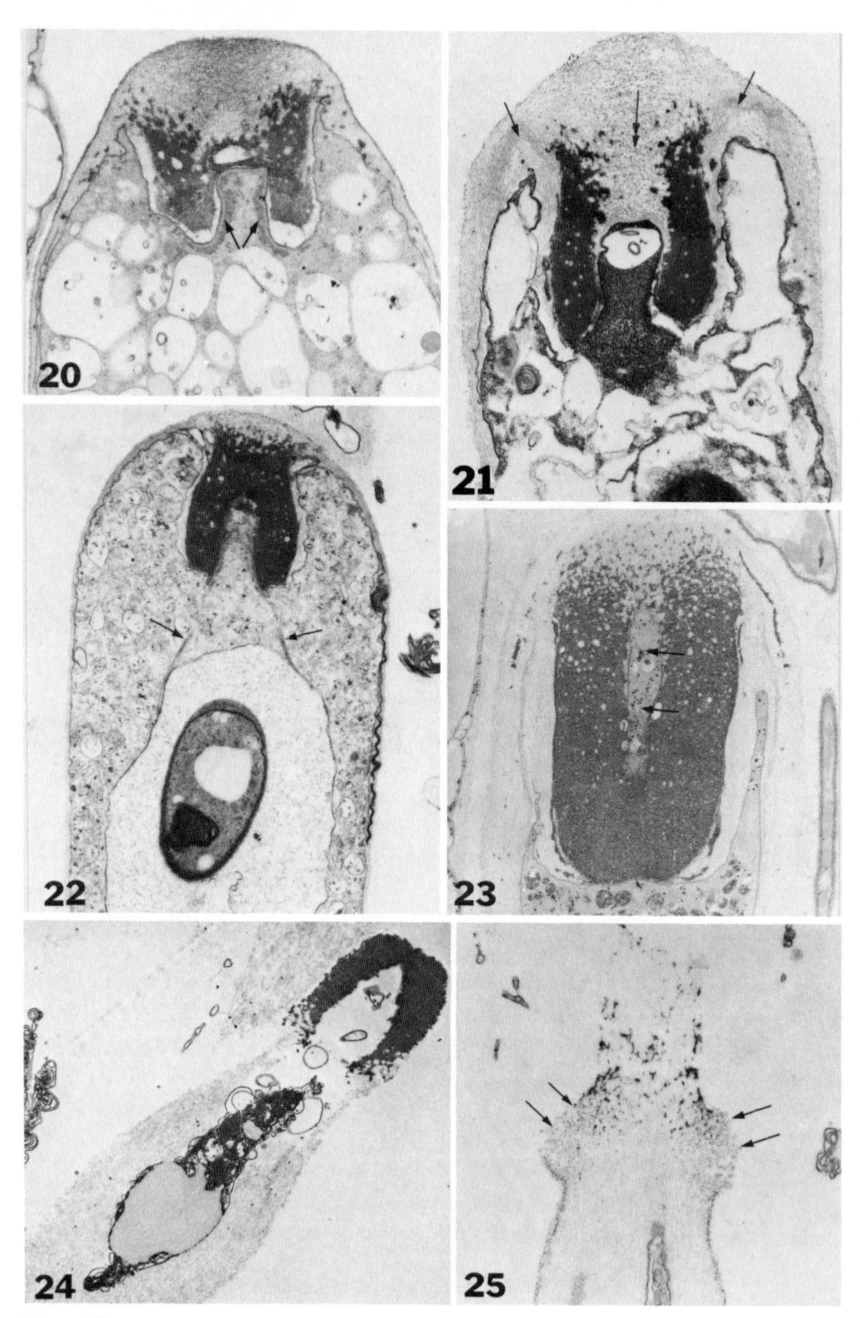

Plate IV

Fig. 2.20. *Xylaria hypoxylon.* Median longitudinal section through apex showing granular material lining the lumen of the apical ring (arrows). Glutaraldehyde/acrolein fixation. ×16 000.

Fig. 2.21. *Xylaria longipes.* Median longitudinal section through apex showing the alignment of the compact fibril-like material (single arrows) on either side of the apical ring. Note the differentiated "plug" region (double arrow), which is the site of rupture when the ascus dehisces. Glutaraldehyde/acrolein fixation. ×15 000.

methods by which internal (hydrostatic) pressure may be generated is not relevant to this discussion but is reviewed in detail elsewhere (Beckett and Read, in preparation). What is relevant is that the assumption of a hydrostatic pressure to provide the motive force for ascospore discharge implies the existence of regulatory mechanisms not only for its generation but also for its maintenance and its release. It is the latter two mechanisms that will be considered here in detail.

There are two ways by which hydrostatic pressure may be released at the apex of the ascus. One is by mechanical rupture of the wall, and the other is by an enzymatic dissolution of wall material. The point at which this pressure is released, irrespective of the method involved, may be governed by the component properties of the apical apparatus. Mechanical rupture could occur at the weakest or least elastic point in the wall, whereas enzymatic dissolution could involve hydrolysis of specific wall components. Both of these, we can speculate, could be localized at the "plug" region of the apical apparatus, which in *Sordaria, Xylaria,* and *Rosellinia* lies at the proximal end of, and to the inside of, the apical ring (Pl. II, Fig. 2.11; Pl. III, Figs. 2.15–2.19; Pl. IV, Figs. 2.20–2.25; Pl. V, Figs. 2.27 and 2.28). It is possible, therefore, that the apical apparatus represents a unique region of the ascus wall, which by virtue of its intrinsically different chemical composition or as a result of its altered wall chemistry, may function as a pressure valve capable of regulating ascus dehiscence. In the case of mechanical rupture of the apical apparatus, regulation could be based on the requirement of a threshold pressure. This would presumably operate over a limited range of pressures for asci with walls of any given chemical composition, although a wider range could occur between genera and species. Alternatively, if hydrolysis of the "plug" region is a requirement for ascus dehiscence, then a more specific, metabolic control can be envisaged that may be directly or indirectly triggered by a stimulus.

The existence of a trigger mechanism involving blue light (470-nm wavelength) is suggested by the work of Ingold and Dring (1957) on *S. fimicola* (Roberge) Cesati and de Notaris. Ingold (1958) showed that, in a mutant strain of *S. fimicola,* which had orange-yellow perithecia, a light stimulus caused an almost immediate increase in spore discharge. This was in contrast to the 1–2 hour lag or "latent period" between the light stimulus and the response in *S. fimicola* wild-

Fig. 2.22. *Xylaria longipes.* Oblique longitudinal section of apex showing the fibrous strand of material (arrows) that links the apical spore with the inside edge of the apical ring. Glutaraldehyde/acrolein fixation. ×10 000.

Fig. 2.23. *Rosellinia aquila.* Near-median longitudinal section of apex stained by the Thiéry technique showing glycogen (arrows) present in the narrow channel within the apical ring. Glutaraldehyde/acrolein fixation. Thiosemicarbazide 24 hours. ×8000.

Fig. 2.24. *Xylaria longipes.* Near-median longitudinal section through the apical region of a dehisced ascus showing the everted apical ring. Glutaraldehyde/acrolein fixation. ×8000.

Fig. 2.25. *Xylaria hypoxylon.* Same as for Fig. 2.24 but stained by the Thiéry technique. Note the everted hinge region of the ascus wall (arrows). Glutaraldehyde/formaldehyde fixation. Thiosemicarbazide 24 hours. ×16 000. All magnifications are approximate.

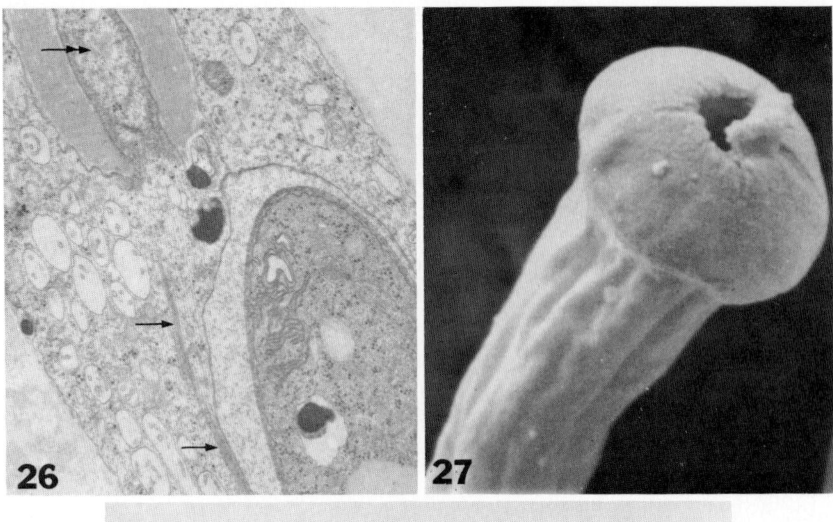

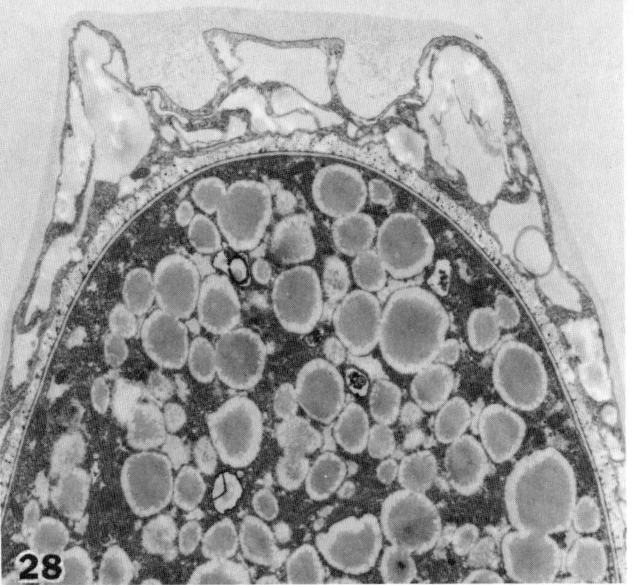

Plate V

Fig. 2.26. *Xylaria longipes.* Median longitudinal section through part of the apical region of an ascus showing the fibrous strand (single arrows) attached to the outer layer of the ascospore wall. Note short fibrous structures (double arrow) within lumen of apical ring. Glutaraldehyde/acrolein fixation. ×22 500.

Fig. 2.27. *Sordaria humana.* Part of the apical region of a dehisced ascus as seen with the stereoscan electron microscope. Note the ruptured "plug" region surrounded by the intact apical ring. ×8000.

Fig. 2.28. *Sordaria humana.* Median longitudinal section of apex of an undehisced ascus showing the relationship between spore diameter and the diameter of the pore region within the apical ring. Glutaraldehyde/formaldehyde fixation. ×8400. All magnifications are approximate.

type (Ingold, 1958) and in *Sordaria macrospora* Auserwald (Walkey and Harvey, 1967). Walkey and Harvey (1968) showed that spore discharge in a variety of pyrenomycetous fungi was also stimulated by rain when in the field and by simulated rain when under laboratory conditions. In contrast to *Sordaria,* spore discharge in *Xylaria longipes* is stimulated by darkness and inhibited by light (Walkey and Harvey, 1966a; Beckett and Crawford, 1973). Furthermore, Walkey and Harvey (1966a) showed that there was a significant number of pyrenomycetous ascomycetes that could be recognized as "nocturnal" in this respect.

The recognition of a possible trigger mechanism, however, has not led to an understanding of the process(es) leading to ascus dehiscence. The "latent period" in *S. fimicola* and *S. macrospora* after a light stimulus is almost certainly related to the time required for what may be a chain of events to occur before the ascus is ready to dehisce. These events are unknown, but it is possible that they include ascus elongation, hydrolysis of "plug" material in the apical ring, and a variety of reactions that may be necessary to increase the hydrostatic pressure within the ascus. The dramatic decrease in the "latent period" seen in the mutant strain of *S. fimicola* (Ingold, 1958) need not necessarily result simply from the removal of a masking effect on light by the dark pigment present in the perithecial wall of the wild type. The genetic nature of the mutant was unknown and could have affected spore discharge in other ways.

From the ultrastructural studies so far made on dehisced asci, it is clear that under normal conditions the region of the wall that is proximal to the apical ring and that overlies the cytoplasmic channel within the lumen of the ring becomes broken down to form a continuous channel connecting the ascus lumen with the outside environment (Pl. IV, Figs. 2.21 and 2.24; Beckett and Crawford, 1973; Codron, 1974; Corlett and Elliott, 1974). Once this has happened, the pressure within the ascus is, of course, released. Therefore, for discharge of ascospores to be efficient, it is essential that they be ejected through the resultant pore rapidly, before the pressure is dissipated. In *Hypoxylon* Builliard ex Fries and in *Xylaria* (Pl. IV, Figs. 2.24 and 2.25), spore discharge is accompanied by the eversion of the apical ring (Ziegenspeck, 1926; Greenhalgh and Evans, 1967; Beckett and Crawford, 1973). A similar, although less pronounced, eversion of apical apparatuses is also seen in *S. sclerotiorum* (Codron, 1974) and in *Ciboria acerina* Whetzel and Buchwald ex Grove and Elliott (Corlett and Elliott, 1974). More recently, Read (1977) has shown by light microscopy that the apical ring in *S. humana* is partially everted, and scanning electron microscopy confirms this observation (Pl. V, Fig. 2.27).

In *Xylaria* and *Rosellinia* (Pl. IV, Fig. 2.21, arrows, and Fig. 2.23), a region of compact fibril-like material can be seen in the ascus wall on either side of, and proximal to, the apical ring. A hinging role has been suggested for this part of the wall (Beckett and Crawford, 1973) because it is here that the apical ring pivots during eversion (Pl. IV, Fig. 2.25, arrows; Beckett and Crawford, 1973). In *Sordaria,* the apical ring does not fully evert but does remain intact after ascus dehiscence. Only the central "plug" region is structurally altered (Pl. V, Fig. 2.27). As with *Xylaria,* the apical ring in *Sordaria* retains its shape and dimension even when the ascus has retracted (compare Figs. 2.27 and 2.28, Pl. V).

Chenantais (1919) showed that in *Lasiosphaeria* de Cesati and *Lasiosordaria* Chenantais the spores were attached to the apical ring by a thin strand of material. Ingold (1933) described a similar attachment of the spores to the tip of the ascus for pyrenomycetes in general and illustrated this in *Podospora curvula*. The nature of these attachments is unknown. They have been variously referred to as mucilaginous and cytoplasmic, but in *Xylaria longipes* (Beckett and Crawford, 1973; Pl. IV, Fig. 2.22, arrows; Pl. V, Fig. 2.26, arrows) an electron-opaque, fibrous strand has been seen that links the inside of the apical ring with both the apical spore and the rest of the spores in what is normally a uniseriate alignment. This strand of material is embedded in a zone of electron-opaque, granular substance (Pl. III, Fig. 2.17, and Pl. IV, Fig. 2.21) which subsequently breaks down to form a uniform cylindrical lining within the apical ring (Pl. III, Fig. 2.16; Pl. IV, Fig. 2.20, arrows, and Fig. 2.22; Pl. V, Fig. 2.26). There is also evidence that the fibrous material of the strand may exist in a partially organized, short-length fibrous form within the lumen of the apical ring (Pl. V, Fig. 2.26, double arrow). That such linkages between the spores and ascus apices have not readily been observed at the ultrastructural level may be because (1) they are transient structures associated with a specific developmental stage, (2) they are artifacts of a particular chemical fixation, or (3) very few detailed electron microscope studies have been made.

What then is the function of the connecting strand? There are at least two possible roles. First, it may serve to align the spores with the apical ring and, more importantly, with the future pore region within the ring. Second, as the ring everts at the time of ascus dehiscence, the strand may cause the whole spore chain to be jerked out as an initial multispored projectile. This would have the further advantage of ensuring that the spores were discharged *before* all the hydrostatic pressure was lost. However, Walkey and Harvey (1966b) found that in *Xylosphaera longipes* (Nitchke) Dennis (= *Xylaria longipes*) there was a marked tendency for the complete breakdown of the spore mass into single-spored projectiles. They further showed that in *X. longipes* the horizontal distance of discharge decreased with an increase in the number of spores per projectile. This was in contrast to their findings with *Sordaria* and to those of Ingold (1961), wherein the average discharge distance increased with the size of the projectiles in accordance with the formula $d = Kr^2$ (Ingold, 1960) where d is distance of discharge, K is a constant related to spore density and the viscosity of the air, and r is the radius of the projectile.

The tendency for the spore mass to break up into single spores is governed by a number of factors depending on the genus and species. In *X. longipes* and other species of this genus, one factor is the length of the apical ring and particularly the diameter of the pore and channel passing through it. Because the spores have a diameter of approximately eight to ten times that of the pore, and because the apical rings of dehisced and undehisced asci do not differ significantly in diameter (Beckett and Crawford, 1973), it is reasonable to assume that either the ring is elastic or the spores are compressible. Observations on *Poronia punctata* (Linneus ex Fries) Fries by Stiers (1974) and on the discomycetous *Pyronema domesticum*

(Sowerby ex Fries) Saccardo by Hung (1977) showed that discharged ascospores had apparently lost their outer wall (perispore) layer. Hung (1977) showed that this layer degenerated immediately prior to discharge. It is possible that this hitherto unsuspected feature is common among ascomycetes, and it may impart a compressible or elastic property to the ascospore wall. Presumably, the wall would later become more rigid on exposure to the atmosphere. Degeneration of the outer wall layer could also provide a lubricant substance to aid the spore in its passage through an elastic apical ring. Furthermore, if the outer layer of the spore wall in *X. longipes* were to degenerate prior to ascus dehiscence, the linkage between the spores and the apical ring could be destroyed or possibly even enhanced depending on the properties of the degeneration product(s). In either event the result is speculative, but it can be assumed that the spores are each gripped, if only momentarily, as they pass through the ring. This is likely to separate them from one another.

Evidence for the hypothesis that at least in some instances ascospores are compressible comes from the observations on *Lophodermella sulcigena* (Rostrup) von Höhnel by Campbell (1973). Here the ascospores are cigar shaped and surrounded by a mucilaginous sheath. As each spore passes through the pore at the tip of the ascus, it is visibly compressed (Campbell, 1973; Beckett et al., 1974). Once the major part of the spore has passed through the pore, the squeezing of the ascus wall against the tapered end of the ascospore aids in throwing the spore clear of the ascus. With the liberation of one spore, the next in line is forced into the pore, thus sealing it and retaining the internal pressure. When all eight spores have been discharged, the vacuolated epiplasm is lost through the open pore (Beckett et al., 1974). This is probably the same procedure as described by Ingold (1933, 1953) for species of *Trichoglossum* Boudier and *Geoglossum* Persoon ex Fries.

The relationship between the shape of the ascospore and the mechanics of its discharge was first discussed by Ingold (1954). He pointed out that in inoperculate ascomycetes successive spore discharge is probably the rule and that in such types, the shape of the spore could determine to some extent the effectiveness of its discharge.

There are two basic categories of ascospore shape. The majority are what is termed bipolar–symmetrical, but some are bipolar–asymmetrical (Ingold, 1954). In the latter type, the widest part of the spore is nearer the apex than the base. With these types, once the widest part of the spore is situated within the ascus pore, the velocity of the spore will be zero or very low, but all subsequent work done by the hydrostatic pressure will be directed toward the discharge of that spore and not toward the stretching of the pore. This work is given by pV, where p is the hydrostatic pressure within the ascus and V is the volume of that part of the spore remaining in the ascus lumen at this stage. Ingold (1954) suggested that for spores of equal mass, the nearer the widest part is to the front end, the greater will be V and, all else being equal, the greater the initial velocity.

In spores such as those of *L. sulcigena* (Campbell, 1973), which are surrounded by a mucilaginous sheath, the friction between spore and pore could presumably be very much reduced, thus favoring this hypothesis. The same might be true for such spores as those of *Poronia* Willdenow ex S. F. Gray (Stiers, 1974), in which the outer wall layer degenerates. It is not clear how these forces would operate however, if, as seems to be the case in *L. sulcigena,* the spore also were to change shape as it passed through the pore (Beckett et al., 1974).

Finally, some consideration must be given to the method(s) by which the hydrostatic pressure within the ascus is maintained until such time as it is utilized during ascus dehiscence. We must assume that this pressure cannot leak back through the base of the ascus into the cushion of pseudoparenchymatous cells at the base of the perithecium and to which the asci are attached.

One possibility is that the pressure is built up by a continual, or sudden, influx of cytoplasm and water into the ascus (Burnett, 1968). Another possibility is that leakage is prevented by the formation, possibly at a critical developmental stage, of an elaborate septal pore apparatus in the cross wall at the base of the ascus. Such an apparatus has been seen in species of *Sordaria* by Furtado (1971) and by Zickler (Beckett et al., 1974). More recently, an extremely complex pore apparatus in the basal septum of the ascus of *S. humana* has been observed (Beckett, in preparation). Cytochemical staining suggests that part of this apparatus is formed from modified ascus plasma membrane and part is from the ascus endoplasmic reticulum.

The role of such septal pore apparatuses is at present unknown, but they may be involved in the regulation of movement of water, solutes, cytoplasm, or ions both into and out of the ascus. Any of these functions, either singly or in combination, could affect the generation or maintenance of the hydrostatic pressure inside the ascus.

General Conclusions

Correlated light and electron microscope studies on the ascus apical apparatus could yield useful information that may aid in the understanding of the different apical types and of the so-called amyloid reaction of certain ascus apices.

However precise the knowledge and understanding of the chemical composition and ultrastructure of the apical apparatuses of different asci become, the fact remains that in the final analysis it may be of limited practical use to the taxonomist who is primarily concerned with those structures or features which are readily visible, at least with the light microscope.

The significance of the differences and similarities (Table 2.1) will only be fully appreciated when they are viewed with a knowledge and understanding of the composition of the structures involved. If an association between morphological types and compositional types can be established, then a more realistic evaluation can be made of the use of specific variations in morphology as an aid to taxonomy. Considerable advances have already been made along these lines with respect to vegetative structure and composition (Bartnicki-Garcia, 1968).

References

Bartnicki-Garcia, S., 1968. Cell wall chemistry, morphogenesis, and taxonomy of fungi. Ann. Rev. Microbiol. 22: 87–108.

Beckett, A. (in preparation). Septal pore apparatuses in asci and ascogenous hyphae of *Sordaria humana.*

Beckett, A., and R. M. Crawford, 1973. The development and fine structure of the ascus apex and its role during spore discharge in *Xylaria longipes.* New Phytol. 72: 357–369.

Beckett, A., I. B. Heath, and D. J. McLaughlin, 1974. An Atlas of Fungal Ultrastructure. Longmans, London.

Beckett, A., and N. D. Read (in preparation). The origin of hydrostatic pressure as the motive force for ascospore discharge.

Bellemére, A., 1969. Quelques observations relatives a l'infrastructure de l'appareil apical des asques du *Bulgaria inquinans* Fr. (discomycète inoperculé). C. R. Acad. Sci. Paris 268: 2252–2255.

Bellemére, A., 1975. Étude ultrastructurale des asques: la paroi, l'appareil apical, la paroi des ascospores chez des discomycètes inoperculés et des Hystériales. Physiol. Veg. 13: 396–406.

Burnett, J. H., 1968. Fundamentals of Mycology. Edward Arnold Ltd., London.

Campbell, R., 1973. Ultrastructure of asci, ascospores and spore release in *Lophodermella sulcigena* (Rostrup) von Hoehnel. Protoplasma 78: 69–80.

Chadefaud, M., 1969. Remarques sur les parois l'appareil et les réserves nutritives des asques. Oesterreich. Bot. Z. 116: 181–202.

Chadefaud, M., 1973. Les asques et la systematique des Ascomycètes. Bull. Soc. Mycol. France 89(2): 127–170.

Chenantais, J. E., 1919. Recherches sur les Pyrenomycètes. Bull. Soc. Mycol. France 35–37: 113–139.

Codron, D., 1974. Étude ultrastructurale de quelques points du développement des asques du *Sclerotinia sclerotiorum* (Lib.) de Bary. Ann. Sci. Nat. Bot. Biol. Veg. 15(3): 255–276.

Corlett, M., and M. E. Elliott, 1974. The ascus apex of *Ciboria acerina.* Can. J. Bot. 52(7): 1459–1463.

Dennis, R. W. G., 1968. British Ascomycetes. Cramer, Stuttgart.

Frei, E. V., and R. D. Preston, 1961. Variants in the structural polysaccharides of algal cell walls. Nature (London) 192: 939–943.

Furtado, J. S., 1971. The septal pore and other ultrastructural features of the pyrenomycete *Sordaria fimicola.* Mycologia 63: 104–113.

Greenhalgh, G. N., and L. V. Evans, 1967. The structure of the ascus apex in *Hypoxylon fragiforme* with reference to ascospore release in this and related species. Trans. Br. Mycol. Soc. 50: 183–188.

Greenhalgh, G. N., and L. V. Evans, 1968. The developing ascospore wall of *Hypoxylon fragiforme.* J. Roy. Microsc. Soc. 88: 545–556.

Griffiths, H. Bronwen, 1971. The structure of the pyrenomycete ascus and its relevance to taxonomy. *In* G. C. Ainsworth and J. Webster (Eds.), First International Mycological Congress Abstracts. Exeter, England, p. 37–38.

Griffiths, H. Bronwen, 1973. Fine structure of seven unitunicate pyrenomycete asci. Trans. Br. Mycol. Soc. 60: 261–271.

Grove, S. N., and C. E. Bracker, 1970. Protoplasmic organization of hyphal tips among fungi: vesicles and Spitzenkorper. J. Bact. 104: 989–1009.

Hung, Ching-Yuan, 1977. Ultrastructural studies of ascospore liberation in *Pyronema domesticum*. Can. J. Bot. 55: 2544–2549.

Ingold, C. T., 1933. Spore discharge in the ascomycetes I. Pyrenomycetes. New Phytol. 32: 175–196.

Ingold, C. T., 1939. Spore Discharge in Land Plants. Oxford, Clarendon Press.

Ingold, C. T., 1953. Dispersal in Fungi. Oxford, Clarendon Press.

Ingold, C. T., 1954. Ascospore form. Trans. Br. Mycol. Soc. 37: 19–21.

Ingold, C. T., 1958. On light-stimulated spore discharge, in *Sordaria*. Ann. Bot. 22: 129–135.

Ingold, C. T., 1960. Dispersal by air and water—the take-off. *In* J. G. Horsefall and A. E. Diamond (Eds.), Plant Pathology Vol. 3. Academic Press, London, p. 137.

Ingold, C. T., 1961. Ballistics in certain ascomycetes. New Phytol. 60: 143–149.

Ingold, C. T., and V. J. Dring, 1957. An analysis of spore discharge in *Sordaria*. Ann. Bot. 21: 465–477.

Jensen, W. A., 1962. Botanical Histochemistry. W. H. Freeman, San Francisco and London.

Kohn, L. M., and R. P. Korf, 1975. Variation in ascomycete iodine reactions: KOH pre-treatment explored. Mycotaxon 3: 165–172.

Luttrell, E. S., 1951. Taxonomy of the pyrenomycetes. Univ. Missouri Studies 24: 1–120.

Maeda, H., and N. Ishida, 1967. Specificity of binding of hexopyranosyl polysaccharides with fluorescent brightener. J. Biochem. 62: 276–278.

Marchant, R., and A. W. Robards, 1968. Membrane systems associated with the plasmalemma of plant cells. Ann. Bot. 32: 457–471.

Nannfeldt, J. A., 1976. Iodine reactions in ascus plugs and their taxonomic significance. Trans. Br. Mycol. Soc. 67: 283–287.

Read, N. D., 1977. A study of the ascus as a highly specialized "spore gun" in *Sordaria humana* using light and fluorescent microscopic techniques. B.Sc. Dissertation, Bristol Univ. Bristol, U.K.

Reeves, F. Brent, 1971. The structure of the ascus apex in *Sordaria fimicola*. Mycologia 63: 204–212.

Schrantz, J.-P., 1970. Étude cytologique, en microscopic optique et électronique, de quelques ascomycètes. II. La paroi. Rev. Cytol. Biol. Veg. 33: 111–168.

Stiers, D. L., 1974. Fine structure of ascospore formation in *Poronia punctata*. Can. J. Bot. 52: 999–1003.

Thiéry, J.-P., 1967. Mise en évidence des polysaccharides sur coupes fines en microscopic électroniques. J. Microsc. 6: 987–1018.

Walkey, D. G. A., and R. Harvey, 1966a. 1. A survey of the periodicity of spore discharge in pyrenomycetes. Trans. Br. Mycol. Soc. 49: 583–592.

Walkey, D. G. A., and R. Harvey, 1966b. Studies of the ballistics of ascospores. New Phytol. 65: 59–74.

Walkey, D. G. A., and R. Harvey, 1967. Spore discharge rhythms in pyrenomycetes. III. Ascospore production and the quantitive and qualitative influences of light on spore discharge in *Sordaria macrospora*. Trans. Br. Mycol. Soc. 50: 241–249.

Walkey, D. G. A., and R. Harvey, 1968. Spore discharge rhythms in pyrenomycetes. IV. The influence of climatic factors. Trans. Br. Mycol. Soc. 51: 779–786.

Ziegenspeck, H., 1926. Schleudermechanismen von Ascomyceten. Bot. Arch. Koenigsberg 13: 341–381.

Chapter 3
The Operculate Ascus and Allied Forms

J. VAN BRUMMELEN

Introduction

When the Crouan brothers (1857, 1858) discovered the operculum at the top of the ascus, they placed all species with this structure in the genus *Ascobolus* Persoon ex Hooker. Boudier (1879, 1885) was the first to recognize the importance of the operculate mode of ascus dehiscence for the classification of the discomycetes. Accordingly, he divided the cup fungi on the basis of this character into two groups: the Operculatae and the Inoperculatae. It is a notable fact that none of Boudier's contemporaries (e.g., Quélet, Karsten, Fuckel, De Notaris, Schroeter, Saccardo, Rehm, Cooke, Phillips, Lindau) accepted this division. It became accepted only much later by Gäumann (1926), Seaver (1927, 1928), Nannfeldt (1932), and Le Gal (1947, 1953). A subdivision of the cup fungi based on the mode of ascus dehiscence is now accepted by most mycologists.

The Ascus Form

The shape and the structure of the asci in Pezizales show a wide range of possibilities. The shape of the asci depends on their number within each fruit body, the number of spores per ascus, and the pressure exerted upon the ascus by the surrounding hymenium and excipulum. In closed (cleistohymenial) fruit bodies with a single ascus, this shape is often subglobular (e.g., in *Thelebolus stercoreus* Tode ex Fries, *Trichobolus zukalii* Heimerl, *Lasiobolus monascus* Kimbrough). Obovoid asci are found in species with only a few asci within each fruit body (e.g., in the genera *Thelebolus* Tode ex Fries, *Leptokalpion* van Brummelen, *Ascodesmis* van Tiegham, *Dennisiopsis* Subramanian and Chandrashekara, and *Saccobolus* Boudier). Clavate and cylindrical asci, however, are most frequently met with in the Pezizales. As a rule, the clavate forms prevail in the smaller fruit bodies which have a less developed lateral excipulum, whereas cylindrical ones occur in the larger fruit bodies.

The ascus demonstrates a clear polarity in the distribution of its contents, which even changes during development. After the formation of ascospore nuclei from the diploid nucleus, two zones can be distinguished: the upper part which becomes the sporogenous zone, and the basal part filled with cytoplasm containing glycogen-like reserves. The structure of the latter was described by Schrantz (1968).

The ultrastructure of the ascoplasm varies considerably. As shown in the studies by Merkus (1975, 1976), in some genera and groups of species it is more complex than in others, whereas the presence of electron-dense globular structures in the epiplasm seems to be correlated with the formation of oil drops in the sporoplasm.

During sporogenesis and the ripening of the ascospores, the epiplasm becomes continuously more vacuolized. Meanwhile, glycogen and organelles are disappearing. Only at the end of this stage is the ascus ripe and ready to discharge its spores.

The Ascospore

In most asci of the Pezizales, one meiosis and two mitoses take place to form eight nuclei, each of which initiates the formation of a unicellular ascospore. In a few species of some unrelated genera, four-spored asci seem to be a constant character (Eckblad, 1968). In some of these, four of the original eight spores degenerate.

Cytological investigations by Berthet (1964) revealed that in most operculate discomycetes the ascospores are uninucleate, except in Sarcoscyphaceae and Morchellaceae where the spores are plurinuclear at maturity or in Helvellaceae where they are tetranuclear. In several genera of coprophilous Thelebolaceae, meiosis is followed by more than two mitoses, resulting in asci having from 16 to over 4000 spores.

Ascosporogenesis of operculate discomycetes was studied with electron microscopy in *Saccobolus kervernii* (Crouan) Boudier (Carroll, 1966, 1967, 1969); in *Ascodesmis sphaerosporus* Obrist (Moore, 1963; Carroll, 1966); in *Ascodesmis nigricans* van Tieghem (Bracker and Williams, 1966); in *Pyronema* Carus (Reeves, 1967; Griffith, 1968); in *Ascobolus immersus* Persoon ex Persoon (Delay, 1966); in *Ascobolus viridulus* Phillips and Plowright (Osio, 1969); in *Ascobolus stercorarius* (Bulliard ex St. Amans) Schroeter (Wells, 1972); and in *Pustularia cupularis* (Linneus ex Fries) Fuckel (Schrantz, 1966, 1967).

Although the ascospores of the Inoperculatae are always smooth, many operculate discomycetes develop a more or less complex spore sculpturing. Le Gal's (1947) detailed studies on ascospore ornamentation—with the aid of light microscopy—have had a far-reaching influence on the taxonomy of the operculate discomycetes. She described a great variation of the patterns of ornamentation, the development of which often seemed to be extremely complex.

A comparative study of the ultrastructure and development of ascospore ornamentation in the Pezizales (Merkus, 1973, 1974, 1975, 1976) has demonstrated that, in principle, the development of the ascospore ornamentation is a single common process with a certain number of variations. In smooth-spored species an initial ornamentation is formed that disappears during further ripening of the ascospores. In a few other smooth-spored species, a permanent, smooth ornamentation is deposited on the ascospore wall (Merkus, 1976).

It is clear from these studies that the ascospore ornamentation should be used with great caution for the delineation of genera or families.

The Ascus Wall

Up to now, over 45 species of Pezizales belonging to about 30 genera have been studied (personal observation) with light and electron microscopy in order to analyze the dehiscence mechanism of their asci. The light microscope has revealed information about the structure and function of the living ascus. Vital and subvital observations have been made in squash mounts in a slightly hypotonic solution of glycose and distilled water. Slides have been examined with phase contrast and Nomarski's interference contrast optics. The observation of unstained asci with polarized light has proved to be of special value. For light microscopy, asci also have been stained with a wide variety of dyes, of which Congo red, acid fuchsin, trypan blue, methyl blue, and methylene blue have given satisfactory results. Also, sections about 0.5 μm thick (fixed in glutaraldehyde and osmium tetroxide and subsequently embedded in epoxy resin) have been stained with toluidine blue—a method proving to be of great value. When Congo red is used as a stain for wall material, it must be remembered that this stain shows two staining mechanisms: one is by chemical linkage (a "true staining"); the other is by physical absorption of the long dipolar molecules in the dye by the surfaces of microfibrillae in the wall where there is sufficient space for them to penetrate.

For electron microscopy, material has been fixed in buffered glutaraldehyde or in $KMnO_4$ and postfixed in 1% OsO_4 (van Brummelen, 1974). For studies of the ascus wall, material fixed in $KMnO_4$ has proved to be best.

The lateral wall of the operculate ascus consists of at least two layers. The outer layer is usually rather thin and strongly birefrigent. It stains red with Congo red and a bluish violet with toluidine blue. The inner wall layer is usually thicker, less rigid, and only weakly anisotropic. It stains reddish with toluidine blue and does not stain with Congo red. In the electron microscope, the inner layer is more electron transparent in permanganate–OsO_4-fixed material.

The outer layer is more variable in appearance. In young asci, its surface is usually sharply delimited, but under other circumstances it may sometimes become swollen and mucilagenous, with a rather diffuse delimitation at the outside. This is in agreement with findings (Reisinger et al., 1977) on the ultrastructure of hyphal walls of ascomycetes and basidiomycetes. In asci with thick walls (e.g., Thelebolaceae) or in heavily swollen layers (e.g., some Sarcosyphaceae), a sublayering or lamellation is often visible. In many operculates, the ascus wall is covered by an extraascan layer (the periascus of Chadefaud) which is usually more apparent in the apical part of the ascus. In genera in which the ascus wall stains blue with iodine, this reaction is strictly confined to the mucilagenous substance of the periascus.

The Ascus Tip

All operculate discomycetes belong to the order Pezizales, but not all fungi arranged within the Pezizales show operculate asci.

More than a century ago, Boudier (1869) described the typical operculate ascus with an apical ring-shaped indentation delimiting the operculum. In *Peziza cunicularia* Boudier (Boudier, 1869), however, he also found asci opening at their tips by a bilabiate split. Shortly afterwards, Renny (1871, 1873) described several such fungi which he placed in *Ascobolus,* section *Ascozonus* Renny. Other non-operculate asci were studied and described in *Thelebolus* Tode ex Fries and allied genera (Zukal, 1886; Ramlow, 1906, 1915; Kimbrough, 1966, 1972; Kimbrough and Korf, 1967). Here mature asci open by an irregular tear in the apical part of the ascus wall.

A special type of operculate ascus has been described by Chadefaud (1946) and Le Gal (1946a, b) as "paraoperculate" and "suboperculate," respectively. They regarded this structure as important enough to distinguish a separate order or suborder between Inoperculatae (Helotiales) and the remainder of the Operculatae. The published terminology for structural details of this type of ascus is rather confusing, as both French authors used mostly different descriptive terms for the same detail and sometimes only slightly different ones for quite different elements (Le Gal, 1946b; van Brummelen, 1975).

The terminology "suboperculate ascus," as introduced by Le Gal, has been generally accepted. The suboperculate ascus was considered the most important character of the Suboperculati. It is defined by the "cousinet apical" which Le Gal described as an interrupted ring located in the "chambre apical" in the top of the ascus. However, both Eckblad (1968, 1972) and van Brummelen (1975) could not find sufficient evidence for the existence of the suboperculate ascus in the original sense. From the drawings published by Le Gal (1946b), it was possible to conclude that the "cousinet apical" represented different structures in the top of the ascus. Although the ascus of *Sarcoscypha* (Fries) Boudier and a few related genera may represent a special type of operculate ascus within the Pezizales, there is little left of the hypothesis that the Suboperculati (or Sarcoscyphineae) represent a taxon intermediate between Inoperculati and Operculati. Samuelson (1975) arrived at about the same conclusion from his study of sarcoscyphyaceous species.

Ascus Function

During ripening, the apical wall of the ascus develops certain structures that form an opening mechanism. The apical part of the wall tends to become more complex and may consist of three to four layers. Certain parts of the apical wall become weakened by the formation of indentation, a fracturing line and weakened zones or by local breakdown or gelatinization of the wall. In other (often adjoining) parts, the wall may become strengthened and more rigid. The moment at which these changes in the wall become manifest differs greatly. Sometimes they are scarcely visible before the moment of ascus dehiscence. The study of emptied asci is always necessary to decide with certainty on the structure and the mechanism of ascus dehiscence.

Ascus Types

The following principal types of asci can be distinguished in the Pezizales.

1. *Ascobolus* type (the traditional standard model of the operculate ascus). The operculum is very large and sharply delimited by a circular internal indentation just at the inner side of a strengthened ring. The operculum and a region under the ascostome are strengthened and rigid. This causes the typical sinuous outline of the apex. The ascostome is smooth. The periascus is of rather uniform thickness and stains blue with iodine. A funiculus is present. Examples: *Ascobolus furfuraceus* Persoon ex Hooker (Pl. I, 3.1), *Ascobolus sacchariferus* van Brummelen, *Saccobolus glaber* (Persoon ex Persoon) Lambotte (Pl. I, Figs. 3.2 and 3.3), *Thecotheus* sp. (Pl. I, Fig. 3.5), *Boudiera echinulata* (Seaver) Seaver (Pl. I, Fig. 3.4), and *Iodophanus carneus* (Persoon ex Persoon) Korf apud Kimbrough and Korf (Pl. I, Fig. 3.6.).
2. *Peziza* type. The operculum is sharply delimited by a circular internal indentation just at the inner side of a weakly developed ring. The operculum and a region under the ascostome are strengthened and rigid. Wall structure is rather complex. The ascostome is smooth. The periascus is strongly developed, with a ring-shaped thickening over the ascostome, tapering toward the base; it stains blue with iodine. A funiculus is present. Examples: *Peziza badia* Persoon ex Mérat, *Peziza succosella* (Le Gal and Romagnesi) Moser (Pl. II, Fig. 3.7), *Peziza depressa* Persoon ex Persoon, *Peziza ammophila* Durieu and Léveillé apud Durieu (Pl. II, Fig. 3.8), and *Peziza cerea* Sowerby ex Mérat.
3. *Ascodesmis* type. The operculum is very large and sharply delimited by a weak indentation and a circular zone of two-sided wall disintegration. The ring is only weakly developed. The operculum is strengthened but not rigid. The ascostome is smooth. The periascus is very thin and is not stained blue with iodine. Funiculus and funnel are absent. Examples: *Ascodesmis nigricans* (Pl. II, Fig. 3.9) and *Ascodesmis microscopica* (Crouan) Seaver (Pl. II, Fig. 3.10).
4. *Octospora* type. The operculum is rather roughly delimited, without indentation or a prominent ring, and sometimes strengthened at its inner side by an electron-transparent layer. The apical wall shows a thimble-shaped, electron-dense lamina interrupted by a thick ring-shaped electron-transparent zone. Dehiscence takes place in a weakened zone between the electron-transparent ring and the strengthened operculum. If the operculum is not strengthened, it may be torn irregularly. Cleavage of wall and operculum along the thimble-shaped lamina above the electron-transparent ring is rather frequent after dehiscence. The ascostome is rough. The periascus is thin if present and is not

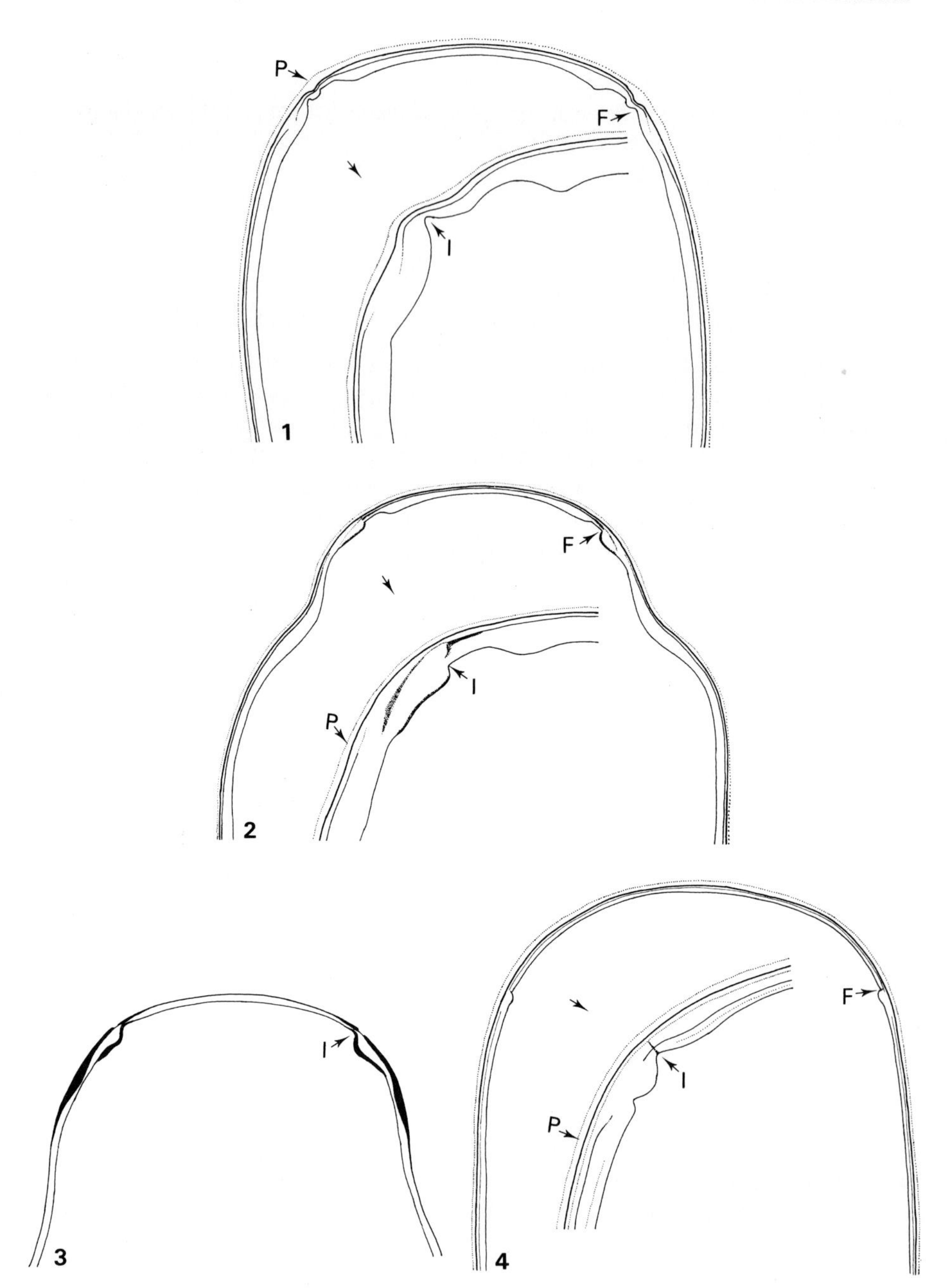
P
F
I
1
F
I
P
2
I
3
F
I
P
4

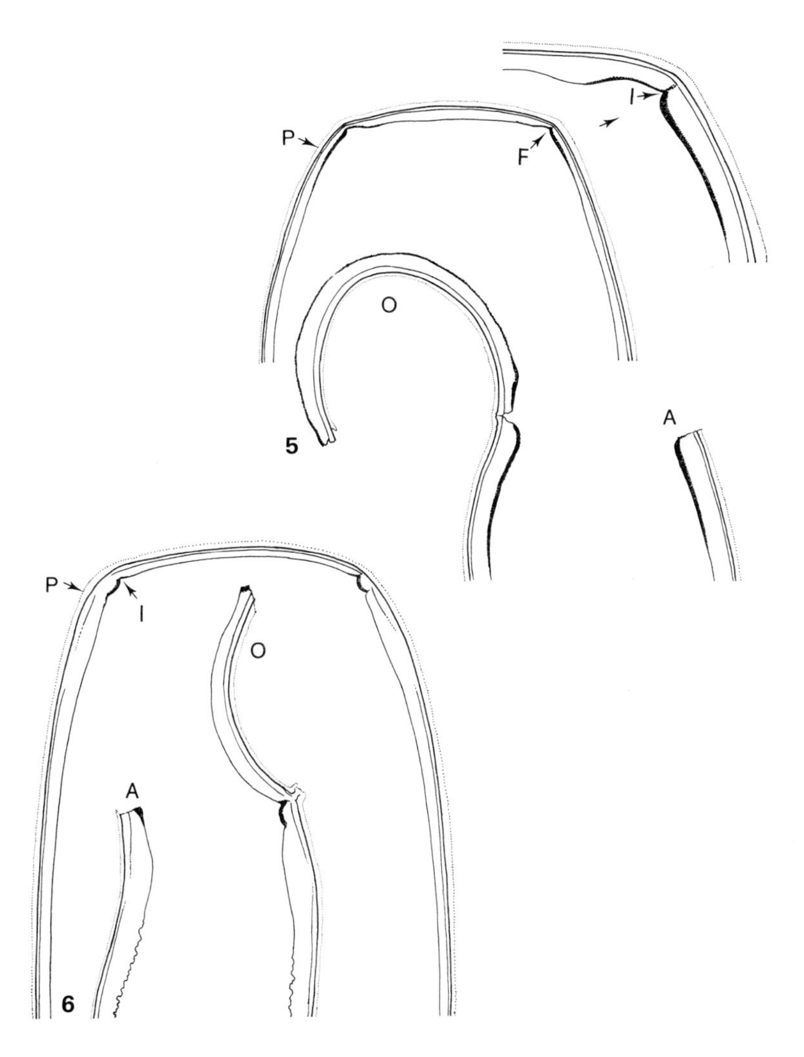

Plate I. Diagrammatic sections of ascus tops as seen with electron microscopy. I, indentation of ascus wall; F, line or zone of fracturing; P, periascus (extraascan layer); O, operculum; A, ascostome.

Fig. 3.1. *Ascobolus furfuraceus.*

Fig. 3.2. *Saccobolus glaber.*

Fig. 3.3. The same, as seen with light microscopy in sections stained with toluidine blue.

Fig. 3.4. *Boudiera echinulata.*

Fig. 3.5. *Thecotheus sp.*

Fig. 3.6. *Iodophanus carneus.*

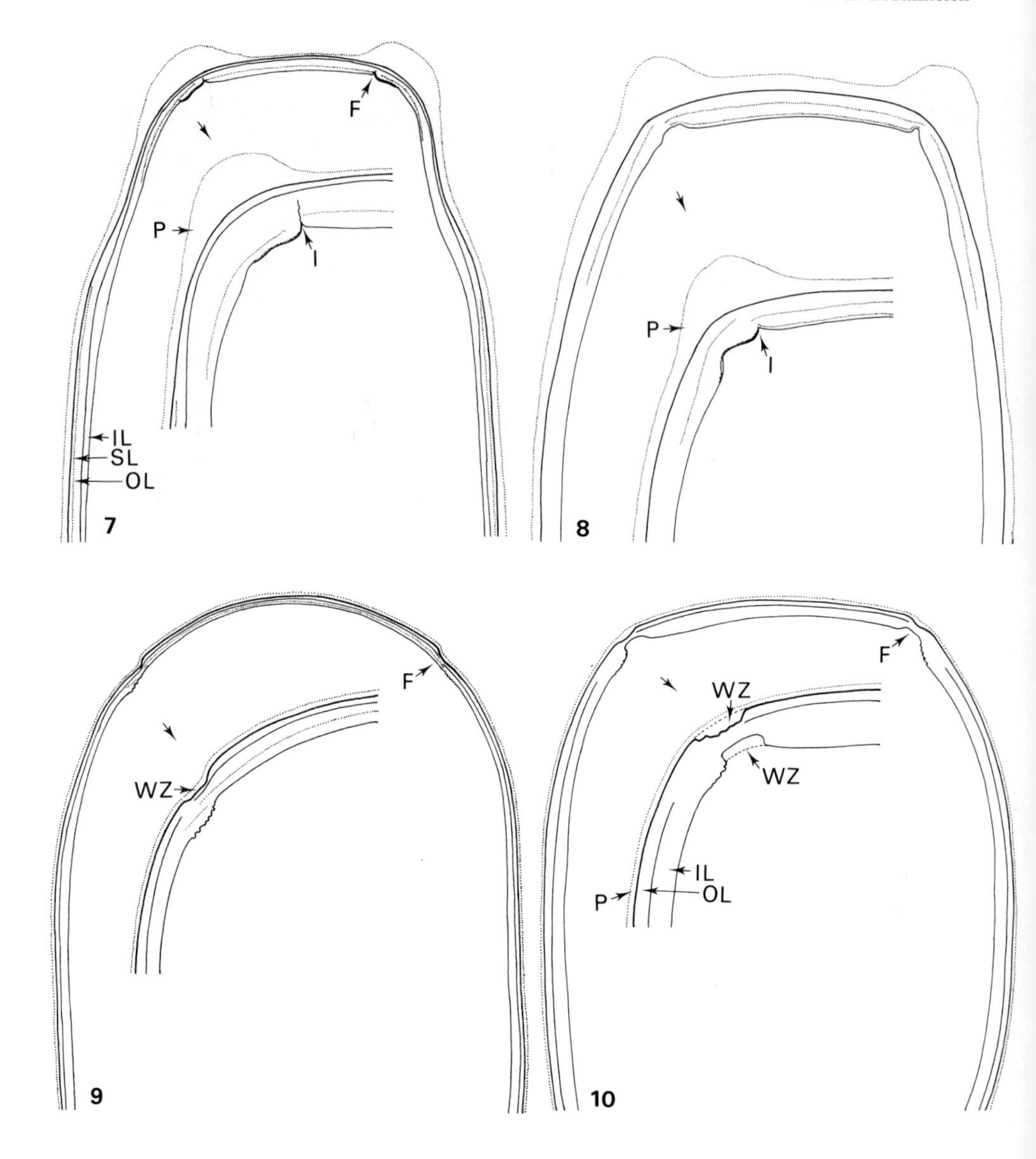

Plate II. Diagrammatic sections of ascus tops as seen with electron microscopy. IL, inner layer of ascus wall; OL, outer layer of ascus wall; SL, sublayering of ascus wall; WZ, weakened zone. Other abbreviations as in Plate I.

Fig. 3.7. *Peziza ammophila.*

Fig. 3.8. *Peziza succosella.*

Fig. 3.9. *Ascodesmis nigricans.*

Fig. 3.10. *Ascodesmis microscopica.*

stained blue with iodine. A funiculus is present. Examples: *Pyronema omphalodes* (Bulliard ex St Amans) Fuckel (Pl. III, Fig. 3.11), *Anthracobia maurilabra* (Cooke) Boudier (Pl. III, Fig. 3.12), *Aleuria aurantia* (Persoon ex Hooker) Fuckel (Pl. III, Figs. 3.13 and 3.14), *Otidea onotica* (Persoon ex S. F. Gray) Fuckel, *Coprobia granulata* (Bulliard ex Mérat) Boudier (Pl. III, Figs. 3.15 and 3.16), *Cheilymenia pulcherrima* (Crouan) Boudier, *Cheilymenia vitellina* (Persoon) Dennis (Pl. IV, Fig. 3.17), *Scutellinia armatospora* Denison (Pl. IV, Fig. 3.18), *Octospora musci-muralis* Graddon (Pl. IV, Fig. 3.19), and *Sowerbyella unicolor* (Gillet) Nannfeldt (Pl. IV, Fig. 3.20).

5. *Helvella* type. The operculum is sharply delimited at its basal side by a ring-shaped breaking line just at the inner side of a narrow, scarcely proliferating ring. The apical wall shows a thimble-shaped, electron-dense lamina which is only interrupted by the proliferating ring. The ascostome is smooth. The periascus is very thin and does not stain blue with iodine. A funiculus is present. Examples: *Helvella crispa* Scopoli ex Fries (Pl. V, Fig. 3.21) and *Geopyxis carbonaria* (Albertini and von Schweinitz ex Persoon) Saccardo (Pl. V, Fig. 3.22).
6. *Urnula* type. The operculum is rather roughly delimited by an external ring-shaped zone of wall disintegration, without indentation or a proliferating ring, and it is not strengthened. In the inner wall layer of the top, a thimble-shaped, electron-dense lamina is formed which is intercepted at the apex by a broad, ring-shaped, electron-transparent zone. Near the upper margin of this zone, the operculum is torn loose. The ascostome is rather rough. The periascus is very thin and does not stain blue with iodine. A funiculus is not yet recorded. Example: *Urnula platensis* Spegazzini (Pl. VI, Figs. 3.23 and 3.24).
7. *Sarcoscypha* type. The operculum is very thick, rather narrow and centrally or obliquely placed; it is sharply delimited in the inner layer and more roughly in the outer layer. The inner layer is strongly swollen, is often stratified and laminated, and forms a thick lenticular body (opercular plug) at the top. The ascostome is smooth. The periascus is clearly developed and does not stain blue with iodine. A funiculus is present. Examples: *Sarcoscypha coccinea* (Scopoli ex S. F. Gray) Lambotte (Pl. VI, Figs. 3.25 and 3.26) (van Brummelen, 1975), *Pithya cupressina* (Batsch ex Persoon) Fuckel (Pl. VI, Fig. 3.27), *Pseudoplectania nigrella* (Persoon ex Persoon) Fuckel (Pl. VI, Fig. 3.28), *Desmazierella acicula* Libert, *Cookeina sulcipes* (Berkeley) O. Kuntze, *Phillipsia domingensis* (Berkeley) Berkeley, and *Wynnea americana* Thaxter have an obliquely placed operculum and belong here.
8. *Thelebolus* type. The operculum is absent in multispored asci or rather roughly delimited in some eight-spored asci. It usually opens by an irregular tear or with a bilabiate split starting from the margin of a small, rigid apical disk. Sometimes external forces produce a more or less circular tear (*Leptokalpion*). A usually pronounced ring in the wall prevents the tear from passing this level. The ascostome is rough (if present). The wall is thick and often strongly stratified or laminated. The periascus and a funiculus are absent. The *Thelebolus* type of ascus covers a wide range of possibilities, many of which have been studied in detail by Kimbrough (1966, 1972) and Kimbrough and Korf (1967).

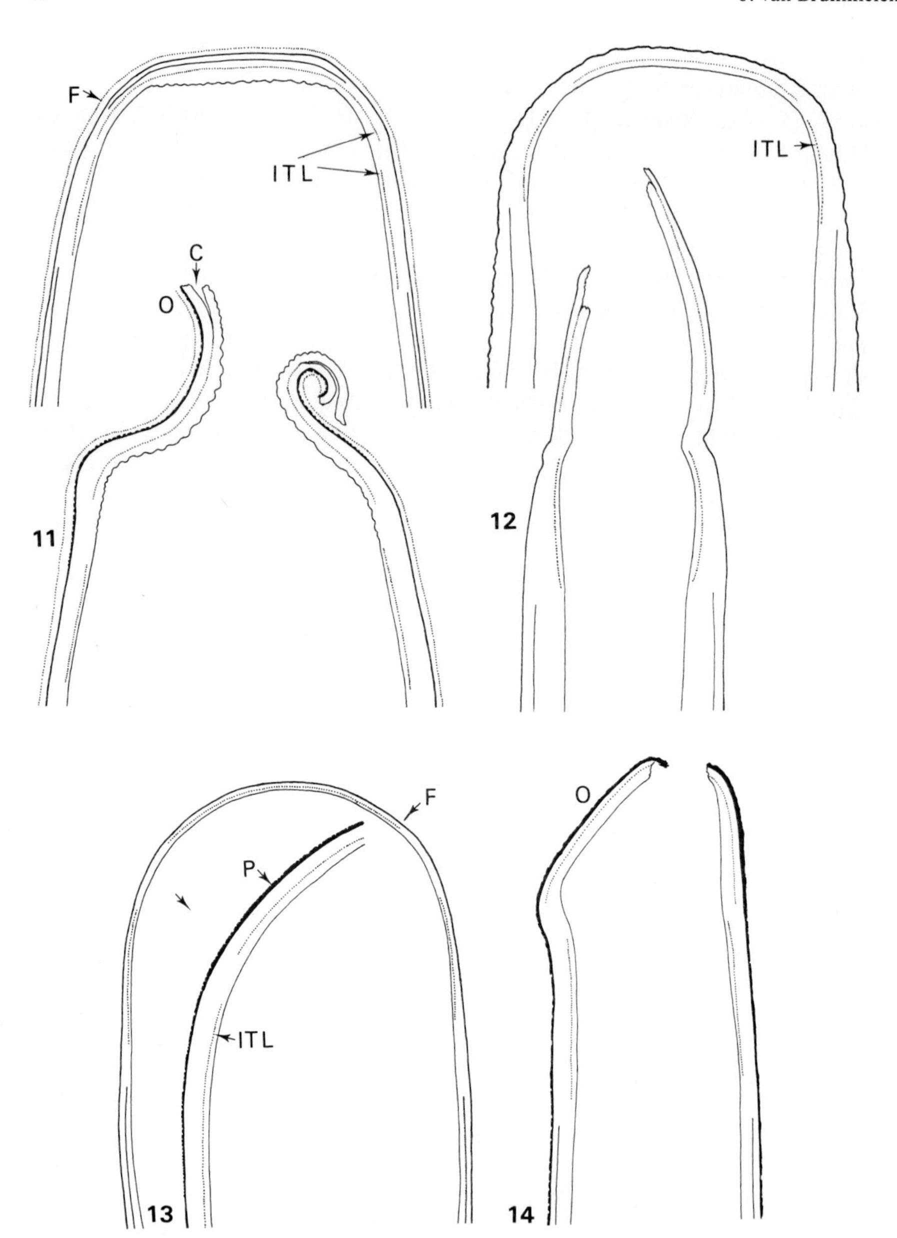
F
ITL
C
O
11
ITL
12
F
P
ITL
13
O
14

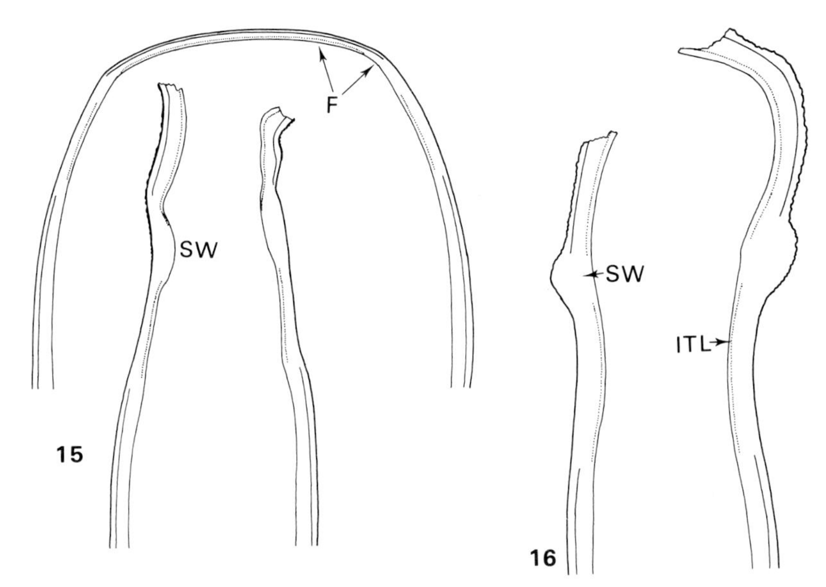

Plate III. Diagrammatic sections of ascus tops, as seen with electron microscopy. C, cleavage of ascus wall; ITL, interrupted thimble-shaped wall; SW, strongly swollen wall region. Other abbreviations as in Plates I and II.
Fig. 3.11. *Pyronema omphalodes.*
Fig. 3.12. *Anthracobia maurilabra.*
Figs. 3.13 and 3.14. *Aleuria aurantia.*
Figs. 3.15 and 3.16. *Coprobia granulata.*

Examples: *Thelebolus stercoreus* (Pl. VII, Figs. 3.29 and 3.32), *"Rhyparobius" myriosporus* (Crouan) Boudier (Pl. VIII, Figs. 3.33–3.38), *"Rhyparobius" crustaceus* (Fuckel) Rehm (Pl. IX, Fig. 3.39), *"Rhyparobius" caninus* (Auerswald) Saccardo (Pl. IX, Fig. 3.40), *"Ascophanus" coemansii* Boudier (Pl. X, Fig. 3.41), *Coprotus* sp., *Lasiobolus monascus* (Pl. X, Fig. 3.42), *Lasiobolus pilosus* (Fries) Saccardo (Pl. X, Fig. 3.43), and *Ascozonus woolhopensis* (Renny) Boudier (Pl. X, Fig. 3.44).

As our knowledge about ascus structure in the genera of the Pezizales is still very incomplete, it is quite possible and even probable that further types of asci will turn up in the future. A more detailed study of the Thelebolaceae will almost certainly lead to a further subdivision of the *Thelebolus* type.

A great deal of information on the structure of the ascus top is still required from many important genera, especially from those with a tropical distribution. Most of the Tuberales should be classified among the Pezizales, but as these fungi have lost their mechanism of ascus dehiscence it is not possible to fit them in on this character.

If we consider the information available and compare the different structures, it is possible to recognize affinities and trends. This has led me first, to distinguish the eight types described above and second, to visualize their possible interrelationships in a scheme (Fig. 3.45).

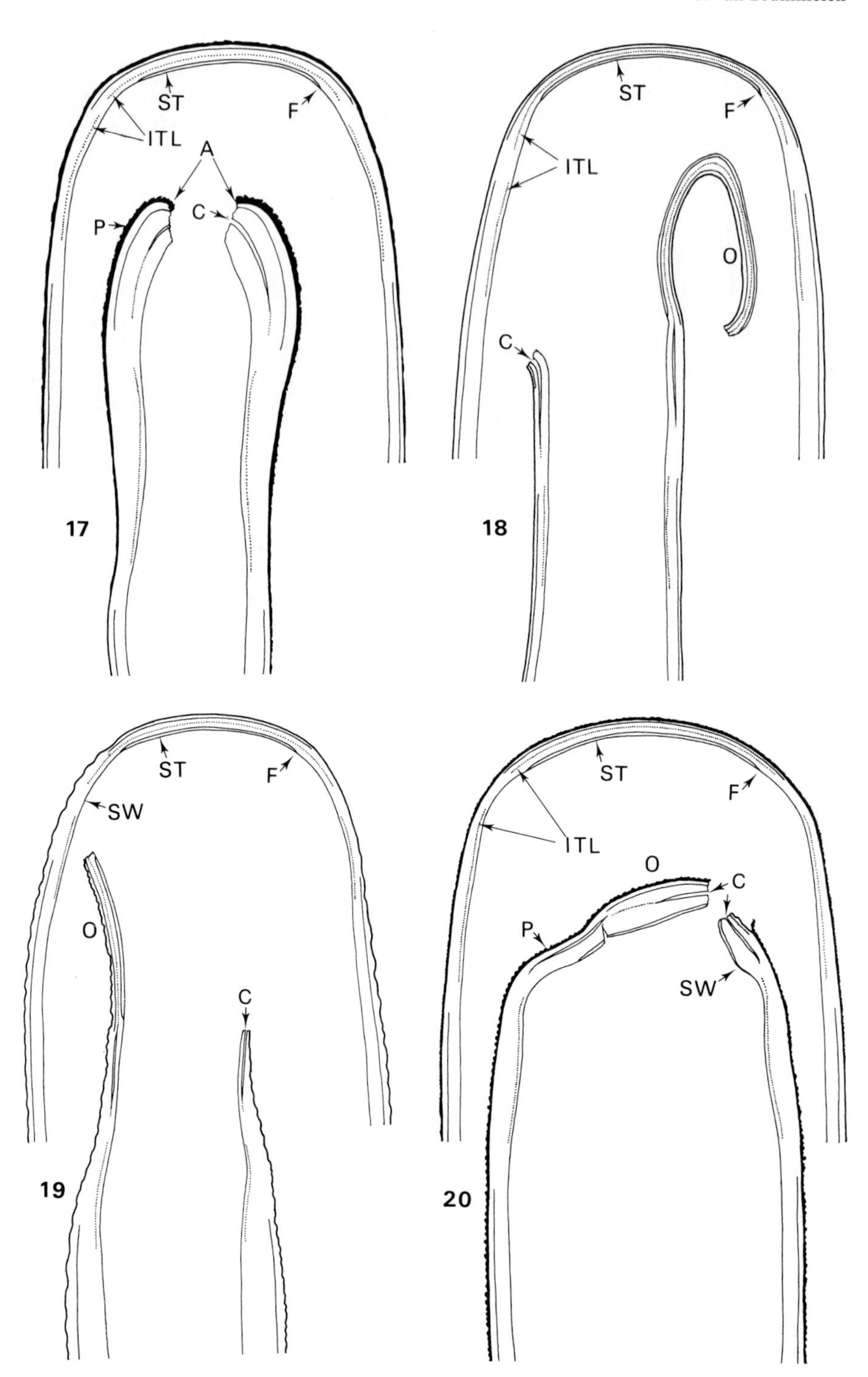
ST
F
ITL
A
C
P
17
ST
F
ITL
O
C
18
ST
F
SW
O
C
19
ST
F
ITL
O
C
P
SW
20

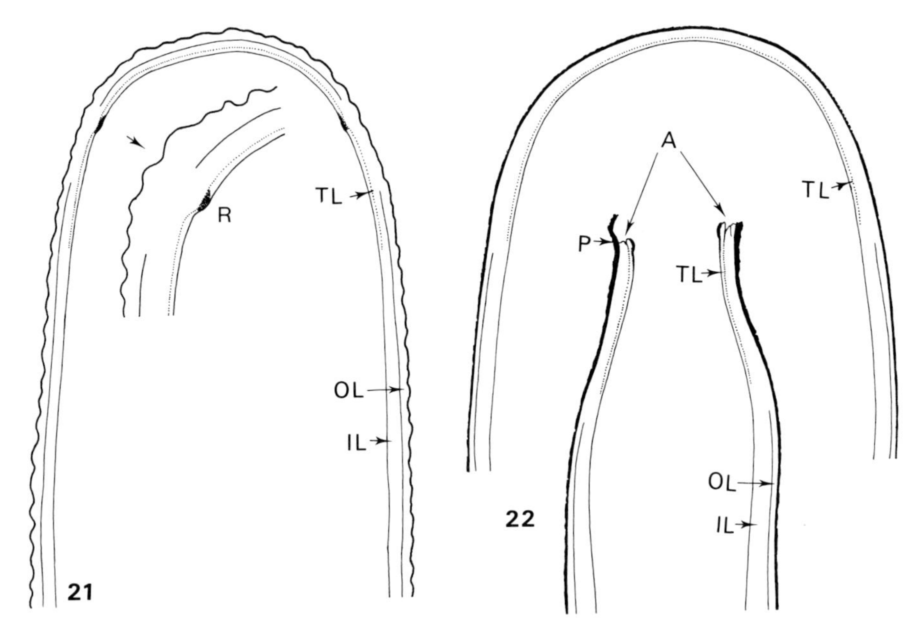

Plate V. Diagrammatic sections of ascus tops as seen with electron microscopy. R, prominent ring; TL, thimble-shaped lamina. Other abbreviations as in Plates I–IV.
Fig. 3.21. *Helvella crispa.*
Fig. 3.22. *Geopyxis carbonaria.*

←

Plate IV. Diagrammatic sections of ascus tops, as seen with electron microscopy. ST, strengthened layer. Other abbreviations as in Plates I–III.
Fig. 3.17. *Cheilymenia vitellina.*
Fig. 3.18. *Scutellinia armatospora.*
Fig. 3.19. *Octospora musci-muralis.*
Fig. 3.20. *Sowerbyella unicolor.*

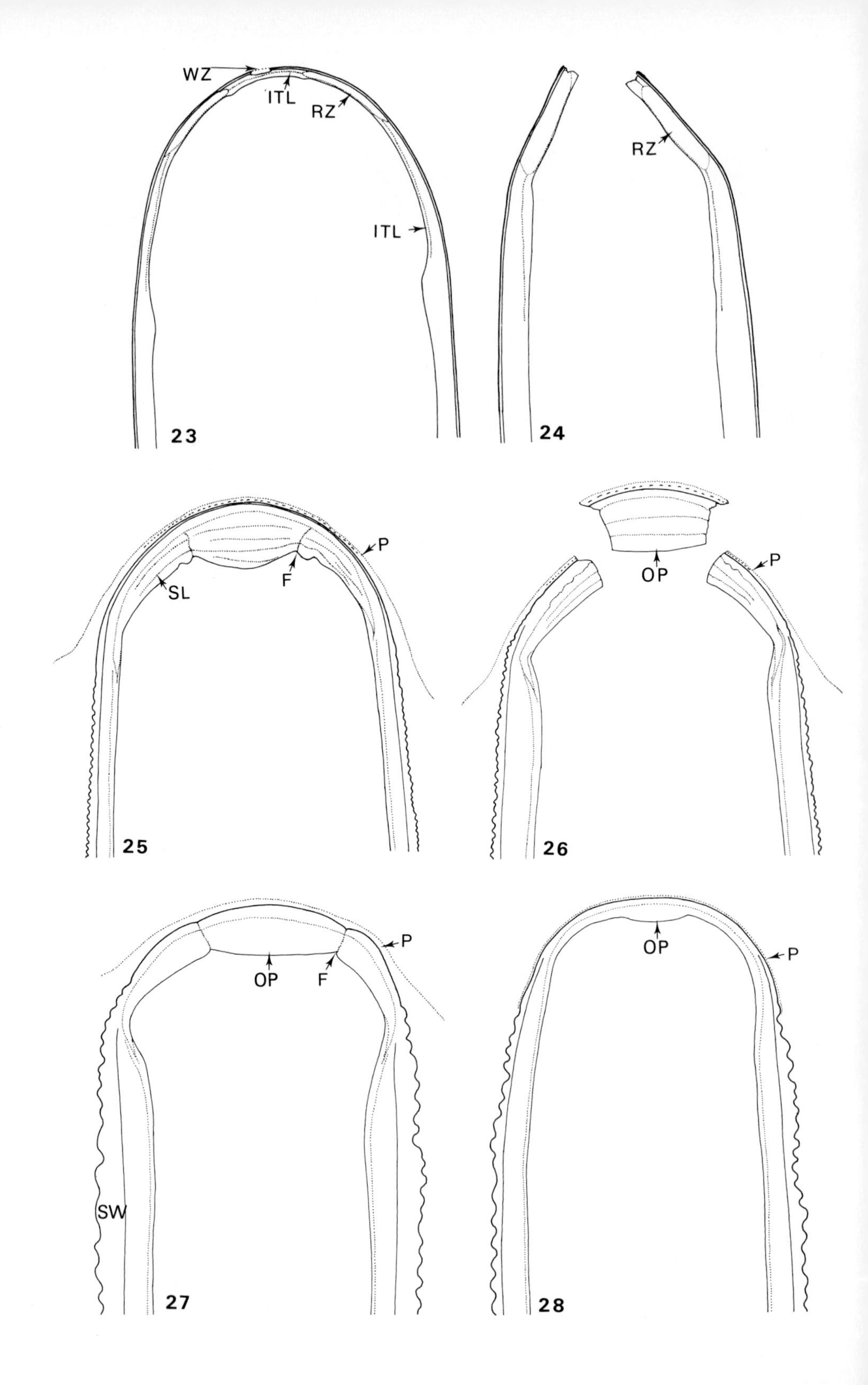
WZ
ITL
RZ
ITL
23
RZ
24
P
SL
F
25
OP
P
26
P
OP
F
SW
27
OP
P
28

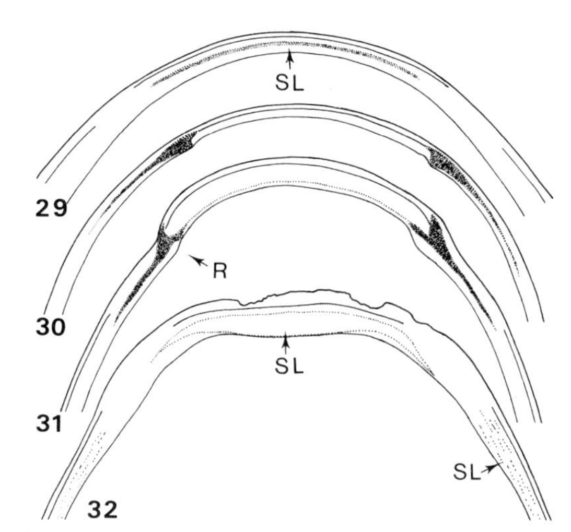

Plate VII. Diagrammatic sections of asci and ascus tops (as seen with electron microscopy, if not otherwise stated). *Thelebolus stercoreus* (with over 1000 spores). Abbreviations as in Plates I–VI.

Fig. 3.29. Very young stage, before ascosporogenesis.

Fig. 3.30. Ripening ascus as seen with light microscopy in sections stained with toluidine blue.

Fig. 3.31. Ripening ascus.

Fig. 3.32. Mature ascus.

←

Plate VI. Diagrammatic sections of ascus tops as seen with electron microscopy. RZ, ring-shaped, electron-transparent zone of ascus wall; OP, opercular plug. Other abbreviations as in Plates I–V.

Figs. 3.23 and 3.24. *Urnula platensis.*

Figs. 3.25 and 3.26. *Sarcoscypha coccinea.*

Fig. 3.27. *Pithya cupressina.*

Fig. 3.28. *Pseudoplectania nigrella.*

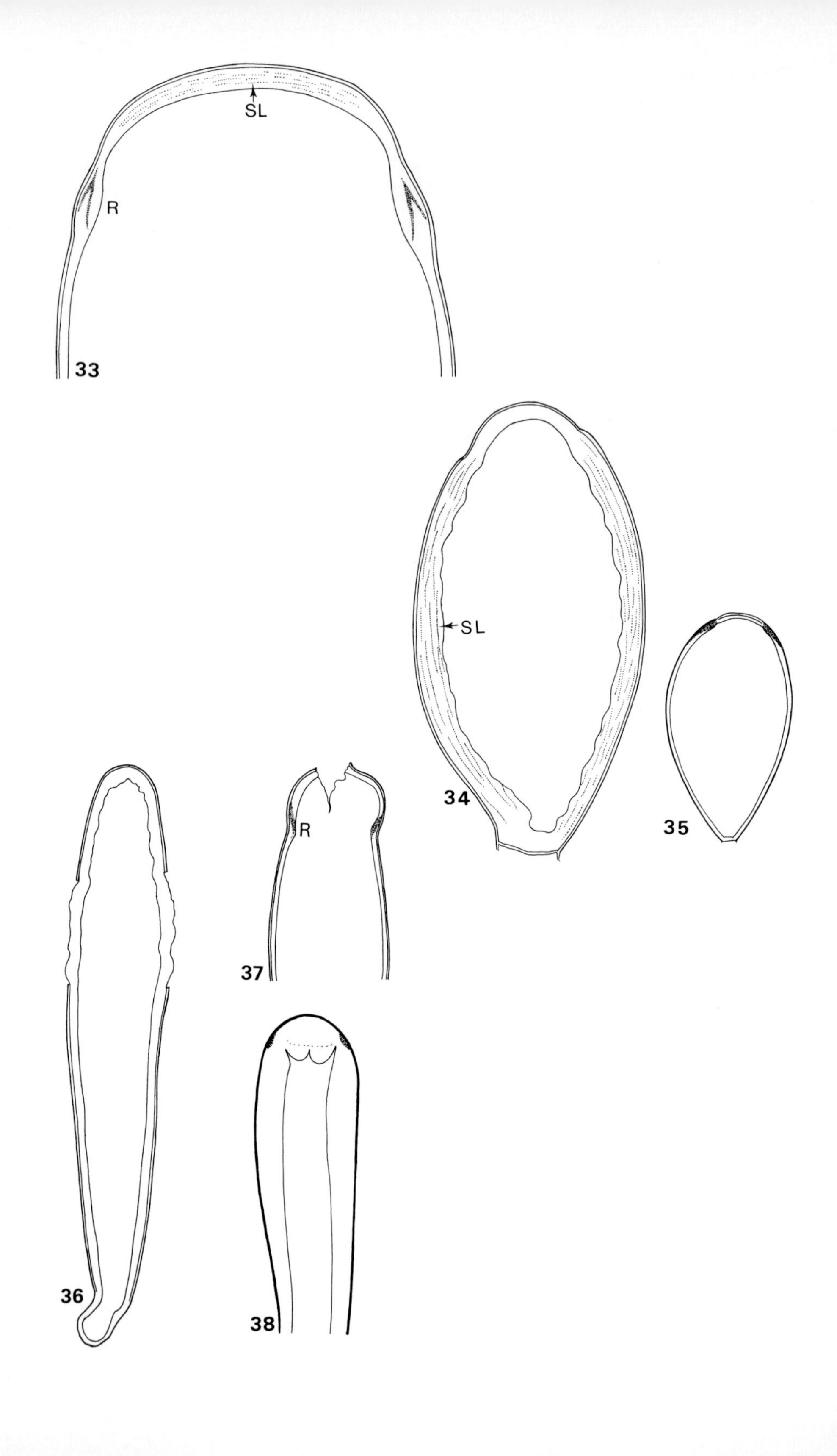
SL
R
33
SL
34
35
R
37
36
38

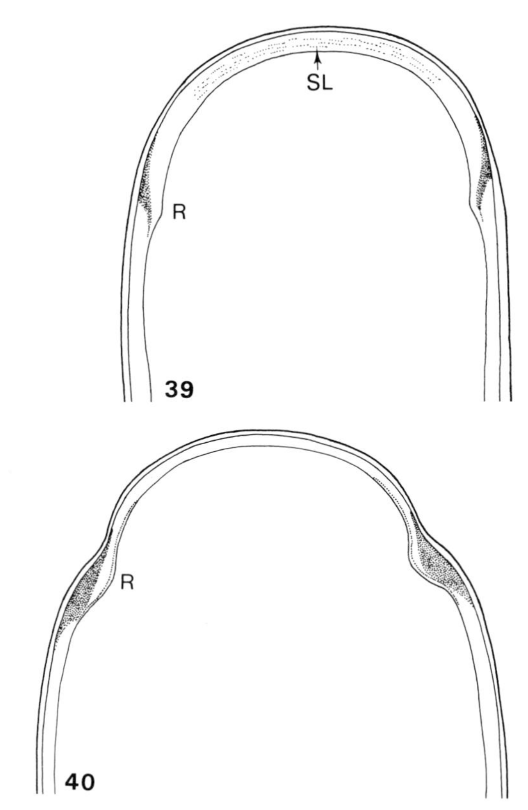

Plate IX. Diagrammatic sections of asci and ascus tops (electron microscopy). Abbreviations as in Plates I–VI.
Fig. 3.39. *"Rhyparobius" crustaceus* (with about 64 spores).
Fig. 3.40. *"Rhyparobius" caninus* (with about 32 spores).

←

Plate VIII. Diagrammatic sections of asci and ascus tops (as seen with electron microscopy unless otherwise noted). *"Rhyparobius" myriosporus* (128–512 spores).
Fig. 3.33. Ripening ascus (with about 256 spores).
Figs. 3.34 and 3.35. Asci (with about 512 spores) as seen with light microscopy, stained with Congo red.
Figs 3.36–3.38. Ripening and mature asci (with about 256 spores), vital observation.

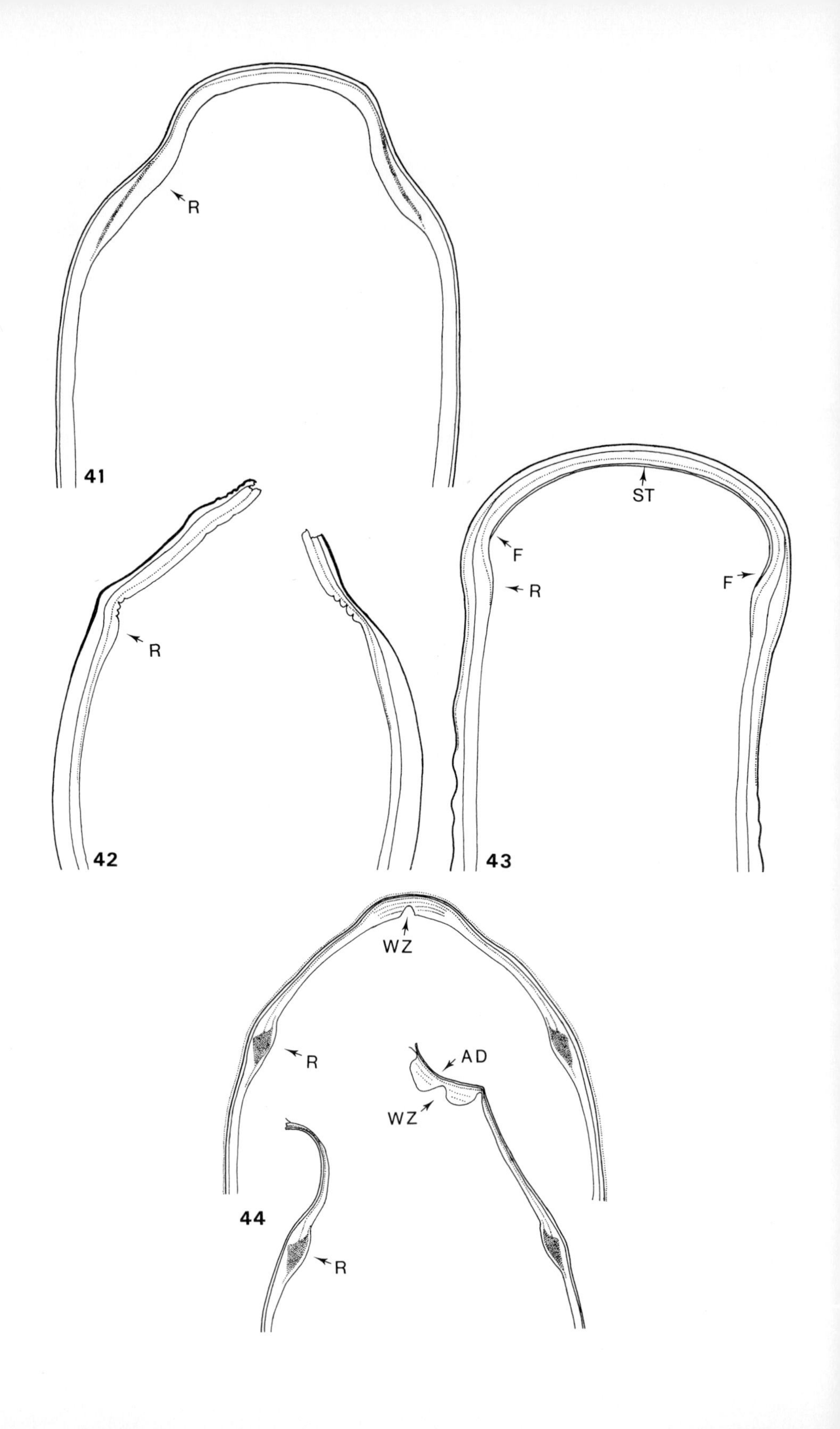
R
41
R
42
ST
F
R
F
43
WZ
R
AD
WZ
44
R

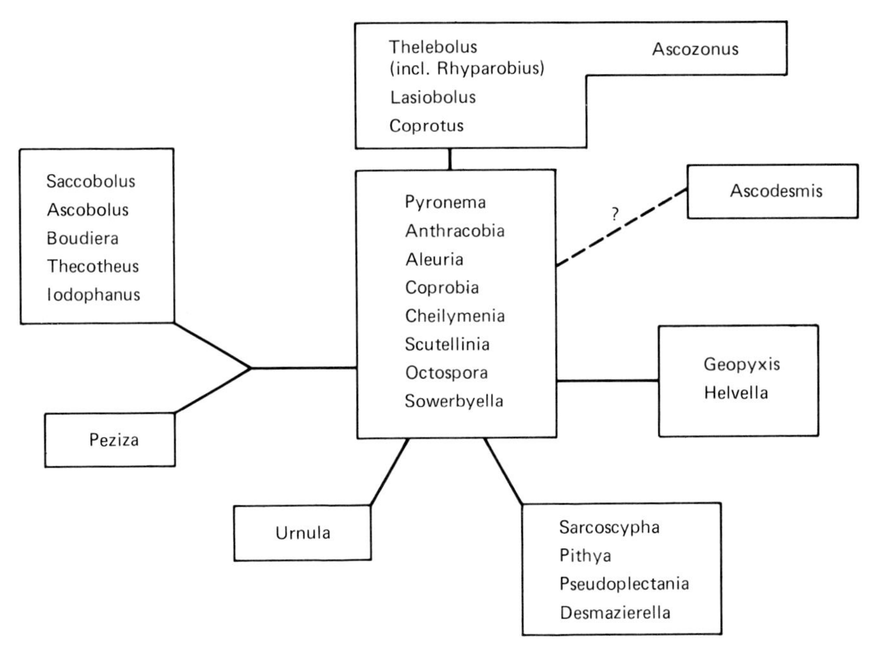

Fig. 3.45. Tentative scheme of possible interrelationships between different types of opening mechanisms of asci in Pezizales.

←

Plate X. Diagrammatic sections of ascus tops as seen with electron microscopy. AD, apical disc. Other abbreviations as in Plates I–VI.

Fig. 3.41. *"Ascophanus" coemansii.*

Fig. 3.42. *Lasiobolus monascus.*

Fig. 3.43. *Lasiobolus pilosus.*

Fig. 3.44. *Ascozonus woolhopensis.*

General Conclusions

The following trends can be recognized:

1. All asci with "amyloid walls" studied so far form a compact and closely related group.
2. The *Ascodesmis* type takes a rather isolated position.
3. There is a rather complete scale of structures ranging from *Thelebolus* via *Lasiobolus* and *Coprotus* via *Pyronema–Aleuria–Coprobia–Cheilymenia–Octospora–Scutellinia–Sowerbyella* to members of the Sarcoscyphaceae.
4. No direct relationship with the asci of Inoperculatae has been detected.

References

Boudier, J. L. E., 1869. Mémoire sur les Ascobolacés. Ann. Sci. Natur. (Bot.) (Séries 5) 10: 191–268.

Boudier, J. L. E., 1879. On the importance that should be attached to the dehiscence of asci in the classification of the discomycetes. Grevillea 8: 45–49.

Boudier, J. L. E., 1885. Nouvelle classification naturelle des discomycètes charnus connus généralement sous le nom de *Pézizes*. Bull. Soc. Mycol. France 1: 91–120.

Bracker, C. E., and C. M. Williams, 1966. Comparative ultrastructure of developing sporangia and asci in fungi. Electron Microscopy, Proc. 6th Internl. Congr. Electron Microscopy, Kyoto. Vol. II, pp. 307–308.

Brummelen, J. van, 1974. Light and electron microscopic studies of the ascus top in *Ascozonus woolhopensis*. Persoonia 8: 23–32.

Brummelen, J. van, 1975. Light and electron microscopic studies of the ascus top in *Sarcoscypha coccinea*. Persoonia 8: 259–271.

Carroll, G. C., 1966. A study of the fine structure of ascosporogenesis in *Saccobolus kerverni* and *Ascodesmis sphaerospora*. Ph.D. Thesis, Univ. of Texas, Austin.

Carroll, G. C., 1967. The ultrastructure of ascospore delimitation in *Saccobolus kerverni*. J. Cell Biol. 33: 218–224.

Carroll, G. C., 1969. A study of the fine structure of ascosporogenesis in *Saccobolus kerverni*. Arch. Mikrobiol. 66: 321–339.

Chadefaud, M., 1946. Les asques para-operculés et la position systématique de la Pézize *Sarcoscypha coccinea* Fries ex Jacquin. C. R. Hebd. Séanc. Acad. Sci. Paris 222: 753–755.

Crouan, P. L., and H. M. Crouan, 1857. Note sur quelques *Ascobolus* nouveaux et sur une espèce nouvelle de *Vibrissea*. Ann. Sci. Natur. (Bot.) (Séries 4) 7: 173–178.

Crouan, P. L., and H. M. Crouan, 1858. Note sur neuf *Ascobolus* nouveaux. Ann. Sci. Natur. (Bot.) (Séries 4) 10: 193–199.

Delay, C., 1966. Étude de l'infrastructure de l'asque d'*Ascobolus immersus* Pers. pendant la maturation des spores. Ann. Sci. Natur. (Bot.) (Séries 12) 7: 361–420.

Eckblad, F. E., 1968. The genera of the operculate discomycetes. A re-evaluation of their taxonomy, phylogeny and nomenclature. Nytt Mag. Bot. 15: 1–191.

Eckblad, F. E., 1972. The suboperculate ascus—a review. Persoonia 6: 439–443.

Gäumann, E. A., 1926. Vergleichende Morphologie der Pilze. G. Fischer, Jena, 626 pp.

Griffith, H. B., 1968. The structure of the pyrenomycete ascus. Ph.D. Thesis. Univ. of Liverpool.

Kimbrough, J. W., 1966. Studies in the Pseudoascoboleae. Can. J. Bot. 44: 685–704.

Kimbrough, J. W., 1972. Ascal structure, ascocarp ontogeny, and a natural classification of the Thelebolaceae. Persoonia 6: 395–404.

Kimbrough, J. W., and R. P. Korf, 1967. A synopsis of the genera and species of the tribe Theleboleae (= Pseudoascoboleae). Am. J. Bot. 54: 9–23.

Le Gal, M., 1946a. Mode de déhiscence des asques chez les *Cookeina* et les *Leotia,* et ses conséquences du point de vue phylogénétique. C. R. Hebd. Séanc. Acad. Sci. Paris 222: 755–757.

Le Gal, M., 1946b. Les discomycètes suboperculés. Bull. Trim. Soc. Mycol. France 62: 218–240.

Le Gal, M., 1947. Recherches sur les ornementations sporales des discomycètes operculés. Ann. Sci. Natur. (Bot.) (Séries 11) 8: 73–297.

Le Gal, M., 1953. Les discomycètes de Madagascar. Prodr. Flore Mycol. Madagascar 4: 1–465.

Merkus, E., 1973. Ultrastructure of the ascospore wall in Pezizales (Ascomycetes)—I. *Ascodesmis microscopica* (Crouan) Seaver and *A. nigricans* van Tiegh. Persoonia 7: 351–366.

Merkus, E., 1974. Ultrastructure of the ascospore wall in Pezizales (Ascomycetes)—II. Pyronemataceae sensu Eckblad. Persoonia 8: 1–22.

Merkus, E., 1975. Ultrastructure of the ascospore wall in Pezizales (Ascomycetes)—III. Otideaceae and Pezizaceae. Persoonia 8: 227–247.

Merkus, E., 1976. Ultrastructure of the ascospore wall in Pezizales (Ascomycetes)—IV. Morchellaceae, Helvellaceae, Rhizinaceae, Thelebolaceae, and Sarcoscyphaceae; general discussion. Persoonia 9: 1–38.

Moore, R. T., 1963. Fine structure of Mycota. I. Electron microscopy of the discomycete *Ascodesmis.* Nova Hedwigia 5: 263–278.

Nannfeldt, J. A., 1932. Studien über die Morphologie und Systematik der nicht-lichenisierten inoperculaten Discomyceten. Nova Acta Roy. Soc. Sci. Upsal. (Series 4) 8(2): 1–368.

Osio, B. A., 1969. Electron microscopy of ascus development in *Ascobolus.* Ann. Bot. 33: 205–209.

Ramlow, G., 1906. Entwicklungsgeschichte von *Thelebolus stercoreus.* Bot. Ztg. 64: 85–99.

Ramlow, G., 1915. Beiträge zur Entwicklungsgeschichte der Ascoboleen. Mycol. Cbl. 5: 177–198.

Reeves, F., 1967. The fine structure of ascospore formation in *Pyronema domesticum.* Mycologia 59: 1018–1033.

Reisinger, O., E. Kiffer, F. Mangenot, and G. M. Olah, 1977. Ultrastructure, cytochimie et microdissection de la paroi des hyphes et des propagules exogènes des ascomycètes et basidiomycètes. Rev. Mycol. 41: 91–117.

Renny, J., 1871. A description of some species of the genus *Ascobolus* new to England. Trans. Woolhope Natur. Field Club 1871: 45–48.

Renny, J., 1873. New species of the genus *Ascobolus.* Trans. Woolhope Natur. Field Club 1872–3: 127–131.

Samuelson, D. A., 1975. The apical apparatus of the suboperculate ascus. Can. J. Bot. 2660–2679.

Schrantz, J. P., 1966. Contribution à l'étude de la formation de la paroi sporale chez *Pustularia cupularis* (L.) Fuckel. C. R. Acad. Sci. Paris 262: 1212–1215.

Schrantz, J. P., 1967. Présence d'un aster au cours des mitoses de l'asque et de la formation des ascospores chez l'Ascomycète *Pustularia cupularis* (Linneus) Fuckel. C. R. Acad. Sci. Paris 264: 1274–1277.

Schrantz, J. P., 1968. Ultrastructure et localisation du glycogène chez l'Ascomycète *Galactinia plebeia* Le Gal. Rev. Cytol. Biol. Vég. 31: 151–157.

Seaver, F. J., 1927. A tentative scheme for the treatment of the genera of Pezizaceae. Mycologia 19: 86–89.

Seaver, F. J., 1928. The North American Cup-fungi (Operculates). Seaver, New York.

Wells, K., 1972. Light and electron microscopic studies of *Ascobolus stercorarius*. II. Ascus and ascospore ontogeny. Univ. Calif. Publ. Bot. 62: 1–93.

Zukal, H. 1886. Mykologische Untersuchungen. Denkschr. (K.) Akad. Wiss. Wien Math. Naturwiss. Kl. 51: 21–36.

Chapter 4
The Bitunicate Ascus

E. Müller

Introduction

The bitunicate ascus as defined by Luttrell (1951) is surrounded by two walls that differ in their ability to extend. The outer, inextensible wall (ectotunica) ruptures when the pressure within the mature ascus increases, whereas the extensible inner wall (endotunica) elongates. The ascospores are pressed out by a vacuole that regulates the inner pressure of the ascus during spore ejaculation.

The term "bitunicate" was introduced and defined by Luttrell (1951) for a type of ascus clearly separated from the type of ascus defined by the term "unitunicate," proposed simultaneously. Luttrell's definition was based on his own experiences; on older reports, including those by Currey (1856) on *Pleospora herbarum* (Fries) Rabenhorst, by Pringsheim (1858) on *Pyrenophora scripi* (Rabenhorst) Wehmeyer, by Hodgetts (1917) on *Leptosphaeria acuta* (Mougeot and Nestler) Karsten, by Hoggan (1927) on *Plowrightia ribesia* (Fries) Saccardo, and by Chadefaud (1942) on several different species; and on a general description of this ascus type given by de Bary as early as 1884:

> The bitunicate ascus is surrounded by two distinct walls: a thin, inextensible outer wall, and a thick, extensible inner wall. Asci containing mature ascospores actively ejaculate them. The outer wall ruptures at the apex and the inner wall expands to form a long sac. The ascospores are pressed successively through an elastic pore in the apex of the expanded inner wall. [Luttrell, 1951]

Luttrell's definition therefore considers the ascus morphology as seen under the light microscope as well as ascus activity during ascospore liberation. Now, more than 25 years later, this definition needs examination in relation to contemporary information. Based on a number of recent contributions to this subject, including those of Bellemére (1971, 1975), Chadefaud (1969a, b, 1973), Funk and Shoemaker (1967), Furtado and Olive (1971), Reynolds (1971), and Schrantz (1970), a more up to date description of the development and the morphology and function of the bitunicate ascus may be undertaken.

Ascus Form

An examination of bitunicate asci with the help of the electron microscope reveals a complex construction of the two walls seen with the light microscope. The ectotunica as well as the endotunica are composed of more than one wall layer, and therefore the bitunicate ascus may be considered to be multilayered. However, the unique behavior of the two walls is confirmed and explained by their characteristic arrangement of wall elements. Therefore, the term "bitunicate" does not seem to be inaccurate.

Similar to the development of asci in the majority of ascomycete species, most bitunicate asci are formed by croziers; the developmental stages do not differ principally from most other ascus types. Before the original nuclear pair fuses, a single primary wall delimits the entire young ascus. Examination with the electron microscope shows that this wall is formed by an amorphous matrix in which microfibrils are distributed. The microfibrils are arranged parallel to each other, parallel to the long axis of the ascus along the sides, and parallel to the protoplast surface at the ascus ends (Reynolds, 1971). The primary wall has a limited expansion capacity during ascus growth. It seems that in most cases the wall is composed of two distinct wall layers, differing in their electron density.

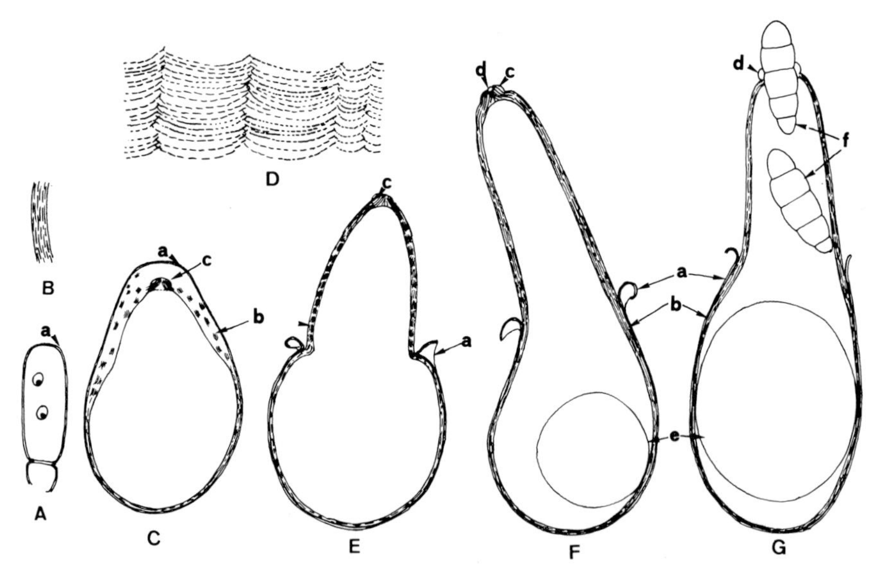

Fig. 4.1. Development of a bitunicate ascus. **A** Young ascus before nuclear fusion, surrounded by a primary wall. **B** Ectotunica with parallel microfibrils within an amorphous matrix. **C** Young ascus before ascospore delimitation. The ectotunica encloses the endotunica, which is comprised of bundled microfibrils that converge in the corona. **D** Bundled microfibrils of the endotunica with typical arrangement. **E** Enlarged mature ascus after the rupture of the ectotunica and with the endotunica extended. Note an apical pore in the corona. **F** Extended ascus ready to ejaculate the ascospores, with a basal vacuole. **G** Ascus during spore ejaculation: a, ectotunica; b, endotunica; c, corona; d, apical pore; e, vacuole. (Modified after Reynolds, 1971.)

The formation of the secondary wall (endotunica) begins before the ascospores are delimited. It develops inside the primary wall (ectotunica) and remains weakly developed at the ascus base; at the lateral and particularly at the upper portion of the ascus, it is well manifested. As in the ectotunica, the secondary wall is composed of two or more layers, as may be seen in electron micrographs. The two layers contain microfibrils, as does the ectotunica. Within the innermost layer of the endotunica these microfibrils are bound together to form larger complexes, generally situated parallel to the contour of the ascus protoplast, giving the innermost part of the wall a banded pattern. However, at certain intervals the fibril orientation turns up from a parallel plane to a direction roughly perpendicular to the protoplast. As seen with the aid of the light microscope, this arrangement may give the impression of vertically arranged striation giving an overall undulating appearance to the innermost surface of the ascus wall. The zone of convergence of the striae at the apex is relieved as a more or less small, often tubular "corona"—the "nasse apicale" of Chadefaud (1973)—which sometimes may appear as spiraled striae in light microscopic mounts (Chadefaud, 1954; Funk and Shoemaker, 1967; Reynolds, 1971). This inner layer sometimes is a complex of more than one sublayers. In contrast to the inner layer, the microfibrils of the outer layer of the endotunica are not bundled, and therefore no marked striation is to be seen.

Ascus Function

The ultrastructured description of the inner endotunica is unique among the ascomycetes (Bellemére, 1971). In addition to the manner of spore ejaculation, it represents the best characteristic for the recognition of the bitunicate ascus. Reynolds (1971) suggested that the banded arrangement of the microfibrils was related to the extension of the endotunica during spore ejaculation.

The development of a bitunicate ascus can be schematized as in Figure 4.1. The primary wall or ectotunica surrounds the young ascus before nuclear fusion. This wall layer is composed of parallel microfibrils within an amorphous matrix. The ectotunica and the secondary wall, the endotunica, are present in the young ascus just before ascospore delimitation. The endotunica contains bundled microfibrils that converge in the "corona" in an arrangement characteristic of this ascus type. The ascus enlarges during ascosporogenesis to accommodate the spores. A vacuole forms at the ascus base, which presses the ascospores toward the apex, and they are ready for ejaculation. During ejaculation, the ectotunica ruptures and the endotunica extends at spore discharge as the basal vacuole enlarges, and the ascocarps are pressed through the apical pore of the "corona."

Water incorporation is an essential condition for ascospore ejaculation. According to Furtado and Olive (1971), mature asci are ready to imbibe larger amounts of water. The inner pressure of the ascus increases by the expansion of gelatinous substances partly manifested as gelatinous sheaths surrounding the ascospores and partly representing the rest of the ascus protoplast not consumed during ascospore development. Under the increasing pressure, the apical portion of the ascus begins to elongate, and the wall of this region becomes uniformly thickened. The thin ectotunica breaks and its delimiting marks are seen at the periphery of the ascus. The elongated asci stretch toward the opening of the ascoma, a vacuole is formed at the base, and the ascospores are pressed successively through an elastic pore in the apex of the "corona." After the ascospores are gone, the ascus collapses and the other asci may ejaculate their mature ascospores.

General Conclusions

If the original definition of the term "bitunicate" and the actual multilayered construction of the ascus wall are compared, it may be concluded that the term "bitunicate" is not accurate. However, the existence of two wall complexes, differing in their behavior during ascospore ejaculation, can be confirmed and more or less explained by their ultrastructure. The term "bitunicate" is therefore acceptable.

In spite of the therefore unequivocal definition of the bitunicate ascus, some doubts have remained about the systematic position of some taxa. The first source of disagreement is the literal interpretation of the term "bitunicate" as "two-layered." A number of taxa not belonging to the bitunicate ascomycetes show two-layered asci under the light microscope at certain stages of development. However, if their kind of ascospore ejaculation is taken into account these may easily be recognized as being unitunicate. In contrast, there are some bitunicate ascomycetes with cylindrical and thin-walled asci in which the two layers are difficult to see under the light microscope. Also in such cases, observation of the mode of ascospore ejaculation may provide good proof.

Other doubtful cases are mentioned by Bellemére (1971). He considered *Agyrium rufum* (Persoon) Fries to be an intermediate between the bitunicate and the unitunicate ascus. Additionally, he mentioned intermediate forms between bitunicate forms and species in the lichen order Lecanorales, especially within the so-called bitunicate discomycetes. To be sure, such intermediate forms have not been confirmed to date. Hafellner and Poelt (1976) studied the genus *Karschia* Korbes, one of the most discussed examples of intermediate forms between Lecanorales and bitunicate ascomycetes. They found that the genus is heterogeneous in reality and contains superficially similar members of the typically lecanoralian genus *Buellia* de Notaris and masked bitunicate ascomycetes, to be included in *Karschia.*

The bitunicate ascus is easy to distinguish from other ascus types. However, within the bitunicate asci, some typical differences exist. These are expressed not only in the form (broadly ellipsoidal, clavate, cylindrical, saccate) and size—traits easily seen under the light microscope—but also in the occurrence of different numbers of wall layers and in slight differences in their construction, observable only with the help of the electron microscope. Final judgment on the significance of such differences is withheld pending more well-investigated cases.

References

Bellemére, A., 1971. Les asques et les apothécies des discomycètes bituniqués. Ann. Sci. Natur. Bot. Biol. Veg. 12: 429–464.

Bellemére, A., 1975. Étude ultrastructurale des asques: la paroi, l'appareil apical, la paroi des ascospores chez des discomycètes inoperculés et les Hystériales. Phys. Veg. 13: 393–406.

Chadefaud, M., 1942. Structure et anotomie comparée de l'appareil apical des asques chez divers discomycetes et pyrenomycetes. Rev. Mycol. 7: 57–88.

Chadefaud, M., 1954. Sur les asques de deux Dothidéales. Bull. Soc. Mycol. Fr. 70: 99–108.

Chadefaud, M., 1969a. Données nouvelles sur la paroi des asques. C. R. Acad. Sci. Paris. (Ser. D) 268: 1041–1044.

Chadefaud, M., 1969b. Remarques sur la paroi, l'appareil apical et les réserves nutritives des asques. Oesterreich. Bot. Z. 116: 181–202.

Chadefaud, M., 1973. Les asques et la systématique des ascomycètes. Bull. Soc. Mycol. France 89: 127–170.

Currey, F., 1856. On the reproductive organs of certain fungi with remarks on germination. J. Microsc. Sci. 4: 198.

De Bary, A., 1884. Vergleichende Morphologie und Biologie der Pilze. W. Engelmann, Leipzig.

Funk, A., and R. A. Shoemaker, 1967. Layered structure in the bitunicate ascus. Can. J. Bot. 45: 1265–1266.

Furtado, J. S., and L. S. Olive, 1971. Ascospore discharge and ultrastructure of the ascus in *Leptosphaerulina australis*. Nowa Hedw. 19: 799–823.

Hafellner, J., and J. Poelt, 1976. Die Gattung Karschia—Bindeglied zwischen bitunicaten Ascomyceten und lecanoralen Flechtenpilzen? Plant Syst. Evol. 126: 243–254.

Hodgetts, W. J., 1917. On the forcible discharge of spores of *Leptosphaeria acuta*. New Phytol. 16: 139–146.

Hoggan, I., 1927. The parasitism of *Plowrightia ribesia* on the currant. Trans. Br. Mycol. Soc. 12: 27–43.

Luttrell, E. S., 1951. Taxonomy of the pyrenomycetes. Univ. Missouri Studies 24: 1–120.

Pringsheim, N., 1858. Ueber das Austreten der Sporen von *Sphaeria scirpi* aus ihren Schläuchen. Jahrb. Wiss. Bot. 1: 189–192.

Reynolds, D. R., 1971. Wall structure of a bitunicate ascus. Planta 98: 244–257.

Schrantz, J. P., 1970. Étude cytologique en microscopie optique et électronique de quelques ascomycètes. II. La Paroi. Rev. Cytol. Biol. Veg. 33: 111–168.

Chapter 5

The Lecanoralean Ascus: An Ultrastructural Preliminary Study

A. BELLEMÈRE AND M. A. LETROUIT-GALINOU

Introduction

The term "Lecanorales" must be precisely defined at the outset because of the diverse concepts of this group held by various authors. Nannfeldt (1932) established a broad definition of the group by uniting under the Lecanorales the ascohymenial discomycetes which are characterized by a perennial ascocarp and an ascus in which the wall stains blue with iodine. As a result, he assigned to this taxon a small number of non-lichen-forming discomycetes, the family Patellariaceae, along with nearly all the discolichens with the exception of the Caliciaceae-like species and some of the paraphysoid graphidiate species. This is nearly the same point of view as that of Dennis (1968). As a result of the work by more recent authors including Richardson (1970), Duncan and James (1970), Luttrell (1973), Poelt (1973), Henssen and Jahns (1974) and Von Arx and Müller (1975), some families or genera have been separated out—generally on the basis of characteristics of the ascus and the ascocarp (e.g., the genera *Patellaria* Fries and relatives, *Baeomyces* Persoon and allies and the families Graphidaceae, Lecanactidaceae, Thelotremataceae and its allies, and Gyalectaceae). These authors differ over only a small number of families. They are in general agreement on a fundamental definition of the order Lecanorales as the ascohymenial discomycetes that have asci of the archaeascé type. This corresponds to the lecanoralean discomycete concept of Chadefaud (1960) with the exclusion of the ascohymenial discomycetes having bitunicate asci.

From a number of light microscope studies (Ziegenspeck, 1926; Magne, 1946; Chadefaud and Galinou, 1953; Galinou, 1955a, b; Chadefaud, 1960, 1964, 1969a, b; Dughi, 1956, 1957; Tomaselli and Hemmeler, 1962; Chadefaud et al., 1963, 1968; Letrouit-Galinou, 1966, 1970, 1971, 1973; Hafellner and Poelt, 1976; Keuck, 1977), the structure of the archaeascé ascus appears to be as follows. The wall in general is formed of two fundamental components—the exoascus and the endoascus. The thin exoascus completely encloses the ascus and is covered with a more or less thickened, generally amyloid, occasionally chitinous gel which covers the ascus entirely, or is partially present, i.e. present only in apical or lateral regions. The endoascus, always thick, is also fundamentally formed of two layers. The outer one is complete and irregular in thickness. It is formed early and is nonamyloid. The inner one usually stains bluish with iodine. It is thickened at the upper part of the ascus where it forms the apical dome. Therefore, in the archaeascé type, the apical thickening is essentially built by the inner part of the endoascus. It may extend more or less along the sides of the ascus wall, forming a subapical pad or may exist only as a plug at the ascus tip. The top of the epiplasm may protrude axially into it. In the resulting ocular chamber, an apical nasse has often been described. The structures constitutive of the apical apparatus exclusively differentiate from the dome and include the axial body, manubrium, pendant, umbilical body, and various amyloid rings.

Chadefaud et al. (1963) distinguished three types of archaeascé asci according to the complexity of the endoascal apical apparatus: (a) in the prearchaeascé ascus the apical thickening of the endoascus is not differentiated; (b) in the euarchaeascé ascus amyloid zones are visible in the apical thickening, sometimes in the form of bells which fit around the ocular chamber and sometimes in the form of superimposed ring-shaped disks surrounding it; (c) in the postarchaeascé ascus a strongly amyloid ring protrudes from the base of the dome into the epiplasm.

As characterized by work with the light microscope, the archaeascé ascus can also be recognized by a synthesis of various features—some belonging to the bitunicate ascus type (i.e., the wall thickening; a nasse apicale), some to the unitunicate or annellasceous ascus type (i.e., amyloid areas in the thickened apex in certain patterns, including that of a ring that projects into the epiplasm, and a unitunicate type of ascus dehiscence).

In the present work, ultrastructural studies on collections of species which developed in their natural habitat have been carried out with the use of comparative light and electron microscope studies and with the use of Thiéry's (1967) polysaccharide stain technique. This technique, which selectively stains certain polysaccharides, sheds light on polysaccharide repartition into the ascus wall, making possible a comparison with the localization of amyloid substances demonstrated with the light microscope.

Previously, this technique has been employed to study nonlichenized discomycetes, leading to the description of four strata in the ascus wall (Bellemère, 1971). The A and B layers probably relate to the exoascus; C and D layers respectively relate to the outer and inner endoascus layers. This terminology is used in the following descriptions.

The selection of the small number of species described here was based on data derived from previous light microscopic studies. There is sufficient evidence to show that the constitutive elements defined by M. Chadefaud with the use of the light microscope can still be identified in the electron microscope. There is also sufficient evidence to be able to state more precisely the nature, structure, and interrelations of these elements. The four ascus types described below also confirm the differences between various ascal structures observed with the light microscope. It is clear from the previous light microscope studies that other ascus types are to be expected in the Lecanorales. Owing to ultrastructural data, the comparableness between the four ascus types recognized here becomes more accurate, and a better comprehension of the archaeascé type is reached. As similar data concerning the nonlichenized ascomycetes become available, comparison with the

Lecanorales becomes possible at the ultrastructural level, leading to a new approach to the problem of their relationships.

Ascus Types

Four major types of apical apparatus can be recognized in the lecanoralean fungi studies here: the Parmelia type, with Lecanora, Physcia and Ramalina variants; the Placynthium type; the Collema type; and the Cladonia type. All types produce an apical dome as a major component of the apical apparatus, as an elaboration of the endoascus D layer. Additionally the epiplasm of the ascus proper extends upward as one to several fingerlike projections into the dome area. The characters that allow comparison of the lecanoralean ascus types are the changes in the endoascus layers. Particularly notable are the simple to complex configurations of the D layer and the associated distribution of polysaccharide components revealed with the Thiéry technique.

The Parmelia type of apical apparatus was shown by Letrouit-Galinou (1970) to have an amyloid apical thickening (apical dome) of the walls, formed by an enlargement of the inner endoascus enclosing the narrow extension of the epiplasm in the apex of the ascus. A nonamyloid projection was described in the axial region of the ascus at the base of the apical dome, which was interpreted as a manubrium. *Parmelia acetabulum* (Necker) Duby represents this type. Along the sides, the ascus wall has four layers, A, B, C and D, which are organized as the exoascus (A and B) and the endoascus (C and D), analogous to those seen in the inoperculate discomycetes (Bellemère, 1977), but with several differences (Fig. 5.1D). The A layer is very thin and externally covered with a gel, which is the ectoascus of Chadefaud (1969b). The gel is composed of a polysaccharide mesh of a strongly inflated and loose texture. The B layer appears as a very thin electron-transparent layer subtending the A layer and is often masked by the polysaccharides that give rise to the external gel. The C layer is well developed, but contains less polysaccharides. The fine texture is apparently due to unoriented subglobose elements. The D layer is hardly distinguishable as a narrow border between the plasmalemma of the ascus protoplast and the C layer. The D layer is differentiated by its rough, coarse constituent elements, stretched more or less parallel to the ascus wall; it is very rich in polysaccharides.

Near the summit of the ascus, the D layer becomes considerably thickened, in contrast to the other three layers, which are not noticeably modified. Three distinguishable regions roughly form concentric domes around a narrow extension of the ascus epiplasm into the apical apparatus (Fig. 5.1). The external region is composed of a very fine, nonfibril network containing occasional polysaccharide fibrils, which become progressively more numerous toward the next layer. The midregion of the apical thickening is filled with short coarse polysaccharide fibrils that are oriented perpendicularly to the external surface of the ascus. These converge around the narrowed epiplasmic projection and form a dense, roughly annular structure. The inner region is quite distinct, with tiny granules that faintly react with Thiéry's stain technique. It projects downward into the epiplasm form-

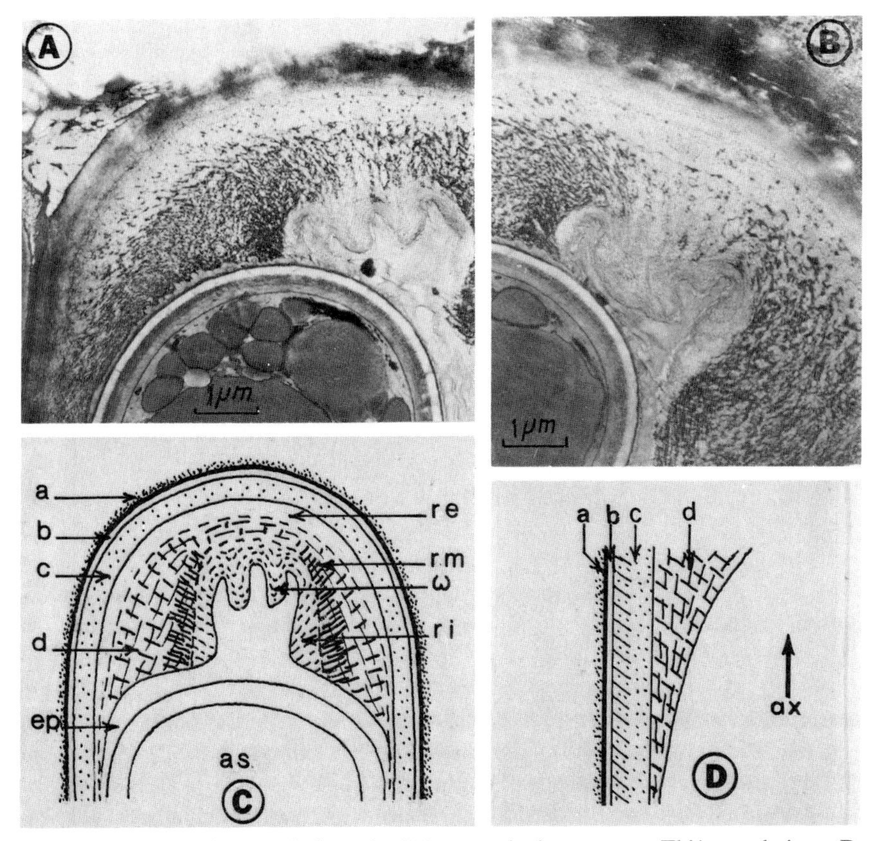

Fig. 5.1A–D. *Parmelia acetabulum*. **A–C** Ascus apical apparatus, Thiéry technique. **D** Ascus lateral wall at the base of the apical apparatus. a,b,A and B layers of exoascus; c,d,C and D layers of endoascus; re, external region of D layer; rm, midregion of D layer; ri, inner region of D layer; WGK omega, omega body; ep, epiplasm; as, ascospore, ax, axis of ascus with arrows directed upward. AD, apical dome; AB, axial body; UB, umbilical body; SP, subapical pad; OC, ocular chamber; P, pendant; a, A layer of exoascus; c, C layer of endoascus; d, D layer of endoascus.

ing a mass which shows several wavy profiles and, in thin section, appears as a lobed omega configuration. This omega body is recognizable under the light microscope because of its location and nonamyloid nature. It was once thought to represent a manubrium.

Several variants of the Parmelia type can be recognized as the result of observations on several species. *Lecanora chlarotera* Nylander (Fig. 5.2) and *Lecanora muralis* (Schreber) Rabenhorst exhibit few differences with *Parmelia acetabulum*. In the ascus wall, the D layer is thinner, but more distinct, and the reduced omega body is not so clearly delimited. The apical apparatus of *Physcia aipolia* (Ehrhart) Hampe and *P. stellaris* (Linnaeus) Nylander (Schoknecht, 1977; Honegger, 1977) demonstrates the basic Parmelia type of composition. However, the D layer shows important differences (Fig. 5.3). The external region is not distinct. The midregion is thinned near the upper center contact with the C layer. The numerous coarse-textured polysaccharide fibrils of the midregion are parallel to the other ascus layers rather than perpendicular as in the Parmelia type. A border area in contact with the plasmalemma lacks them (Fig. 5.3B). Two polysaccharide-rich bands traverse the upper portion of the inner region, and the omega body is not found at its base. Modifications of the Physcia variant worth mentioning are to be found in *Buellia punctata* (Hoffman) Massal (Fig. 5.4) where the polysaccharide fibrils of the midregion of the apical thickening look like those of the Lecanora variant, and in *Ramalina fraxinea* (Linnaeus) Archarius (Fig. 5.5) where at a certain developmental stage, the midregion of the dome protrudes downward around the top of the epiplasm like an annular pendant.

The Placynthium type of apical apparatus recognizably differs from the Parmelia type. In *Placynthium nigrum* (Hudson) Gray (Fig. 5.6), first, in the thick lateral walls of the ascus, the two endoascus layers are clearly distinguishable.

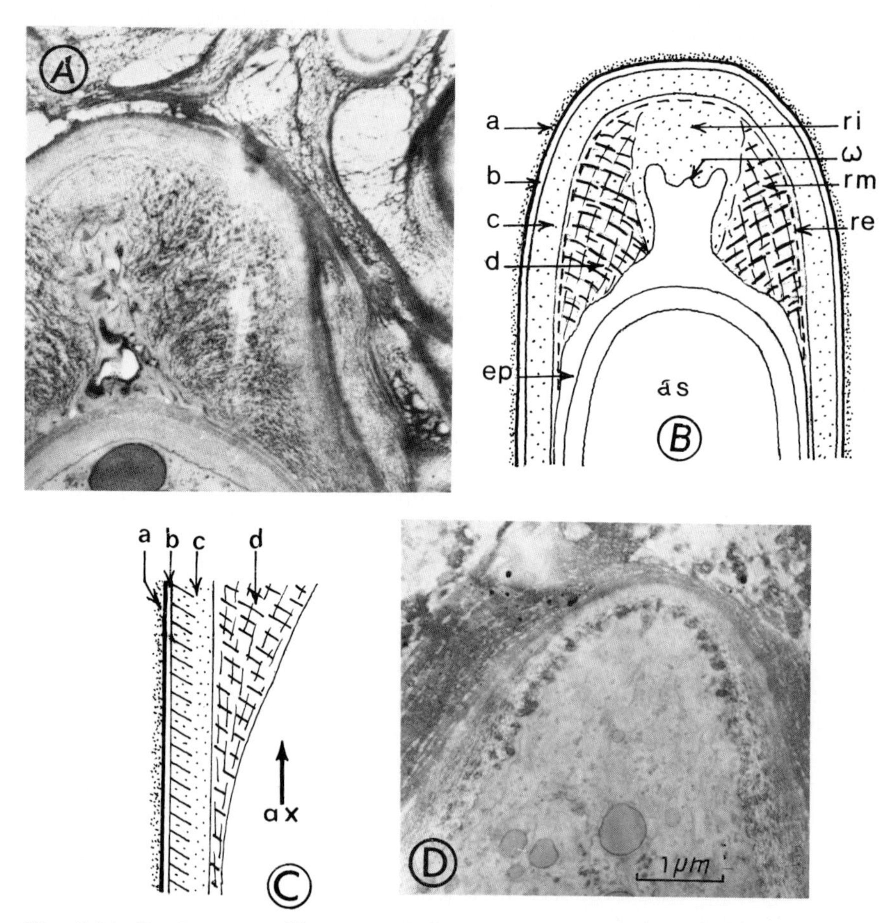

Fig. 5.2A–D. *Lecanora chlarotea.* **A,B** Ascus apical apparatus, Thiéry technique. **C** Ascus lateral wall at base of apical apparatus. **D** An early stage in formation of the apical apparatus. Abbreviations as in Fig. 5.1.

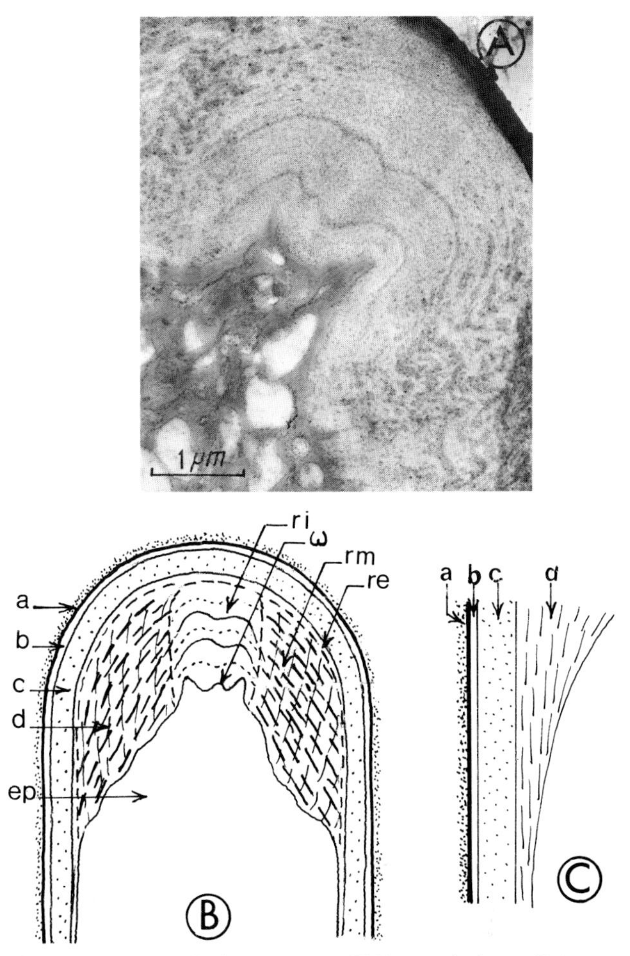

Fig. 5.3A–C. *Physica aipolia.* **A,B** Ascus apical apparatus, Thiéry technique; **C** Ascus lateral wall at base of the apical apparatus. Abbreviations as in Fig. 5.1.

Second, the D layer progressively becomes more pronounced toward the ascus apex so that the apical dome extends downward into a subapical pad. Third, the structure of the dome shows interesting peculiarities. Three regions that fit one inside the other and over the epiplasmic extension into the apical dome are distinguishable in the D layer. The thin, finely granular external region is most visible at the apex of the D layer as a domelike region. The peripheral part of the midregion (Fig. 5.6B), corresponding to the subapical pad, is relatively poor in polysaccharide fibrils in contrast with its deep region which is bell-shaped and rich in coarse, irregularly shaped polysaccharide fibrils. The fibrils are oriented generally parallel to the ascus axis at the base of the dome and become tangentially oriented toward the ascus apex. They are concentrated near the border with the inner region. The inner region contains few polysaccharide components; fine granules appear in lines parallel to the ascus axis. A structure like an omega body can be

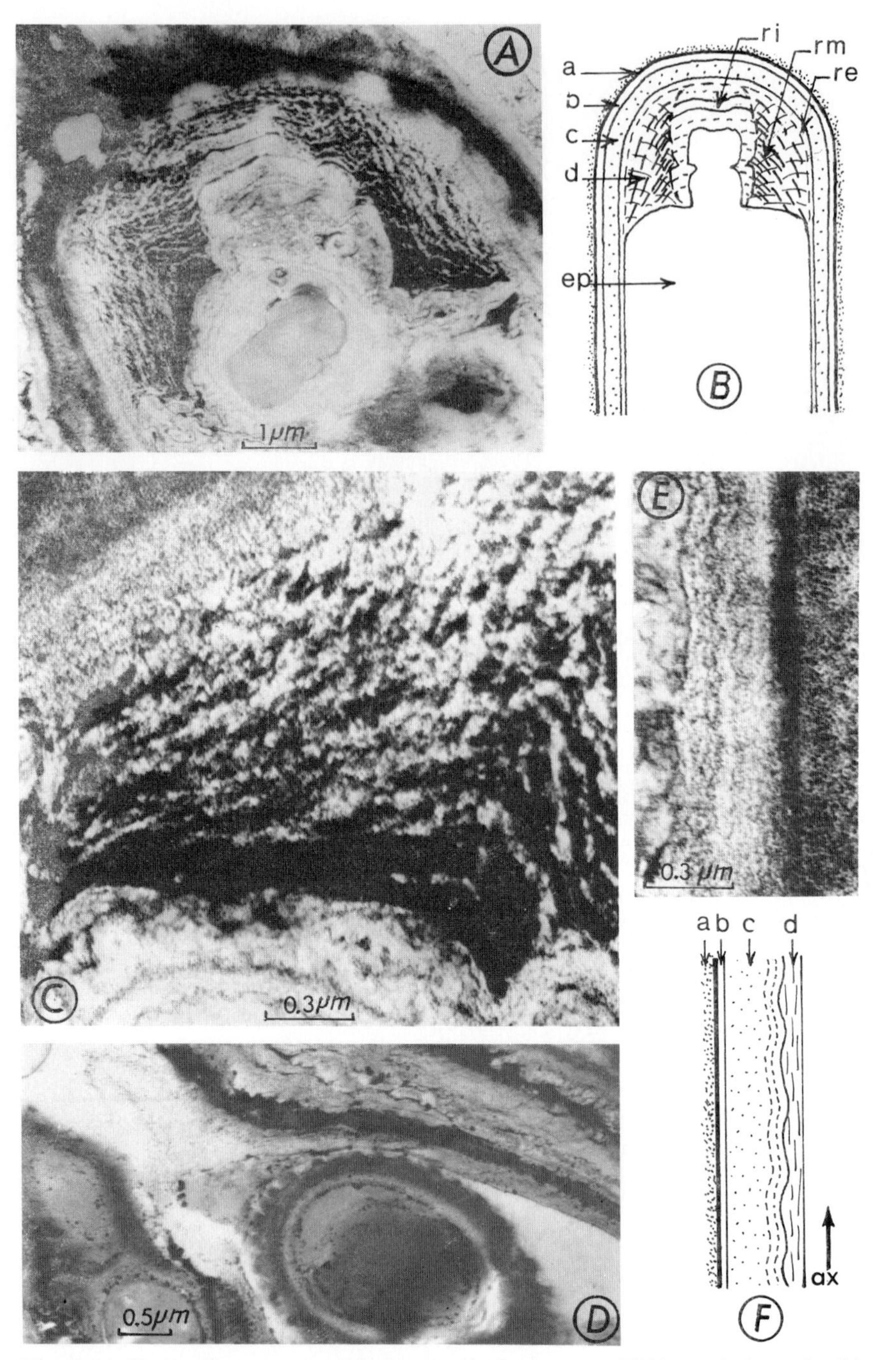

Fig. 5.4A–F. *Buellia punctata.* **A,B** Ascus apical apparatus, Thiéry technique; **C** Midregion of D layer. Note polysaccharide appearance, Thiéry technique. **D** Ascus dehiscence. An ascospore is positioned between the epiplasm and thin components of the apical apparatus. **E,F** Ascus lateral wall. The C layer and the D layer of the endoascus are separated by a different layer. Abbreviations as in Fig. 5.1.

recognized at the base, but it is surrounded with more irregular projections (Fig. 5.7). The base of the bell-shaped polysaccharide-rich portion of the midregion extends into the outermost one.

The Collema type (Fig. 5.7) is particularly distinguished by the distribution of polysaccharides in the apical region. In the lateral ascus layers of *Collema polycarpon* Hoffman, the D layer is barely distinct but it thickens considerably in the apical region of the ascus, forming a subapical pad under the dome. Only two

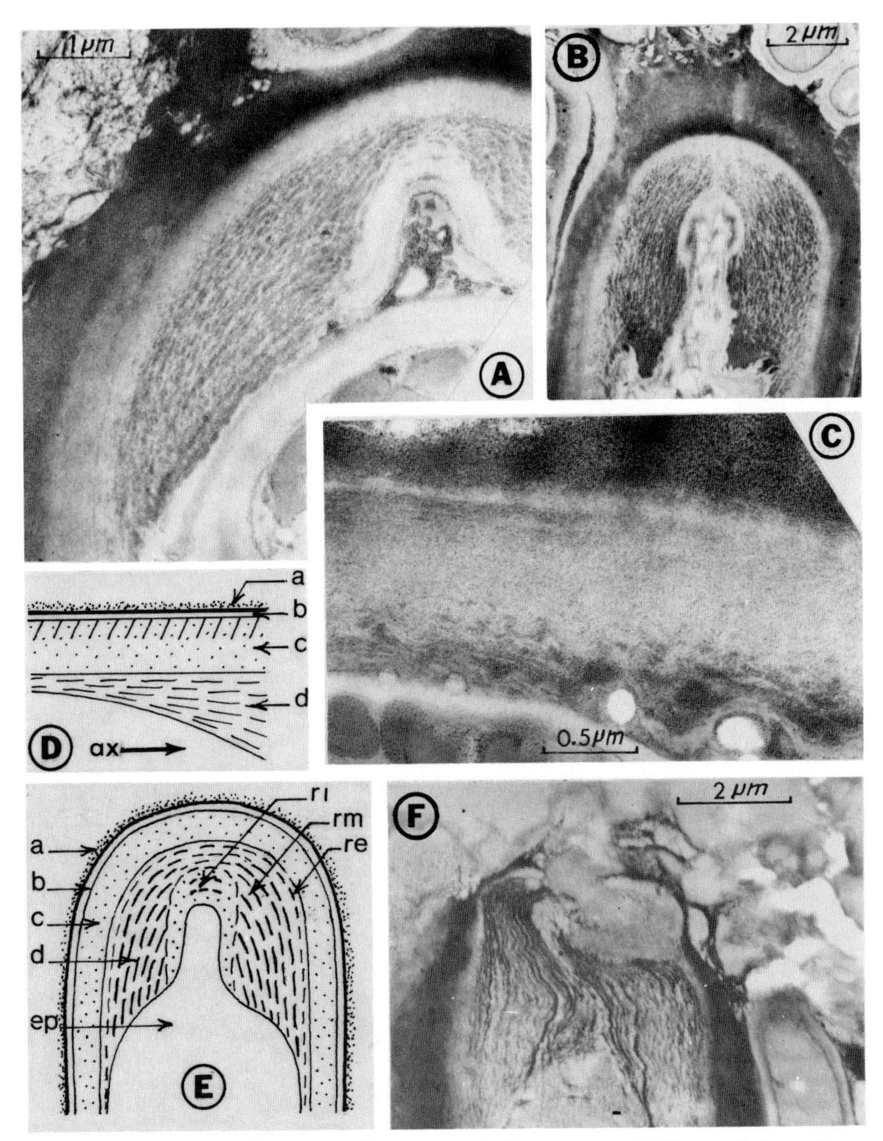

Fig. 5.5A–F. *Ramalina fraxinea.* **A,B** Ascus apical apparatus, Thiéry technique. Two different stages of development. **C** Ascus wall with young ascospore in the epiplasm. **D** Ascus lateral wall at base of apical apparatus. **E** Apical apparatus. **F** "en rostre" dehiscence. Abbreviations as in Fig. 5.1.

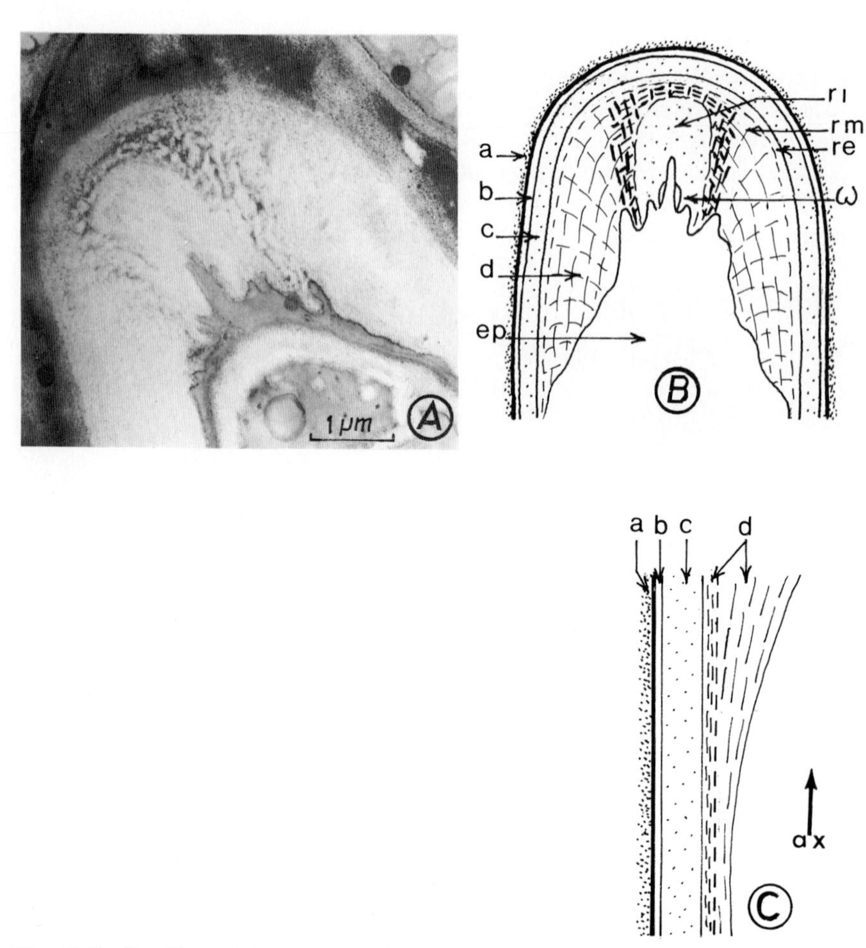

Fig. 5.6A–D. *Placynthium nigrum.* **A–B** Ascus apical apparatus, Thiéry technique. **C** Ascus lateral wall below apical apparatus. Abbreviations as in Fig. 5.1.

regions are present in the dome, corresponding to the mid- and inner regions. The inner region is a narrow and axial column, finely granular, apparently divided into two parts: an upper one and a lower one. The midregion as in the Placynthium type has two parts. The peripheral one corresponds to the subapical pad; the deep one which surrounds the inner region contains short, strongly angular polysaccharide elements, perpendicular to the ascus axis. Its lower part where the fibrils become parallel to the axis protrudes downward into the epiplasm, like the annular formation surrounding the omega body in the Placynthium type. The absence at the top of the epiplasmic chamber of such an omega body is noticeable.

The Cladonia type is original. The ultrastructure of *Cladonia flabeliformis* (Floerke) Vain var. *tubaeformis* (Mudd) Vain (Fig. 5.8) ascus reveals a thin D layer in the lateral wall that abruptly becomes thickened in the form of a plug at the ascus apex. The base of this plug largely protrudes into the top of the epiplasm, and is surrounded by a narrow sinus. Of the three regions of the D layer

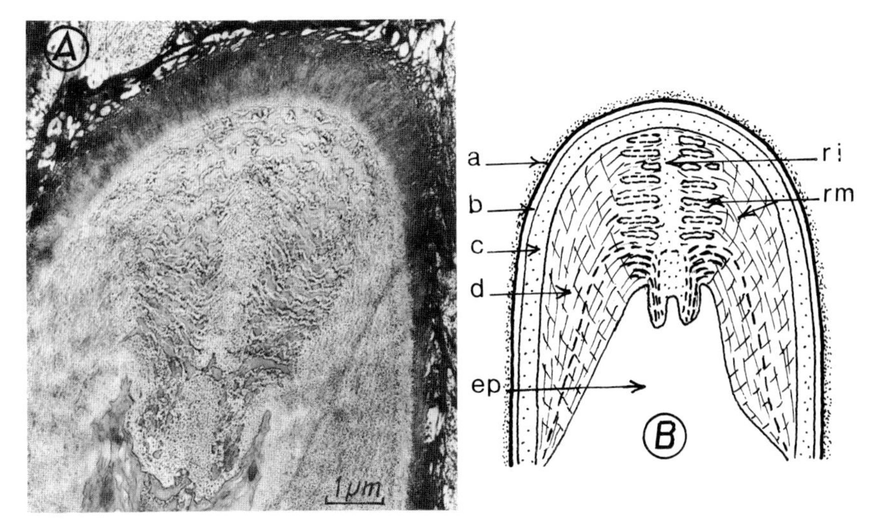

Fig. 5.7A,B. *Collema polycarpon.* **A,B** Apical apparatus, Thiéry technique. Abbreviations as in Fig. 5.1.

of the dome recognizable in the other lecanoralean types, the external one appears to be missing and the inner region is probably reduced to a narrow clear column supported by the upper epiplasmic projection (not visible in the nonaxial Fig. 5.9). The part of the D layer comparable to a midregion of other lecanoralean ascus types therefore nearly comprises the whole plug. The thick, coarse-textured polysaccharide fibrils are arranged in a complicated pattern that matches the amyloid reaction zones detected with light microscope techniques. Near the apex, the fibrils become layered and nearly parallel to the other ascus wall layers.

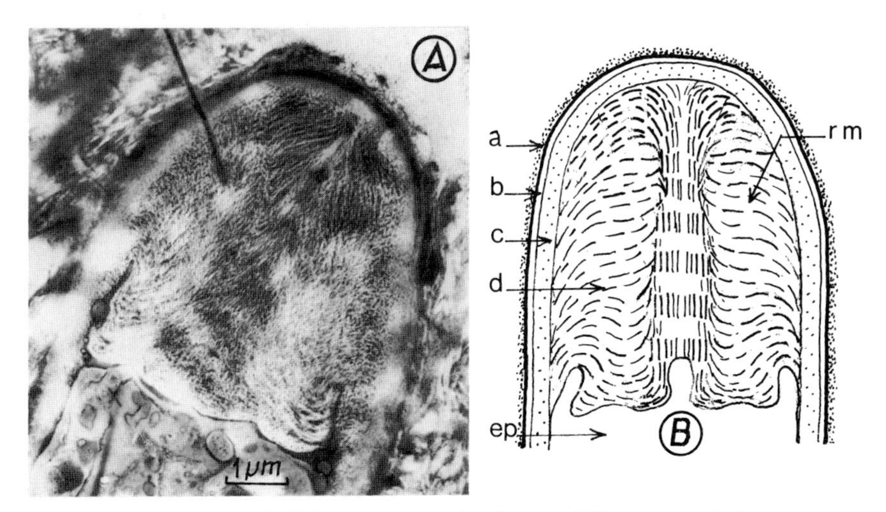

Fig. 5.8A,B. *Cladonia flabelliformus* var. *tubaeformis.* **A,B** Ascus apical apparatus, Thiéry technique. Abbreviations as in Fig. 5.1.

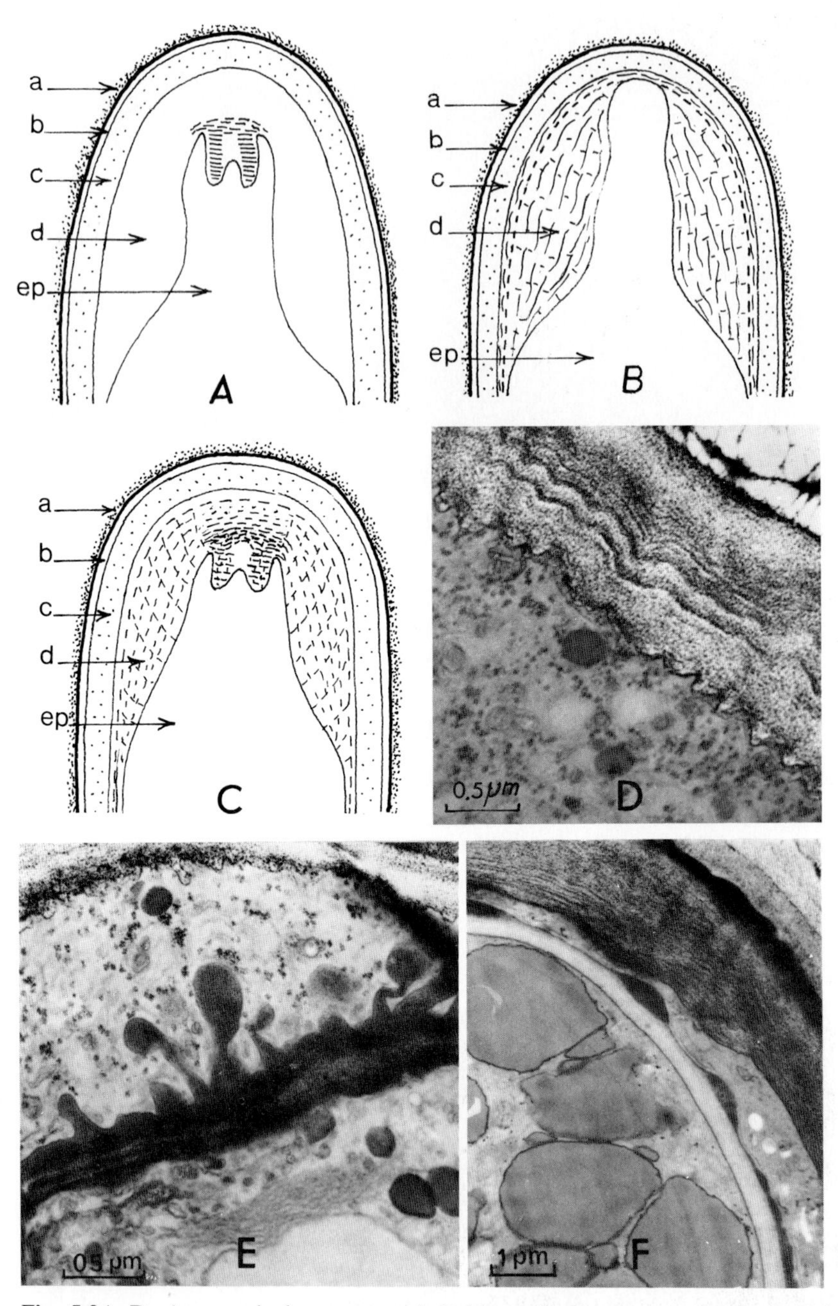

Fig. 5.9A–D. Ascus apical apparatus. **A** *Peltigera.* **B** *Xanthora parietina.* **C** *Solorina crocea.* **D** Ascus wall of *Diploschistes scriuposus* var. *bryophilus,* Thiéry technique. The multilayered D layer resembles a bitunicate-nassacsé type of configuration. **E** Ornamentation of the ascospore wall in *Diploschistes scruposus* var. *bryophyllus,* Thiéry technique. **F** Wall of the ascospore and wall of the ascus in *Pannaria pezizoides.* Abbreviations as in Fig. 5.1.

Besides the structure of the ascus wall, some other cytological features have been observed: concentrical bodies in the ascus epiplasm in *Ramalina fraxinea*, the ascus vesicle in *Placynthium nigrum* and *Collema polycarpon*, and an ascospore wall structure which is generally more complex than known in nonlichenized inoperculate discomycetes (Fig. 5.9E,F).

General Conclusions

Concerning the basic structure of the apical thickening, the present work shows that in the apical dome there exists not only a radial disposition (from axial to lateral parts) but also an unsuspected zonation (with a deep region, a midregion and an outer region). Otherwise, a careful examination of the different types of asci—especially those of the Parmelia type—suggests that the deposits of polysaccharide fibrils are laid into the meshes of a tridimensional net of Thiéry nonreactive substances. It is tempting to consider the structural elements of the apical apparatus as the result of the peculiarities of these deposits along or in between the meshes of this single fundamental net according to the radial or vertical directions. With such a model, some odd structures like the amyloid bell-shaped deposit in the Placynthium type can be explained without too much difficulty.

A correlation of the observations from the electron microscope on the asci of species discussed here and the scheme proposed from light microscope work for similar asci (Parguey-Leduc and Chadefaud, 1963) can clarify some terminology. The present study confirms the existence of four layers in the ascus wall like those of nonlichenized ascomycetes. The A, B, C layers surround the ascus completely and are relatively unchanged structurally all over it. Opposingly, the D layer (=inner endoascus) which laterally may be very thin (Parmelia) or sometimes well developed (Placynthium) always thickens at the top, forming either a subapical pad and a dome (Placynthium type; Collema type) or a single plug (Parmelia type; Cladonia type). With the exception of the apical nasse which has not been clearly observed in this work, all the constitutive elements of the apical apparatus are, in the lecanoralean asci, included in the D layer of the apical thickening.

The tentative clarification of other terms used in the scheme proposed by Parguey-Leduc and Chadefaud (1963) for the ascus apex structure (pendant, umbilicate body, manubrium, axial body) needs the comparison between the Collema type and the Placynthium type. The first type will help achieve a better understanding of the pendant while the second one will lead to an interpretation of the umbilicate body.

In the Collema type, a distinct amyloid ring is included in a basal protrusion of the midregion of the dome. According to the scheme proposed by Parguey-Leduc and Chadefaud (1963), this protrusion may be identified as a pendant. The present electron microscope studies show that this pendant derives from the deep part of the midregion. Additionally, it appears to be distinct from the subapical pad which derives from the peripheral part of the midregion and has a different structure. The term "ocular chamber" (=chambre oculaire) could be kept for the epiplasmic area extended in the axial part of the dome. It is noticeable that this ocular chamber is not laterally in contact with the midregion, but with the inner region of the dome. This peculiarity has been observed in all the species studied here and might be a general characteristic.

In the Placynthium type, the omega body which protrudes downward into the top of the ocular chamber has the position of the umbilicate body in the scheme proposed by Parguey-Leduc and Chadefaud (1963). Therefore, it may be identified with the latter. Consequently, the umbilicate body (=omega body) appears to be an annex of the inner region of the dome by opposition with the pendant, described above, which is suspended into the midregion. A striking point is that in the Placynthium type, both the pendant and the umbilicate body are present. So, at least for those elements, the synthetic model elaborated by Parguey-Leduc and Chadefaud (1963) actually exists.

The ocular chamber as defined above is present in all the species studied here, even in the Cladonia type, though it is rather indistinct in the nonaxial (Fig. 5.8). At the top of the ocular chamber, the umbilicate body is especially well developed in *Parmelia acetabulum;* probably various species lack it like *Collema polycarpon,* or *Ramalina fraxinea.*

In the scheme of Parguey-Leduc and Chadefaud (1963), the umbilical body is said to be hung from a manubrium, which itself is contained in a hollow space or oculus in the center part of the apical dome. The results from electron microscopic studies by Griffith (1973) of the pyrenomycete asci and by Bellemère (1975) of discomycete asci indicate that the axial part of the apical dome is solid and has lateral continuation. There is neither an oculus nor a manubrium and the axis of the apical dome ought to be called the axial body (=pseudomanubrium, =*piéce axiale*). The present observations show that the axial body is made of the axial part of the midregion plus the part of the inner region of the dome overlaying the umbilicate body. Therefore, the term "axial body" would have a topographical meaning, not a structural one.

The relative position of the pendant, the umbilicate body, the ocular chamber and the axial body are indicated in Fig. 5.8.

As a result of correlative light and electron microscope studies, the archaeascé type of ascus recognized by Chadefaud and Galinou (1953) must be divided into many subtypes. Several evolutionary pathways can be formulated to show interrelationships of the lecanoralean fungi by ascus structure (Fig. 5.10). As stated above, the Placynthium type seems to be synthetic. The Collema type, with a subapical pad and a developed pendant, may be close to it, but differs in lacking an umbilicate body and having a different structure of the apical dome, a more developed pendant and a reduced inner region. The ascus of Peltigera possesses an amyloid ring that projects into the epiplasm in the ascus apex according to Magne (1946) and J. C. Boissiére (personal communication). This structure is homologous to the pendant of the Collema type (Fig. 5.11F). The apical apparatus found in the ascus of *Solorina crocea* (Fig. 5.9C) connects the Peltigera type (Figs. 5.9A and 5.11C) and the Placynthium type (Fig. 5.11B) through the appearance of the omega body and the formation of an amyloid ring in a prominent dome. Probably, these two genera are in the neighborhood of the two pre-

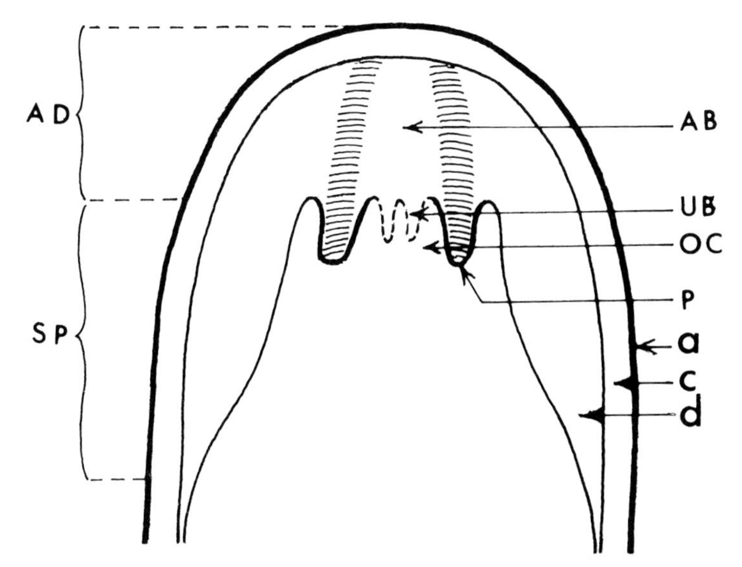

Fig. 5.10. Diagrammatic representation of the lecanoralean apical apparatus. Abbreviations as in Fig. 5.1.

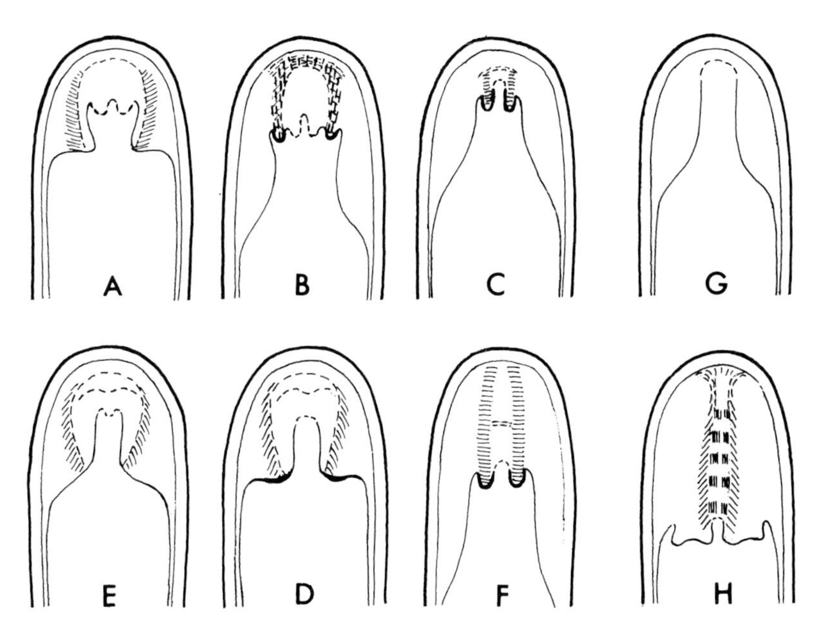

Fig. 5.11A–H. Comparative diagrammatic representations of the lecanoralean apical apparatus. **A** Parmelia type. **B** Placynthium type. **C–E** Parmelia type variants: **C** Peltigera variant, **D** Ramalina variant, **E** Physcia variant. **F** Collema type. **G** Xanthoria type. **H** Cladonia type.

ceding types. All these genera also have in common a preferential lichenization with cyanophytes or Coccomyxa species.

The Parmelia type differs by the absence of a subapical pad, the presence of an apical dome developed into a plug with three concentric regions, and the absence or extreme reduction of the pendant. In this type, the D layer of the ascus lateral wall is reduced in opposition with the preceding types. In the Parmelia type, the variations of the umbilicate body and the inner region of the dome have probably an evolutionary significance.

In the Cladonia type, as previously suggested by light microscope studies, the apical thickening would be interpreted as an enlarged pendant. Though probably related to the Parmelia type, it suggests an original evolutionary pathway.

The Xanthoria type (Honegger, 1977) of apical apparatus (Fig. 5.9B) could represent another mode of evolution leading to a type where the pendant and amyloid ring were not differentiated and the D layer-originated apical dome was reduced to the position of the two-layered subapical pad (Chadefaud, 1960, 1969a).

Because of the uniqueness of the Parmelia type of lecanoralean apical apparatus, along with the related variations, it is to be considered as the mainstay of the lecanoralean fungi.

Some features of the lecanoralean asci, as seen by light and electron microscopic techniques, evoke the bitunicate-nassascé ascus; others the annellasceous ascus. Therefore, the question of the affinities between these types arises.

A few characters of various lecanoralean asci are in common with the bitunicate-nassascé ascus (i.e. a thick D layer, a well developed subapical pad, an apical nasse). But a thick D layer can also be found in certain annellascé asci such as those of *Pezicula* Tulasne and *Rhabdocline* Sydow (Bellemère, 1977). Moreover, the thick D layer of the bitunicate ascus and its apical thickening form a distinct structure (Reynolds, 1971) which has not been found in the lecanoralean species. In the discolichen *Diploschistes scruposus* (Schreber) Norman (Fig. 5.9D) placed in the Thelotremaceae, the many-layered endoascus more closely resembles the bitunicate-nassascé ascus. Concerning the apical nasse, use of the electron microscope yields no data, though some reticular elements visible in the ocular chamber of *Physica aipolia* might be constitutive of a nasse.

The apical apparatus of the lecanoralean ascus differs from the annellaceous ascus in the elaboration of the D layer in the former and the B layer and largely the C layer of the latter. The amyloid ring of the lecanoralean apical apparatus corresponds only to the lower ring of the annellascé ascus. Therefore, the apical apparatus of the Parmelia type and related forms does not appear to be in any way allied to that of the annellascé ascus type. However, the upper amyloid ring of the C layer of certain annellaceous species, such as those of *Sclerotinia* Fuckel (Codron, 1974), is diminished, the apical apparatus suggesting a simplified Cladonia type.

It is tempting to make a correlation between the Collema and Peltigera types and that known in *Vibrissea* Fries (Bellemère, 1977). The differences lie in the reaction to the Thiéry technique and polysaccharide fibril distribution in the apical dome, the subapical pad, and the pendant. A comparison of the ascus apical apparatus of *Placynthium* and *Stictis* Persoon ex. Gray can be made. The *Stictis* ascus possesses an apical dome without an amyloid ring, has a nonamyloid two-layered subapical pad, and is without a pendant (Bellemère, 1977).

Therefore, at the present time, in spite of some annellasceous or bitunicate-nassasceous traits, the lecanorales with archaeascé asci may be considered as forming an original group.

References

Bellemére, A., 1971. Les asques et les apothécies des discomycètes bituniqués. Ann. Sci. Natur. Bot. Biol. Veg. 12: 429–464.

Bellemére, A., 1975. Étude ultrastructurale des asques: la paroi, l'appareil apical, la paroi des ascospores chez les discomycètes inoperculés et des hystériales. Physiol. Veg. 13: 273–286.

Bellemére, A., 1977. L'appareil apical de l'asque chez quelques discomycètes: Etude ultrastructurale comparative. Rev. Mycol. 41: 233–264.

Chadefaud, M., 1960. Les végétaux non vasculaires (cryptogamie). *In* M. Chadefaud and L. Emberger (Eds.), Traité de Botanique Systématique, Vol. 1. Masson, Paris.

Chadefaud, M., 1964. Sur l'origine et la structure des asques du type annellascé. C. R. Acad. Sci. Paris (Sér. D) 258: 299–301.

Chadefaud, M., 1969a. Une interprétation de la paroi des ascospores septées, notamment celle des *Aglaospora* et des *Pleospora*. Bull. Soc. Mycol. Fr., 85: 145–157.

Chadefaud, M., 1969b. Données nouvelles sur la paroi des asques. C. R. Acad. Sci. Paris (Sér. D) 268: 1041–1044.

Chadefaud, M., and M. A. Galinou, 1953. Sur l'asque des lichens du genre *Pertusaria* et son importance phylogénétique. C. R. Acad. Sci. Paris (Ser. D) 237: 1178–1180.

Chadefaud, M., M. A. Letrouit-Galinou, and M. C. Favre, 1963. Sur l'évolution des asques et du type archaeascé chez les discomycètes de l'ordre des Lécanorales. C. R. Acad. Sci. Paris (Sér. D) 257: 4003–4005.

Chadefaud, M., M. A. Letrouit-Galinou, and M. C. Janex-Favre, 1968. Sur l'origine phylogénétique et l'évolution des ascomycètes des lichens. Mem. Soc. Bot. Fr. Colloque sur les Lichens, 79–111.

Dennis, R. W. G., 1968. British ascomycetes.

Dughi, R., 1956. Appareils apicaux des asques et taxonomie des *Collema*. C. R. Ac. Sci. Paris (Sér. D) 243: 1911.

Dughi, R., 1957. Membrane ascale et reviviscence chez les champignons lichéniques discocarpes inoperculés. Ann. Fac. Sci. Marseille 26: 3–20.

Duncan, U. K., and P. W. James, 1970. Introduction to British Lichens. T. Buncle Co.

Galinou, M. A., 1955a. Le système apical des asques chez différentes espèces de lichens des g. *Nephroma, Solorina* et *Peltigera* (Peltigéracées). C. R. Acad. Sci. Paris (Sér. D) 241: 99–101.

Galinou, M. A., 1955b. Recherches sur la flore et la végétation des lichens épiphytes en forêt de Mayenne, suivies de remarques anatomiques et physiologiques sur quelques espèces. Dissertation, Facultés des Sciences de Rennes.

Griffiths, H. B., 1973. Fine structure of seven unitunicate pyrenomycete asci. Trans. Br. Mycol. Soc. 60: 261–271.

Hafellner, J., and J. Poelt, 1976. Die Gattung Karschia. Bindeglied zwischen bitunicaten Ascomyceten und lecanoralen Flechtenpilzen? Plant Syst. Evol. 126: 243–254.

Henssen, A., and M. Jahns, 1974. Lichenes. Geo. Thieme Verlag, Stuttgart.

Honegger, R., 1977. Development and function of the ascus apex in some Lecanorales (lichenized fungi). *In* H. E. Bigelow and E. G. Simmons (Eds.), Second International Mycological Congress Abstracts, Vol. A–L, pp. 302–303. IMC-2, Inc., Tampa.

Keuck, G., 1977. Ontogenetisch-systematische Studien über Erioderma im Vergleich mit anderen cyanophilen Flechtengattungen. Bibliotheca Lichenologica 6: 1–176.

Letrouit-Galinou, M. A., 1966. Recherches sur l'ontogenie et anatomie comparées des apothécies de quelques discolichens. Rev. Bryol. Lichénol. 34: 413–588.

Letrouit-Galinou, M. A., 1970. Les apothécies et les asques du *Parmelia conspersa* (Discolichen, Parméliacées). Bryologist 73: 39–58.

Letrouit-Galinou, M. A., 1971. Etude sur le *Lobaria laetevirens* (Light.) Zahlb. (Discolichen, Stictacée), I. Le thalle, les apothécies, les asques. Le Botaniste 54: 189–234.

Letrouit-Galinou, M. A., 1973. Les asques des lichens et le type archaeascé. Bryologist 76: 30–47.

Luttrell, E. S., 1973. Loculoascomycetes. *In* G. C. Ainsworth, F. K. Sparrow, and A. S. Sussman (Eds.), The Fungi, Vol. 4a. Academic Press. New York, pp. 135–219.

Magne, F., 1946. Anatomie et morphologie des asques de quelques Lichens. Rev. Bryol. Lichenol. 15: 203–209.

Nannfeldt, J. A., 1932. Studien über die Morphologie und Systematik der nichtlichenisierten inoperculaten Discomyceten. Nov. Act. Reg. Soc. Sci. Upsala 4(8): 1–368.

Parguey-Leduc, A., and M. Chadefaud, 1963. Les asques du *Cainia incarcerata* (Desm.) von Arx et Müller et la position systématique du genre *Cainia*. Rev. Mycol. 28(3/4).

Poelt, J., 1973. Classification. Appendix A. *In* V. Ahmadjian and M. E. Hale (Eds.), The Lichens. Academic Press, New York.

Reynolds, D. R., 1971. Wall structure of a bitunicate ascus. Planta 98: 244–257.

Richardson, D. H. S., 1970. Ascus and ascocarp structure in lichens. Lichenologist 4: 350–361.

Schoknecht, J. D., 1977. The ascus apex in discolichens. *In* H. E. Bigelow and E. G. Simmons (Eds.), Second International Mycological Congress Abstracts. Vol. 2: 597, p. 597. IMC-2, Inc., Tampa.

Thiéry, J. P., 1967. Mise en évidence des polysaccharides sur coupes fines en microscopie électronique. J. Microsc. 6: 987–1018.

Tomaselli, R., and M. Hemmeler, 1962. Nuove osservazioni sulla structura dell'asco in *Peltigera polydactyla* Hoffman. Boll. Ist. Bot. Univ. Cataina 3(3): 14–21.

von Arx, J. A., and E. Müller, 1975. A re-evaluation of the bitunicate ascomycetes with keys to families and genera. Stud. Mycol. 9: 1–159.

Ziegenspeck, H., 1926. Schleudermechanismen von Ascomyceten. Bot. Arch. Koenigsberg 13: 341–381.

The Centrum

Chapter 6

The Plectomycete Centrum

D. Malloch

Introduction

Over the last 75 years, studies on ascocarp development have played an increasingly important role in ascomycete taxonomy. The classic studies by Dangeard (1907), Nannfeldt (1932), Luttrell (1951, 1954), and several others are so well known and so fundamental that they are often quoted without literature citations. Although these studies have contributed greatly to an understanding of relationships among certain groups of ascomycetes, the plectomycetes or cleistothecial forms have remained relatively unstudied. The reasons for this neglect are several but stem mainly from traditional taxonomic concepts.

The term "plectomycete" generally refers to ascomycetes that have their asci borne within an entirely enclosed ascocarp or cleistothecium. The asci in this group are evanescent and do not forcibly discharge their ascospores, so that the cleistothecial cavity or centrum becomes filled with ascospores at maturity. Many authors, for example Luttrell (1951) and Nannfeldt (1932), included in the plectomycetes certain forms with ostiolate ascocarps having evanescent asci (such as species of *Ceratocystis* Ellis and Halstead and *Microascus* Zukal).

Until relatively recently, most taxonomists have considered the plectomycetes to be a group of primitive and related ascomycetes that were the ancestors of the more advanced ostiolate and apothecial forms (von Arx and Müller, 1954; Gaumann, 1964). Because this group was believed to be relatively homogeneous, mycologists generally recognized only two types of centrum structure. The first of these, represented by the order Plectascales, was characterized by ascocarps in which the asci were irregularly arranged in the centrum and usually broadly clavate to spherical. The second group, the Perisporiales, included species having clavate to cylindrical asci arranged in a parallel of fasciculate basal hymenium. Most mycologists accepted the view that the two groups of plectomycetes were natural and homogeneous ones and did not develop much interest in either their taxonomy or their development. The first group to be questioned was the Perisporiales. In 1951, Luttrell wrote that the Perisporiales were "an obviously heterogeneous group of fungi . . ." and suggested that its members should be distributed in a number of different groups. This group has all but disappeared in recent systematic treatments.

The disuse of the order Plectascales has been much slower, although its first critics appeared more than 40 years ago. Bessey (1935) observed that "This group is perhaps heterogeneous as regards certain of the included families."

In the middle and late 1950s, a new trend in taxonomic thought appeared that would strongly influence future work on the plectomycetes. Doguet (1955, 1956) questioned the traditional placement of the genus *Thielavia* Zopf in the Plectascales and suggested that it was a closely related descendant of the pyrenomycetous genus *Melanospora* Corda. Simultaneously, Cain (1956a), cautiously at first, suggested that the Aspergillaceae were descended from the pyrenomycetous Hypocreales and then (Cain, 1956b) boldly stated that he believed the Plectascales to represent the termination in a large number of unrelated and highly evolved taxa. His argument was that these fungi were adapted to passive spore dispersal in very specific ecologic niches and were not merely generalized forms occurring ubiquitously. Cain concluded the latter paper by stating that "as long as the mycologist avoids tackling the problem of the distinction between parallel evolution and homologous development, his classification of the fungi will remain purely artificial." True relationships among the Plectascales, according to Cain, would be known only when the individual fungi were studied in detail, including "all stages of development of the ascocarps." In subsequent papers, Cain (1956c, 1959, 1961a–c, 1972) and Malloch and Cain (1970a, b, 1971a–d, 1972a, b, 1973a, b) expanded upon these ideas.

In the 1960s, a renewed interest in these fungi occurred, resulting in a number of new taxa being described and in several important studies on ascocarp development. As ideas changed and developmental studies accumulated, it is logical that the subject of centrum structure in the plectomycetes should be reassessed. If Cain and others are correct in their belief that this group is polyphyletic, then a number of different centrum types may be found. If centrum development is a conservative feature in the ascomycetes, it may aid in determining relationships between perithecial and cleistothecial species.

A Taxonomic Scheme

Centrum structure can be considered family by family. However, the taxonomy of the plectomycetes is still highly unsettled and ranges from one system with the single order Eurotiales (Webster, 1970) to others with several orders. Consequently, a unique placement of families in orders is used here (Fig. 6.1).

Pleosporales

Sporormiaceae

The basis or fundamental centrum type for the family is found in the genus *Sporormiella* Ellis and Everhart, a genus characterized by ostiolate ascocarps and bitunicate, forcibly discharging asci. The development of six species, *Sporormiella intermedia* (Auerswald) Ahmed and Cain (Dangeard, 1907; Satina, 1918), *Sporormiella leporina* (Niessl) Ahmed and Cain (Arnold, 1928), *Sporormiella aus-*

PLEOSPORALES	HYPOCREALES
Eremomycetaceae	Nectriaceae
Phaeotrichaceae	Trichocomaceae
Sporormiaceae	
Zopfiaceae	
DIAPORTHALES	PEZIZALES
Cephalothecaceae	Ascobolaceae
Chaetomiaceae	Monascaceae
Coniochaetaceae	Onygenaceae
Endomycetaceae	Pyronemataceae
Melanosporaceae	
Microascaceae	
Pseudeurotiaceae	
Sordariaceae	
Xylariaceae	

Fig. 6.1. A taxonomic scheme for the plectomycetes.

tralis (Spegazzini) Ahmed and Cain (Blanchard, 1972), *Sporormiella cylindrospora* Ahmed and Cain (Ahmed, 1964), *Sporormiella minima* (Auerswald) Ahmed and Cain (Ahmed, 1964), and *Sporormiella minimoides* Ahmed and Cain [Ahmed, 1964 (as *Sporormiella pseudominima* unpubl.)] has been studied. Based on the studies of Arnold (1928) and Dangeard (1907), Luttrell (1951) assigned *Sporormiella* to his *Pleospora* Developmental Type, characterized by a centrum composed of pseudoparaphyses and basally arranged asci. According to Arnold (1928) and Blanchard (1972), the sterile tissues of the centrum in *S. leporina* and *S. australis* originate at the centrum apex and grow downward and remain free from contact at the base or become attached [hence they are pseudoparaphyses in the restricted sense (Kowalski, 1965a)]. In contrast, Ahmed (1964) found the sterile centrum tissues of *S. cylindrospora, S. minima,* and *S. minimoides* to be attached at the top and bottom of the centrum from the beginning [paraphysoids (Kowalski, 1965a)].

The genus *Preussia* Fuckel is considered by many authors (Cain, 1961b; Chadefaud et al., 1966) to be a close relative to *Sporormiella*. Species of *Preussia* have multiseptate ascospores with germ slits similar to those of *Sporormiella* species but differ in having cleistothecia and evanescent asci. The development of several species of *Preussia* has been studied, including *Preussia flanaganii* Boylan (Boylan, 1970), *Preussia funiculata* (Preuss) Fuckel (Beatus, 1938; Kowalski, 1966), *Preussia isomera* Cain (Kowalski, 1968), *Preussia multilocularis* Maciejowska and Williams (Maciejowska and Williams, 1963), *Preussia nigra* (Routien) Cain (Routien, 1956), and *Preussia typharum* (Saccardo) Cain (Kowalski, 1965). As was the case in *Sporormiella,* there seem to be some developmental differences among the species. In both *P. funiculata* and *P. typharum,* the centrum is composed of a basal layer of asci interspersed among vertically oriented sterile tissues that are attached at the top and the bottom of the centrum from the beginning (paraphysoids). Asci in both species arise from croziers.

In *P. nigra* and *P. multilocularis,* the sterile filaments grow down from the apex of the centrum and become embedded at the base (pseudoparaphyses). Unlike *P. funiculata* and *P. typharum,* which are very similar to species of *Sporormiella, P. nigra* and *P. multilocularis* exhibit further modifications. In *P. nigra,* the asci line the base and sides of the centrum wall and radiate toward the center of the cavity. *Preussia multilocularis,* although considered by Maciejowska and Williams (1963) to have a *Pleospora*-type centrum, sometimes develops ascocarps that lack sterile tissues in the centrum and in which the asci become rather irregularly disposed.

Preussia flanaganii and *P. isomera* differ and from the preceding species in lacking pseudoparaphyses *and* paraphysoids. In both of these species, hyphal threads arise from the cells lining the locule and grow irregularly throughout the centrum. In *P. isomera,* the asci are reported to grow out from the ascogenous hyphae at the center of the locule in an irregularly radiating group. The asci in *P. flanaganii* are thought to arise from the ingrowing threads and are irregularly arranged.

Species of *Pycnidiophora* Clum are closely related to some of those of *Preussia* and differ in having subcylindrical to spherical, irregularly arranged asci and ascospores without germ slits. Cain (1961b), Chadefaud et al. (1966), and others consider *Preussia* and *Pycnidiophora* to be synonymous. The development of two species of *Pycnidiophora, Pycnidiophora multispora* (Saito and Minoura ex Cain) Thompson and Backus (Chadefaud et al., 1966) and *Pycnidiophora dispersa* Clum (Kowalski, 1964), has been examined. In *P. multispora,* the centrum becomes filled with sterile tissues arising from the locule walls. Ascogenous hyphae are reported to arise from several scattered helical ascogonia and to grow among the sterile threads producing irregularly disposed asci. In *P. dispersa,* sterile tissues are said to be lacking. Ascogenous hyphae arise from the walls of the locule *and* from the central cells and grow irregularly to fill the centrum.

Phaeotrichaceae

There are two genera in the family Phaeotrichaceae: *Phaeotrichum* Cain and Barr and *Trichodelitschia* Munk. *Trichodelitschia* has ostiolate ascocarps and basally arranged asci having forcible ascospore discharge, whereas *Phaeotrichum* has cleistothecia and evanescent asci. The development of *Trichodelitschia munkii* Lundquist and *Trichodelitschia bisporula* (Cr.) Lundquist was studied by Parguey-Leduc (1974), who demonstrated the centrum to be composed of pseudoparaphyses and a basal cluster of asci and who assigned it to the *Pleospora* developmental type.

Barr (1956) made a study of the development of *Phaeotrichum hystricinum* Cain and Barr and found no evidence of sterile filaments in the centrum. Instead, certain cells at the center of the developing ascocarp give rise to ascogenous hyphae, which in turn produce irregularly disposed clusters of asci. The locule seems to form by the disintegration of the central tissues, aided by the expanding ascogenous hyphae. Barr (1956) labeled this type of development the *Phaeotrichum* developmental type and compared it to that of *Pseudeurotium zonatum* van

Beyma, *Pseudeurotium ovalis* Stolk (Pseudeurotiaceae) and *Westerdykella ornata* Stolk (Sporormiaceae). It differed from the first two mainly in the early stages of ascocarp initiation. As is shown below, the Pseudeurotiaceae have their ascocarps initiated by ascogonia, whereas ascocarps of the Pleosporales have the pseudoparenchymatous initials typical of most loculoascomycetes.

Zopfiaceae

The only member of the Zopfiaceae that has been studied developmentally is *Lepidosphaeria nicotiae* Parguey-Leduc (placed in Testudinaceae by von Arx, 1971). Its development was studied by Parguey-Leduc (1970), who determined it to be a loculoascomycete with a radially arranged centrum. Pseudoparaphyses arise from the walls of the locule and radiate toward the center. The ascogenous elements arise at the center and produce a centrifugally radiating fascicle of asci.

Eremomycetaceae

The development of *Eremomyces* Malloch and Cain and *Rhexothecium* Samson and Mouchacca, the only members of this family, has not been studied.

Diaporthales

Xylariaceae

According to Luttrell (1951), members of this family have the base and sides of the centrum lined with a hymenium of asci and apically free paraphyses. Although this is a very large family, the species in only one genus, *Pulveria* Malloch and Rogerson, are known to produce cleistothecia. Although they did not attempt a detailed developmental study, Malloch and Rogerson (1977) were able to determine that the ascogenous hyphae are produced centrifugally from a central ascogonium and that the paraphyses radiate inward from the walls of the locule. Asci are produced sympodially along the radiating ascogenous hyphae and are uniformly distributed throughout the centrum.

Sordariaceae

Although the development of a number of ostiolate members of the Sordariaceae has been studied, that of the cleistothecial forms has been neglected. In general, most cleistothecial Sordariaceae closely resemble perithecial forms, having basally arranged asci with prominent apical rings. The genus *Anixiella* Saito and Minoura ex Cain is a good example, and *Anixiella indica* Rai, Wadhwani and Tewari, the one species that has had its development studied (Rai and Wadhwani, 1970), seems to have a typical *Sordaria*-type centrum, as defined by Huang (1976). In this type, paraphyses, when present, arise from the centrum pseudoparenchyma rather from the inner wall cells.

Chaetomiaceae

The genus *Chaetomium* Kunze ex Fries, a pyrenomycete with evanescent asci, is usually said to be characterized by a *Xylaria*-type centrum (Luttrell, 1951; Whiteside, 1961). As Huang (1976) points out, however, the centrum structure tends to differ from species to species, with lateral paraphyses originating from pseudoparenchyma cells in some, hymenial paraphyses in others, and with paraphyses entirely lacking in at least four species.

There has been some debate over the number of cleistothecial members of this family. Malloch and Cain (1973b), taking a fairly conservative approach, have placed all of the cleistothecial species in the Chaetomiaceae in the genus *Thielavia.* von Arx (1975), however, studied the same species as Malloch and Cain and assigned them to five genera, only one of which, *Chaetomidium* (Zopf) Saccardo, was considered to be a member of the Chaetomiaceae. He believed *Thielavia,* the largest of his five genera, to belong in the Sordariaceae.

The genus *Thielavia* (sensu Malloch and Cain, 1973b) has attracted the attention of a number of developmental morphologists. *Thielavia (Chaetomidium) fimeti* (Fuckel) Malloch and Cain is perhaps the best place to begin, as most recent authors would agree on its placement in the Chaetomiaceae. Whiteside (1962) studied the development of this species and found it to differ from that of the *Chaetomium* species he had examined. According to Whiteside, the young centrum is composed of a rather closely packed pseudoparenchymatous tissue formed by the centripetal growth of septate strands originating from the inner layers of the cavity. The asci later develop in a basal layer and grow up into the pseudoparenchymatous core, which apparently becomes crushed. No paraphyses were observed, unless the strands forming the core pseudoparenchyma are considered to be such.

Thielavia sepedonium Emmons and *Thielavia setosa* Dade, both placed in the genus *Corynascus* von Arx by von Arx (1975), are species characterized by ascospores having two germpores. The development of the former was studied by Emmons (1932) and the latter by Douget (1956). In both cases, the asci arise without croziers directly from the ascogenous hyphae and grow into the pseudoparenchymatous centrum tissue. In *T. setosa,* the asci develop centrifugally in a central fascicle, whereas in *T. sepedonium,* the ascogenous hyphae penetrate the entire cavity and bear asci lacking a regular orientation.

Thielavia terricola (Gilman and Abbott) Emmons was studied by both Emmons (1930, 1932) and Douget (1956), who described the centrum as pseudoparenchymatous at first. The ascogenous hyphae radiate throughout the centrum and bear irregularly disposed asci from croziers, crushing the pseudoparenchyma in the process. No paraphyses are involved.

Nicot and Longis (1961) reported two new species, *Thielavia coactilis* Nicot and Longis and *Thielavia hyrcaniae* Nicot and Longis, and illustrated the immature ascocarps. In *T. coactilis* there is a radiating central fascicle of asci, similar to that of *T. setosa,* whereas in *T. hyrcaniae* the asci are borne in a basal hymenium. In both cases, the centrum is pseudoparenchymatous at first and the cavity is created by the developing asci. Paraphyses are lacking.

Coniochaetaceae

The only cleistothecial members of this family are *Coniochaetidium* Malloch and Cain and *Ephemeroascus* van Emden. The developmental morphology of neither has been reported.

Melanosporaceae

The development of several species of the pyrenomycetous genus *Melanospora* Corda has been studied (Nichols, 1896; Vincens, 1917; Cookson, 1928; Douget, 1955; Kowalski, 1965b). In condensing this research, Douget observed that the centrum in all species studied is pseudoparenchymatous at first and is crushed by the developing asci. The origin of the asci, however, varies from species to species, ranging from basal to parietal to centrally fasciculate. Paraphyses are lacking.

The development of the two cleistothecial members of the Melanosporaceae, *Microthecium* Corda and *Rhytidospora* Jeng and Cain, has not been investigated. In species of *Microthecium,* the asci are arranged in a hymenium, and there is really little more than the lack of an ostiole in the ascocarps to separate them from those of *Melanospora.* Species of *Rhytidospora* have irregularly disposed asci and are quite distinct from those of *Microthecium.*

Pseudeurotiaceae

The Pseudeurotiaceae are all plectomycetous forms and are not obviously related to any of the families of perithecial or apothecial fungi. Development is always preceded by a coiled ascogonium, indicating that the family belongs with the Hymenoascomycetes. Of the nine genera of Pseudeurotiaceae recognized by Malloch and Cain (1970b), only one, *Fragosphaeria* Shear, has been studied developmentally. *Albertiniella polypoicola* (Jaczevski) Malloch and Cain undoubtedly also belongs in this family and was studied developmentally by Gusseva (1925).

Chesters (1935) studied the development of *Fragosphaeria reniformis* (Saccardo and Therry) Malloch and Cain and *Fragosphaeria purpurea* Shear (as species of *Cephalotheca* Fuckel). In both species, the centrum is pseudoparenchymatous at first, and the ascogenous hyphae grow throughout the centrum, bearing the spherical asci uniformly and obliterating the pseudoparenchyma. Apparently the same development occurs in *Albertiniella* Kirschten (Gusseva, 1925), although a pseudoparenchymatous centrum was not reported.

Mycogala marginata Crooks (? = *Xylogone sphaerospora* von Arx and Nilsson) may belong with the Pseudeurotiaceae. Its development was studied by Crooks (1935), who reported a pseudoparenchymatous centrum similar to that of *Fragosphaeria* species.

All species of Pseudeurotiaceae so far known lack paraphyses and have uniformly distributed, subspherical to spherical asci. Croziers are known in *Cryptendoxyla hypophloia* Malloch and Cain (Malloch and Cain, 1970b) but have not been reported in other members of the family.

Endomycetaceae

As defined by Redhead and Malloch (1977), the Endomycetaceae contains an assortment of species ranging from those with perithecial ascocarps (*Ceratocystis* Ellis and Halstead, *Ophiostoma* Sydow) to cleistothecial ascocarps (*Amorphotheca* Parberry, *Europhium* Parker) to forms lacking ascocarps altogether (*Endomyces* Reess, *Cephaloascus* Hanawa). Developmental studies have been made on the ostiolate members by a number of authors, and the results have been summarized by Luttrell (1951) as the *Ophiostoma* developmental type. In this type, the ascocarp is initiated by an ascogonium. The centrum is pseudoparenchymatous at first, and the ascogonium lies near the apex. Chains of ascogenous hyphae extend down from the apex, causing the pseudoparenchyma to disintegrate, while the asci mature along the chains from apex to base. The asci are evanescent and dissolve very quickly after ascospore formation. Paraphyses seem to be lacking.

The development of the two cleistothecial members of the family has not been studied. *Europhium* may differ very little in its development from *Ceratocystis;* in fact, Upadhyay and Kendrick (1975) consider the two genera to be synonymous. *Amorphotheca* does not have cleistothecia in the accepted sense; instead, there is an irregular mass of ascogenous elements and asci surrounded by a funnel-shaped amorphous and noncellular mass of material.

Microascaceae

This family, as defined by Malloch (1970a), contains five genera, three of which are ostiolate and two cleistothecial. The development of ostiolate members of the family was studied by Emmons and Dodge (1931), Moreau and Moreau (1953), Douget (1955), and Corlett (1966). In *Microascus* Zukal and *Petriella* Curzi species, the centrum is composed at first of mycelial elements that radiate inward from the sides and base of the locule. In species of *Microascus,* the ascogonium may be central or apical, and the ascogenous elements radiate out among the sterile filaments to fill the locule. Asci develop terminally, laterally, and intercalarily on the ascogenous hyphae and so come to uniformly occupy the centrum. In *Petriella* species studied by Corlett (1966), the ascogonium is central, and the ascogenous hyphae radiate centrifugally, producing asci only at the tips. As the asci enlarge and expand, they eventually fill the locule. Ascocarp development in *Lophotrichus ampullus* Benjamin and *Lophotrichus martinii* Benjamin was studied by Whiteside (1962) who found a filamentous centrum structure similar to that of species of *Microascus* and *Petriella.* Although he was unable to discuss the origin of the ascogenous hyphae, these were seen to eventually fill the centrum and to produce terminal, lateral, and intercalary asci.

The cleistothecial Microascaceae have been more poorly studied. Only *Kernia nitida* (Saccardo) Nieuw has been examined in detail (Satina, 1923; Moreau and Moreau, 1953). In this species, centrum development proceeds in a manner similar to that of the ostiolate Microascaceae. Sterile filaments radiate inward from the walls to fill the locule. The ascogenous hyphae arise at the center and penetrate throughout, producing asci terminally, laterally, and intercalarily. Satina

(1923) reported the asci to arise from croziers, but Moreau and Moreau (1953) were unable to confirm this. Malloch (1970a) was unable to find croziers in any of the Microascaceae.

Cephalothecaceae

The Cephalothecaceae family name is presently used only for a single species, *Cephalotheca sulfurea* Fuckel, although it had been used in the past for all plectomycetous forms with a cephalothecoid peridium (see Malloch and Cain, 1970b). In spite of the fact that the developmental morphology of *C. sulfurea* has been studied in detail, other taxonomically important data are lacking; notably a precise description of the ascospores.

According to Chesters (1934), the centrum of *C. sulfurea* is filled at first with a pseudoparenchymatous tissue. Ascogenous hyphae arise near the center and spread throughout the centrum, producing the subspherical asci uniformly and obliterating the pseudoparenchyma. Paraphyses are lacking.

Hypocreales

Nectriaceae

According to Luttrell (1951), the *Nectria*-type development is characterized by the production of perithecia within a stroma and by a centrum composed of pseudoparaphyses interspersed with basally arranged unitunicate asci.

There are five genera of cleistothecial forms that probably belong in the Nectriaceae: *Amylocarpus* Currey, *Battarrina* (Saccardo) Clements and Shear, *Heleococcum* Jorgensen, *Mycorhynchidium* Malloch and Cain, and *Roumegueriella* Speg. Of these, only *Amylocarpus* has been studied developmentally (Lindau, 1899; Cavaliere, 1966). Although no information is available on the very early stages of centrum development, Lindau reported the young ascocarps to be filled with narrow hyphae. Later, the ascogenous hyphae grow throughout the sterile tissues and produce asci uniformly. Unfortunately, Lindau was unable to determine the origin of the sterile threads.

Trichocomaceae

This large and important family is composed entirely of plectomycetous forms. As recently monographed by Malloch and Cain (1972b) and Subramanian (1972), the family contains 20 genera. Most species are characterized by complex phialidic structures assignable to *Aspergillus* Micheli ex Link, *Paecilomyces* Bainer, or *Penicillium* Link ex Fr.

Cain (1956a) and Malloch and Cain (1972b) have suggested that the Trichocomaceae have evolved from a hypocrealean ancestor and have undergone progressive stages of reduction in ascocarp complexity (Fig. 6.2).

Developmental studies of species having stromata with cleistothecia have been

Group I. Stromata with cleistothecia
- *Dichlaena* Montagne and Durand
- *Emericella* Berkeley and Broome
- *Fennellia* Wiley and Simmons
- *Petromyces* Malloch and Cain
- *Trichocoma* Junghuhn

Group II. Stromata with cleistothecia lacking
- *Dendrosphaera* Patouillard
- *Eupenicillium* Ludwig
- *Hemicarpenteles* Sarbhoy and Elphick
- *Penicilliopsis* Solms-Laubach

Group III. Stromata lacking, cleistothecia pseudoparenchymatous
- *Chaetosartorya* Subramanian
- *Dactylomyces* Sopp
- *Eurotium* Link ex Fries
- *Neosartorya* Malloch and Cain
- *Sclerocleista* Subramanian

Group IV. Stromata lacking, cleistothecia (gymnothecia) hyphal
- *Hamigera* Stolk and Samson
- *Sagenoma* Stolk and Orr
- *Talaromyces* C. R. Benjamin
- *Warcupiella* Subramanian

Group V. Stromata and cleistothecia lacking
- *Byssochlamys* Westling
- *Edyuillia* Subramanian
- *Saccharomycopsis* Schiönning

Fig. 6.2. Stages of reduction in ascocarp complexity.

attempted only with *Emericella* species. The tissue of the stroma is reduced to masses of individually free cells, called "hulle cells," and the ascocarps are borne within these. Benjamin (1955) reported on the development of four species of *Emericella* Berkeley and Broome. The asci are produced uniformly throughout the centrum, apparently without paraphyses or similar structures. Ascocarps are initiated by a thin-walled cell similar to the young hulle cells that becomes surrounded by a tight hyphal coil (Malloch, 1970b). The most complete account of ascocarp development in *Emericella* is that of Borzi (1885) who studied *Emericella variecolor* Berkeley and Broome (as *Inzengaea erythrospora* Borzi). Borzi also reported thin-walled ascocarp initials and was able to follow the centrum development in considerable detail. The centrum is reported to be filled with a mass of hyphae from a very early stage of development. Eventually these hyphae give rise to the asci, which are uniformly distributed in the locule. Sterile centrum tissues are evidently lacking.

The striking species *Trichocoma paradoxa* Junghuhn has attracted the attention of two developmental morphologists (Fischer, 1890; Boedijn, 1935). At first, the ascocarps are composed of a rather woody cuplike base containing a single large ascocarp with a hyphal peridium. Inside the peridium is a dense network of vertically oriented parallel hyphae that completely fills the centrum. At a fairly early stage of development, the peridium exposed at the apex of the stromatic base ruptures, and the hyphal interior (called "the gleba" by Boedijn) expands upward and then frays out. The whole mature structure resembles the tip of a camel's-hair brush. The ascogenous hyphae are produced at the base of the cleistothecium where it is in contact with the cup. These hyphae radiate a short distance up into the gleba and produce uniformly distributed spherical asci. The oldest asci are at the tips of the ascogenous hyphae, and the spores released through their dissolution are pushed up through the gleba by the developing asci below. Eventually, the entire gleba fills with spores.

The best studied genus of fungi with stromatic ascocarps without cleistothecia is *Eupenicillium* Ludwig (Brefeld, 1874; Dodge, 1933; Emmons, 1935; Benjamin, 1955). The ascocarps in *Eupenicillium* are usually described as sclerotioid and are not initiated by ascogonia. Instead, a highly branched mass of mycelium arises that eventually forms at its center the hard sclerotioid bodies. Within these bodies (stromata), the ascogenous hyphae arise (it has not yet been established how) and grow out into the disintegrating tissue of the stroma. Asci are produced uniformly throughout the locule and may occur in helicoid chains or from croziers. There is often a considerable lag in the time between the maturation of the stromata and initiation of the ascogenous hyphae. Certain species of *Penicillium,* such as *Penicillium thomii* Maire, produce stromata that have never been found to contain asci. It is quite possible that these structures require specific stimuli to initiate formation of the ascigerous stage, as in many Sclerotiniaceae.

The developmental morphology of the astromatic pseudoparenchymatous cleistothecia in the genera *Neosartorya* Malloch and Cain and *Eurotium* has been examined in some detail. *Eurotium* species, studied first by De Bary (1854) and later by Dangeard (1907), H. C. I. Fraser and Chambers (1907), Dale (1909), Moreau and Moreau (1953), and Benjamin (1955), begin development with a tightly coiled ascogonium. The ascocarps, even at maturity, are very small and consist at first of the ascogonial coil surrounded by two or three layers of pseudoparenchymatous tissue. Ascogenous hyphae arise from the coil and displace all but the outer layer of pseudoparenchyma, producing asci uniformly from croziers. Sterile hyphae are absent from the centrum.

The development of *Neosartorya fischeri* (Wehmer) Malloch and Cain was studied by Domaradsky (1908) and Olive (1944). Development in this species resembles that of *Eurotium* species, except that the ascocarps are larger and the tissue of the centrum tends to remain more hyphal in nature until late in development. The ascogenous hyphae obliterate the sterile centrum tissues and produce abundant, uniformly distributed asci from croziers.

Astromatic *Talaromyces* species are typical of cleistothecia which are hyphal in nature and lack pseudoparenchyma altogether. They have been studied by Dangeard (1907), Emmons (1935), and Benjamin (1955), who reported the simple development of ascogenous hyphae and asci within the loose hyphal envelope. At this stage of development, the use of the term centrum becomes pointless.

A final group lacks ascocarps, and hence centra are lacking.

Pezizales

Ascobolaceae

The development of the one cleistothecial member of this family, *Guilliermondia saccoboloides* Boudier, has not been studied in detail. However, Boudier (1904) illustrated the asci as radiating from a basal cushion among apically free paraphyses. The asci consist of spherical spore-bearing portions borne on a long stripe, and they are evanescent. The purplish ascospores remain in a group as in *Saccobolus*.

Monascaceae

The genus *Monascus* von Tiegham has attracted the attention of a number of developmental morphologists (Barker, 1903; Kuyper, 1905; Dangeard, 1907; Young, 1931). The ascogonium in *Monascus* species is extremely characteristic, being borne at the apex of a hyphal stalk and subtended by a filamentous antheridium. After fertilization, the ascogonium becomes surrounded by hyphae arising from the cells below it to form the ascocarp primordium. The ascogenous hyphae arise from the division of the ascogonium and produce the asci directly along their length, without croziers. There are no sterile filaments, and the centrum tissues merely disintegrate as the asci develop. The ascocarps are very small and may contain as few as one or two asci.

Onygenaceae

This family, as treated by Malloch and Cain (1971a), contained only fungi having cleistothecia with a distinct pseudoparenchymatous peridium. The Gymnoascaceae should be included with this group as it differs only in having hyphal peridia.

The hyphal members of the Onygenaceae (Gymnoascaceae) have been studied developmentally by a number of authors. Because the ascocarp is only a hyphal network, there is no centrum in the strict sense. The ascogenous hyphae and asci merely develop at the center of the hyphal mass.

The pseudoparenchymatous members of the family, however, do form a true centrum; unfortunately, they have not been adequately studied. Dangeard's (1907) study of *Aphanoascus cinnabarinus* Zukal (not *A. cinnabarinus* sensu Udagawa and Takada, 1973) is the most complete for the family and is probably a fairly typical representative. The ascogonium becomes surrounded by a dense hyphal network, which enlarges to the size of the mature ascocarp. The ascogenous system spreads throughout this network, producing asci. When the asci are nearly mature, the outer tissues of the ascocarp become more compact and form a pseudoparenchymatous peridium. The sterile filaments in the centrum finally disintegrate, leaving only the dry mass of spores.

The centrum, composed at first of a dense mass of sterile filaments, appears to be general throughout the Onygenaceae (Malloch and Cain, 1970a, 1971a) and helps to distinguish them from members of other families. In some species, these centrum elements have become modified. In *Ascocalvatia alveolata* Malloch and Cain, they are irregularly thickened and probably aid in spore dispersal (Malloch and Cain, 1971a), whereas in *Pleuroascus nicholsonii* Massee and Salmon, they disarticulate to form conidia (Malloch and Benny, 1973).

The genus *Onygena* Persoon ex Fries presents an unusual situation in having what appears to be more than one ascogonium per ascocarp (personal observation).

Pyronemataceae

Most members of this family are apothecial with ascocarps containing a basal layer of operculate asci and paraphyses. A few species, however, have become reduced to the plectomycetous form.

Warcupia Paden and Cameron, a fairly representative cleistothecial member of the Pyronemataceae, has been studied by Paden and Cameron (1972). The centrum in young ascocarps is packed with filamentous tissues that are free at one end and probably represent paraphyses. The asci appear in small, randomly oriented clusters along the wall of the locule. The asci are deliquescent, and the ascocarp opens irregularly. The paraphyses are persistent at maturity.

Xeromyces Fraser, studied by L. Fraser (1953), is a much simpler form in which no paraphyses or sterile cells occur. The ascogonium is three celled and becomes surrounded by hyphae arising from its base to form the young cleistothecium. The two-spored asci are produced directly from the cells of the ascogenous hyphae.

The development of *Microeurotium albidum* Ghatak is even simpler (Ghatak, 1936). Here, the ascogonium itself becomes the single ascus and is loosely surrounded by hyphal peridial tissue.

Although it has not been studied developmentally, *Orbicula parietina* (Schrader ex Fries) Hughes serves as an example of a more discomycete-like form. The cylindrical evenescent asci dissolve, and the locule becomes filled with a dry mass of ascospores.

General Conclusions

From a strictly developmental point of view some of the families used above will appear to be extremely heterogeneous. The reason for this is that they have been defined largely on taxonomic characters other than developmental ones. What interests the taxonomist is not the presence of a single character in all members of a taxon; instead he looks for consistent *character groups,* where any one character may be lacking but many others are present. Thus, to a taxonomist, a dog born with three legs is still a dog because he has most of the characters of a dog, in spite of not being a quadruped.

If, then, you accept the taxonomic soundness of the groups outlined (Fig. 6.1) (and many do not) it is possible to assess what is happening developmentally. In doing this I offer the following hypothesis: *in each family there is a basic developmental type from which all the plectomycetous forms in that family can be derived, and that this evolution proceeds in the direction of increasing simplification.*

In many of the families there have been just too few developmental studies to test this hypothesis, but in some such a pattern seems to emerge. Most notable in this respect is the Sporormiaceae.

If we accept the works cited above, it appears that in the genus *Sporormiella* there are at least two forms of the Pleospora-type centrum; one with pseudoparaphyses and one with paraphysoids. In the cleistocarpous species *Preussia funiculata* and *P. typharum* paraphysoids occur while in *P. nigra* and *P. multilocularis* pseudoparaphyses are found. Thus the two types seem to have been carried into the first plectomycetous level of evolution. Within the genus *Preussia,* we have observed above that there are species having neither paraphysoids nor pseudoparaphyses but, instead, a mass of hyphal filaments radiating into the locule from the inner walls. The arrangement of the asci in this genus varies from strictly basal to a more or less loosely parietal arrangement to irregular. The important thing to note here is that *disorganization of the sterile tissue goes hand in hand with the loss of hymenial layers.* What appears to be happening here is that once the ostiole is lost, forcible discharge of ascospores is lost, followed by a more space-saving arrangement of asci. It is possible, of course, to argue that evolution occurred in the other direction, but a little thought will show this to be a rather teleological explanation. The gradual evolution of the centrum toward forcible discharge of ascospores through an ostiole, before the ostiole has evolved, is extremely unlikely. Evolutionary mechanisms just do not plan for the future!

The final (or latest) stage in the sporormiacean cleistothecial line is seen in *Pycnidiophora* species. *Pycnidiophora multispora* still has sterile filaments in the centrum but has uniformly distributed spherical asci. *Pycnidiophora dispersa* has lost the filaments altogether and economically uses the entire locule for the production of closely packed, small, spherical asci.

A similar pattern of evolution can be observed in the Chaetomiaceae, where we can follow, in the genus *Thielavia,* the gradual loss of an ostiole-oriented hymenium. In this family the sterile filaments of the hymenium seem to be on the wane in *Chaetomium* and to be already lacking in most species of *Thielavia* (although Malloch & Cain (1973b) reported paraphyses in *Thielavia spirotricha* (R. K. Benj.) Malloch & Cain). The evolution toward a more compact arrangement of asci proceeds from the basal arrangement in *T. fimeti* and *T. hyrcaniae* through the centrally fasciculate asci of *T. setosa* and *T. coactilis* to the uniform arrangement of *T. sepedonium* and *T. terricola. Thielavia terrestris* (Apinis) Malloch & Cain, with its uniformly distributed, subglobose asci is representative of the most highly evolved species.

Further evolutionary patterns can be drawn from the Nectriaceae and Pyronemataceae, but in these families developmental studies are few.

A different situation arises in the large and highly evolved families Pseudeurotiaceae, Endomycetaceae, Microascaceae, Trichocomaceae, and Onygenaceae.

Here the ancestral groups are not easily determined and the direction of evolution is often unclear. As noted above, the Trichocomaceae is possibly derived from the Hypocreaceae, but connecting links are obscure. By comparing what few taxonomic characters are left to us in these highly reduced families we can speculate that the Pseudeurotiaceae and Endomycetaceae are derived from forms resembling members of the Diaporthaceae and that the Microascaceae are similar to some Chaetomiaceae. The Onygenaceae were thought by Cain (1959) to be derived from certain Pezizales.

If the above five families leave us guessing as to their origins, their high degree of evolution compensates in a fairly uniform centrum structure. The Pseudeurotiaceae, Endomycetaceae and many Trichocomaceae, with their pseudoparenchymatous centra, lack of sterile filaments and uniformly dispersed asci form a developmental type that, in a somewhat expanded sense, conforms to Luttrell's (1951) *Ophiostoma* type.

The Microascaceae form a second type, called here the *Microascus* developmental type, whereby the walls lining the locule produce short tapering filaments, among which the ascogenous hyphae develop. The asci are always uniformly distributed at maturity.

Many Onygenaceae produce a developmental type in which the centrum is entirely filamentous until late in development. The ascogenous hyphae and asci simply develop among the filaments. When the asci are nearly mature, the cleistothecial wall becomes pseudoparenchymatous (or at least solid). In those species where the ascocarp becomes pseudoparenchymatous at an earlier stage of development (such as *Arachnomyces* Mass. & Salm.) the centrum remains filamentous. Because the most detailed study of this type was on *Aphanoascus* Zukal (Dangeard, 1907), I prefer to call it the *Aphanoascus* developmental type.

The filamentous ascocarps in many Onygenaceae and some Trichocomaceae are reduced to the point that there is really no centrum at all. Ascocarps of this type, called "gymnothecia" by Novak & Galcozy (1966), occur in the end lines of the Onygenaceae, Trichocomaceae and possibly Pseudeurotiaceae (*Myxotrichum* Kunze ex Fr.) and are not of taxonomic significance at the family level.

Finally, recognition should be extended to the *Phaeotrichum* developmental type of Barr (1956). This type occurs in cleistothecial forms with pseudoparenchymatous ascocarp initials and centra characterized by a lack of sterile filaments and uniformly distributed asci. It is known in *Phaeotrichum hystricinum* and *Pycnidiophora dispersa* but probably occurs in other Pleosporales as well. It is the loculoascomycetous counterpart of the *Ophiostoma* developmental type.

In summary, then, we can say that the plectomycete centrum is a more complex subject than it was originally thought to be. It appears that in each family the most "primitive" cleistothecial forms have a centrum type similar to that of its ancestral "conventional" form and that progressive simplification occurs. In the most highly evolved families there is a more or less stable centrum morphology that can be assigned to one of at least four types. Convergent evolution is common in these fungi, however, and taxonomists should be cautious in using centrum structure at the expense of other evidence.

References

Ahmed, S.I., 1964. Classification of coprophilous ascomycetes. The genus *Sporormia*. Ph.D. Thesis, University of Toronto, Ontario, Canada.

Arnold, C.A., 1928. The development of the perithecia and spermogonium of *Sporomia leporina* Niessl. Amer. J. Bot. 15: 241–245.

Barker, B.T.P., 1903. The morphology and development of the ascocarp in *Monascus*. Ann. Bot. 17: 167–236.

Barr, M.E., 1956. The development of the ascocarp in *Phaeotrichum hystricinum*. Can. J. Bot. 34: 563–568.

Beatus, R., 1938. Entwicklungsgeschichte und zytologische Untersuchungen der Ascomyceten. *Perisporium funiculatum*. Jahrb. Wiss. Bot. 87: 301–323.

Benjamin, C.R., 1955. Ascocarps of *Aspergillis* and *Penicillium*. Mycologia 47: 669–687.

Bessey, E.A., 1935. A Text-Book of Mycology. Blakiston's, Philadelphia.

Blanchard, R.O., 1972. Origin and development of ascogenous hyphae and pseudoparaphyses in *Sporormia australis*. Can. J. Bot. 50: 1725–1729.

Boedijn, K.B., 1935. On the morphology and cytology of *Trichocoma paradoxa*. Ann. Jard. Bot. Buitenzorg 44: 243–266.

Borzi, A., 1885. *Inzengaea,* ein neuer Askomycet. Jahb. Wiss. Bot. 16: 450–463.

Boudier, E., 1904. Sur un nouveau genre et une nouvelle espece de Myriangiacees, le *Guillermondia saccoboloides*. Bull. Soc. Mycol. France 20: 19–22.

Boylan, B.V., 1970. The cytology and development of *Preussia flanaganii* sp. nov. Can. J. Bot. 48: 163–166.

Brefeld, O., 1874. Botanische Untersuchungen über Schimmelpilze. II. Die Entwicklungsgeschichte von *Penicillium*. Felix Verlag, Leipzig.

Cain, R.F., 1956a. Studies of soil fungi. *Saturnomyces,* a new genus of the Aspergillaceae. Can. J. Bot. 34: 135–141.

Cain, R.F., 1956b. Studies of coprophilous ascomycetes. II. *Phaeotrichum,* a new cleistocarpous genus in a new family, and its relationships. Can. J. Bot. 34: 675–687.

Cain, R.F., 1956c. Studies of coprophilous ascomycetes. IV. *Tripterospora,* a new cleistocarpous genus in a new family. Can. J. Bot. 34: 699–710.

Cain, R.F., 1959. The Plectascales and Perisporiales in relation to the evolution of the ascomycetes. *In:* Ninth International Bot. Congr. Abstracts, Vol. 2, p. 56.

Cain, R.F., 1961a. Studies of soil fungi. III. New species of *Coniochaeta, Chaetomidium,* and *Thielavia*. Can. J. Bot. 39: 1231–1237.

Cain, R.F., 1961b. Studies of coprophilous ascomycetes. VII. *Preussia*. Can. J. Bot. 39: 1633–1666.

Cain, R.F., 1961c. *Anixiella* and *Diplogelasinospora,* two genera with cleistothecia and pitted ascospores. Can. J. Bot. 39: 1667–1677.

Cain, R.F., 1972. Evolution of the fungi. Mycologia 64: 1–14.

Cavaliere, A.R., 1966. Marine ascomycetes: Ascocarp morphology and its application to taxonomy. I. *Amylocarpus, Ceriosporella* gen. nov., *Lindra*. Nova Hedw. 10: 387–398.

Chadefaud, M., A. Parguey-Leduc, and M. Boudin, 1966. Sur les périthèces et les asques du *Preussia*. Bull. Soc. Mycol. France 82: 93–122.

Chesters, C.G.C., 1935. Studies on British pyrenomycetes. I. The life histories of three species of *Cephalotheca*. Trans. Br. Mycol. Soc. 19: 261–279.

Cookson, I., 1928. The structure and development of the perithecium in *Melanospora zamiae*. Ann. Bot. 42: 255–269.

Corlett, M., 1966. Developmental studies in the Microascaceae. Can. J. Bot. 44: 79–88.

Crooks, K.M., 1935. An account of the cultural and cytological characteristics of a new species of *Mycogala*. Proc. Roy. Soc. Victoria New Ser. 47: 352–364.
Dale, E., 1909. On the morphology and cytology of *Aspergillus repens* de Bary. Ann. Mycol. 7: 215–225.
Dangeard, P.A., 1907. L'origine du périthèce chez les ascomycetes. Le Botaniste 10: 1–385.
De Bary, A., 1854. Ueber die Entwicklung und den Zusammenhang von *Aspergillus glaucus* und *Eurotium*. Bot. Z. 12: 425–471.
Dodge, B.O., 1933. The perithecium and ascus of *Penicillium*. Mycologia 25: 90–104.
Doguet, G., 1955. Le genre *Melanospora:* Biologie, morphologie, développement, systématique. Le Botaniste 39: 1–313.
Doguet, G., 1956. Le genre *Thielavia* Zopf. Rev. Mycol. 21, Suppl. v. 1: 1–21.
Domaradsky, M., 1908. Zur Fruchtkörperentwicklung von *Aspergillus fischeri* Wehmer. Ber. Deutsch. Bot. Ges. 26: 14–16.
Emmons, C.W., 1930. *Coniothyrium terricola* proves to be a species of *Thielavia*. Bull. Torrey Bot. Club 57: 123–126.
Emmons, C.W., 1932. The development of the ascocarp in two species of *Thielavia*. Bull. Torrey Bot. Club 59: 415–422.
Emmons, C.W., 1935. The ascocarps in species of *Penicillium*. Mycologia 27: 128–150.
Emmons, C.W., and B.O. Dodge, 1931. The ascosporic stage of species of *Scopulariopsis*. Mycologia 23: 313–331.
Fischer, E., 1890. Beiträge zur Kenntnis exotischen Pilze, *Trichocoma paradoxa* Jungh. Hedwigia 29: 161–171.
Fraser, H.C.I., and H.S. Chambers, 1907. The morphology of *Aspergillus herbariorum*. Ann. Mycol. 5: 419–431.
Fraser, L., 1953. A new genus of the Plectascales. Proc. Linn. Soc. N.S.W. 78: 241–246.
Gaumann, E., 1964. Die Pilze. Birkhauser Verlag, Basel.
Ghatak, P.N., 1936. On the development of the perithecium of *Microeurotim albidum*. Ann. Bot. 50: 849–861.
Gusseva, K., 1925. Zur Entwicklungsgeschichte von *Cephalotheca polyporicola* Jaczevski. 1. Soc. Bot. Russ. 10: 229–238.
Huang, L.H., 1976. Developmental morphology of *Triangularia backusii* (Sordariaceae). Can. J. Bot. 54: 250–267.
Kowalski, D.T., 1964. The development and cytology of *Pycnidiophora dispersa*. Amer. J. Bot. 51: 1076–1082.
Kowalski, D.T., 1965a. Development and cytology of *Preussia typharum*. Bot. Gaz. 126: 123–130.
Kowalski, D.T., 1965b. The development and cytology of *Melanospora tiffanii*. Mycologia 57: 279–290.
Kowalski, D.T., 1966. The morphology and cytology of *Preussia funiculata*. Amer. J. Bot. 54: 1036–1041.
Kowalski, D.T., 1968. Morphology and cytology of *Preussia isomera*. Bot. Gaz. 129: 121–125.
Kuyper, H.P., 1905. Die Perithecien-Entwicklung von *Monascus purpureus* Went und *Monascus parkeri* Dangeard und die systematische Stellung dieser Pilze. Ann. Mycol. 3: 32–71.
Lindau, G., 1899. Ueber Entwicklung und Ernährung von *Amylocarpus encephaloides* Currey. Hedwigia 38: 1–19.
Luttrell, E.S., 1951. Taxonomy of the pyrenomycetes. Univ. Missouri Studies, No. 24, 1–120.

Luttrell, E.S., 1955. The ascostromatic ascomycetes. Mycologia 47: 511–532.

Maciejowska, Z., and E.B. Williams, 1963. Studies on a multiloculate species of *Preussia*. Mycologia 55: 257–270.

Malloch, D., 1970a. New concepts in the Microascaceae illustrated by two new species. Mycologia 62: 727–740.

Malloch, D., 1970b. The genera of cleistothecial Ascomycota. Ph.D. Thesis, University of Toronto.

Malloch, D., and G.L. Benny, 1973. California ascomycetes: four new species and a new record. Mycologia 65: 648–660.

Malloch, D., and R.F. Cain, 1970a. The genus *Arachnomyces*. Can. J. Bot. 48: 839–845.

Malloch, D., and R.F. Cain, 1970b. Five new genera in the new family Pseudeurotiaceae. Can. J. Bot. 48: 1815–1825.

Malloch, D., and R.F. Cain, 1971a. New genera of Onygenaceae. Can. J. Bot. 49: 839–846.

Malloch, D., and R.F. Cain, 1971b. Four new genera of cleistothecial ascomycetes with hyaline ascospores. Can. J. Bot. 49: 847–854.

Malloch, D., and R.F. Cain, 1971c. The genus *Kernia*. Can. J. Bot. 49: 855–867.

Malloch, D., and R.F. Cain, 1971d. New cleistothecial Sordariaceae and a new family Coniochaetaceae. Can. J. Bot. 49: 869–880.

Malloch, D., and R.F. Cain, 1972a. New species and combinations of cleistothecial ascomycetes. Can. J. Bot. 50: 61–72.

Malloch, D., and R.F. Cain, 1972b. The Trichocomataceae: Ascomycetes with *Aspergillus, Paecilomyces* and *Penicillium* imperfect states. Can. J. Bot. 50: 2613–2628.

Malloch, D., and R.F. Cain, 1973a. The Trichocomaceae (ascomycetes): Synonyms in recent publications. Can. J. Bot. 51: 1647–1648.

Malloch, D., and R.F. Cain, 1973b. The genus *Thielavia*. Mycologia 65: 1055–1077.

Malloch, D., and C.T. Rogerson, 1977. *Pulveria,* a new genus of Xylariaceae (ascomycetes). Can. J. Bot. 55: 1505–1509.

Moreau, F., and M. Moreau, 1953. Étude du développement de quelques Aspergillacees. Rev. Mycol. 18: 165–180.

Nannfeldt, J.A., 1932. Studien über die Morphologie und Systematik der nicht-lichenisierten inoperculaten Discomyceten. Nova Acta. Reg. Soc. Sci. Upsala, Ser. IV 8(2): 1–368.

Nichols, M.A., 1896. The morphology and development of certain pyrenomycetous fungi. Bot. Gaz. 22: 301–328.

Nicot, J., and D. Longis, 1961. Structure des spores et organisation des périthèces de deux *Thielavia* du sol. C. R. Acad. Sci. Paris 253: 304–306.

Novak, E., and J. Galcozy, 1966. Notes on dermatophytes of soil origin. Mycopathol. Mycol. Appl. 28: 289–296.

Olive, L.S., 1944. Development of the perithecium in *Aspergillus fischeri* Wehmer, with a description of crozier formation. Mycologia 36: 266–275.

Paden, J.W., and J.V. Cameron, 1972. Morphology of *Warcupia terrestris,* a new ascomycete genus and species from soil. Can. J. Bot. 50: 999–1001.

Parguey-Leduc, A., 1970. Le développement des périthèces du *Lepidosphaeria nicotiae.* Bull. Soc. Mycol. France 86: 715–724.

Parguey-Leduc, A., 1974. Les asques et l'ontogenie des périthèces chez les *Trichodelitschia.* Bull. Soc. Mycol. France 90: 101–120.

Rai, J.N., and K. Wadhwani, 1970. *Anixiella indica*—Studies in ascocarp development and ascus cytology. J. Gen. Appl. Microbiol. (Tokyo) 16: 251–258.

Redhead, S.A., and D. Malloch, 1977. The Endomycetaceae: New concepts, new taxa. Can. J. Bot. 55: 1701–1711.

Routien, J.B., 1956. A new species of *Muellerella* and its development. Bull. Torrey Bot. Club 83: 403–409.

Samson, R.A., and J. Mouchacca, 1975. Two new soil-borne cleistothecial ascomycetes. Can. J. Bot. 53: 1634–1639.

Satina, S., 1918. Studies in the development of certain Sordariaceae. Bull. Nat. Moscow 1918: 166–142.

Satina, S., 1923. Beiträge zur Kenntnis des Ascomyceten *Magnusia nitida* Sacc. I. Befruchtung und Entwicklungsgeschichte des Peritheciums; Nebenfruchtform des Pilzes. Bot. Ark. 3: 273–381.

Subramanian, C.V., 1972. The perfect states of *Aspergillus*. Curr. Sci. 41: 755–761.

Udagawa, S., and M. Takada, 1973. The rediscovery of *Aphanoascus cinnabarinus*. J. Jap. Bot. 48: 21–26.

Upadhyay, H.P., and W.B. Kendrick, 1975. Prodromus for a revision of *Ceratocystis* (Microascales, ascomycetes) and its conidial states. Mycologia 67: 798–805.

Vincens, F., 1917. Une nouvelle espece de *Melanospora: M. mangini*. Bull. Soc. Mycol. France 33: 67–69.

von Arx, J.A., 1971. Testudinaceae, a new family of ascomycetes. Persoonia 6: 365–369.

von Arx, J.A., 1975. On *Thielavia* and some similar genera of ascomycetes. Stud. Mycol. 8: 1–29.

von Arx, J.A., and E. Müller, 1954. Die Gattungen der amerosporen Pyrenomyceten. Beitr. Krypt. Fl. Schweiz 11: 1–434.

Webster, J., 1970. Introduction to Fungi. Cambridge Univ. Press, London.

Whiteside, W.C., 1961. Morphological studies in the Chaetomiaceae. I. Mycologia 53: 512–523.

Whiteside, W.C., 1962. Morphological studies in the Chaetomiaceae. III. Mycologia 54: 611–620.

Young, E.M., 1931. The morphology and cytology of *Monascus ruber*. Amer. J. Bot. 18: 499–517.

Chapter 7
The Discomycete Centrum

J. W. KIMBROUGH

Introduction

Although the term "centrum" has not been generally used to describe apothetical development in the discomycetes, the concept includes the ascogenous system and sterile tissues occupying cleistothecial, perithecial, and pseudothecial cavities of other groups of ascomycetes. "Centrum" applies equally as well to discomycetes. Certainly a large number of taxa currently considered among the discomycetes display patterns of centrum development that are similar to, or often almost indistinguishable from, types of development described in other euascomycetes. In the discomycetes, any attempt to recognize centrum patterns must also take into account the role that the sterile tissue of the ascocarps plays in displaying the hymenium.

The Ascogonium

Among the discomycetes there appear to be some distinct patterns of development associated with the appearance and types of ascogonia. One very striking observation is that among the inoperculate discomycetes, ascogonia—as a general rule—develop within a primordium. As was well documented by Drayton (1934) in his study of the Sclerotiniaceae, the ascogonium differentiates within a cellular primordium. Bellemère (1967), in his outstanding work on close to 50 taxa of inoperculate discomycetes, shows that in all species studied, ascogonia or "sporophytic cells" develop within a primordium. Carpenter (1976) has shown more recently that ascogonia and trichogynes also develop within preexisting primordia of *Gelatinodiscus*.

In contrast, the ascogonium plays a dominant role in the initiation of apothecia in most of the operculate discomycetes. In numerous cytologic studies of operculate species (summarized by Kimbrough, 1970), the vegetative hyphae give rise to variously shaped ascogonia around which the ascocarp develops. There appear to be at least four types of ascogonia within the operculate discomycetes: the clustered, intertwined type, as found in *Ascodesmis* Van Teigham (O'Donnell et al.,

1976a); clusters of paired ascogonia and antheridia, as in *Pyronema* Carus (Harper, 1900); stalked ascogonia, as in species of *Scutellinia* (Cooke) Lambotte, *Coprobia* Boudier (Blackman and Fraser, 1906), and *Lasiobolus* Saccardo; beaded chains or ascogonial coils as in *Ascobolus* Persoon ex Hooker (O'Donnell et al., 1974); and several other Pezizales (O'Donnell and Hooper, 1974; O'Donnell et al., 1976a). In a large number of discomycetes, however, information on the ascocarp initials is lacking.

The development of a functional hymenium from an ascogonium in association with the formation of a stalk presents additional problems. The most thorough study of the types of stipitate discomycetes was provided by Corner (1929b, 1930a). In all types, the ascogonium is initially ensheathed. Then considerable growth of the excipular tissues of the apothecium occurs before hymenial differentiation. When excipular growth is active below the ascogonium (Fig. 7.1, B and C), the ascogonium becomes elevated somewhere within the stipe and often near the hymenium. In other taxa, the excipular growth is more active above the ascogonium, resulting in the latter remaining near the base of the stipe (Fig. 7.1, A and D). Through parallel growth of the excipular and ascogenous hyphae, the hymenium eventually differentiates in the terminal area.

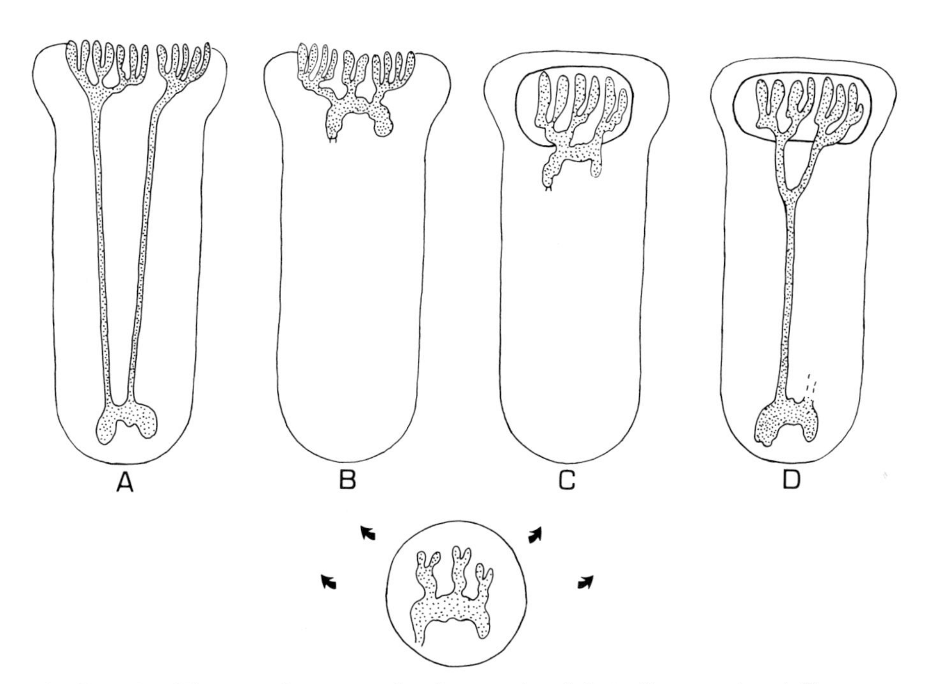

Fig. 7.1A–D. Diagram of ascocarp development in stipitate discomycetes. **A** Paragymnohymenial development with ascogonium remaining below; delayed hymenial development. **B** Paragymnohymenial, with excipular expansion below the ascogonium; delayed production of hymenium. **C** Cleistohymenial, with ascogonium and subsequent asci developing at the top of the stalk. **D** Cleistohymenial, with basal ascogonium; excipular cells interspersed with ascogenous hyphae.

The Hymenium

In cleistothecial and perithecial ascomycetes, the expansion of the ascocarp and creation of the hymenium are brought about, at least in part, by internal centrifugal forces resulting from the growth of the centrum. In the discomycetes, however, the growth and differentiation of the ascocarpic sterile tissues play a dominant role in determining the extreme variation in the display of the hymenium (Corner, 1929a, b, 1930a, b).

The hymenium in the discomycete fungus is analyzed in terms of the relationship to the formation of the ascocarp wall tissue. Two extremes of a continuum are the hymenium development within a complete closure and exposed at maturity or the hymenium development as an exposed superficial layer from inception. These two ends of the developmental spectrum as well as modifications, such as Corner's hemiangiocarpic type, show variance in aspects of the ascocarp wall.

The enclosed hymenium was called angiocarpic by Corner (Figure 3, 1929b). van Brummelen (1967) has proposed the term "cleistohymenial" for species in which the hymenium is enclosed, at least through early development. The cleistohymenial forms have been organized further according to the hymenium when asci are exposed. These include: (1) the archihymenial phase, before the initials of the hymenium are formed; (2) the prohymenial phase, when paraphyses are present but not croziers; (3) the mesohymenial phase, varying from crozier development into young asci until ripening asci appear; (4) the telohymenial phase, with mature asci present and spores discharged; and (5) the posthymenial phase, when the hymenium becomes obsolete (Fig. 7.2).

Most of both operculate and inoperculate discomycetes possess a cleistohymenial type of development. [A few, such as *Saccobolus versicolor* (Karsten) Karsten, open in the prohymenial phase (van Brummelen, 1967). Most, however, open during the mesohymenial phase (Paden and Stanlake, 1973).] All 50 taxa of inoperculate discomycetes studied by Bellemère (1967) appear cleistohymenial, opening in the mesohymenial phase. Several species of *Ascobolus* typify this type of development in the Pezizales (O'Donnell et al., 1974; Paden and Stanlake, 1973). Among the coprophilous discomycetes, there are many species that open in the telohymenial phase. *Ascobolus immersus* Persoon ex Persoon (van Brummelen, 1967), *Thelebolus crustaceous* (Fuckel) Kimbrough (Fig. 7.3, A) and *Lasiobolus monascus* Kimbrough (Fig. 7.3, B) are examples of species whose asci push through the excipulum at the time of spore liberation. Stipitate members of families of discomycetes, other than those mentioned below, appear to be cleistohymenial.

In gymnohymenial development, the hymenium is exposed from the first through the maturation of asci. Two groups of species can be recognized, according to the extent of investing hyphae, as eugymnohymenial and parahymenial types.

Eugymnohymenial development—in which the ascogonium is naked from the beginning—is typified by *Ascodesmis* species in which there is little or no excipular growth (O'Donnell et al., 1976a). *Pyronema* (Fig. 7 4, A) and *Coprotus* Korf and Kimbrough (Fig. 7.4, B) have a scant to well-developed excipulum.

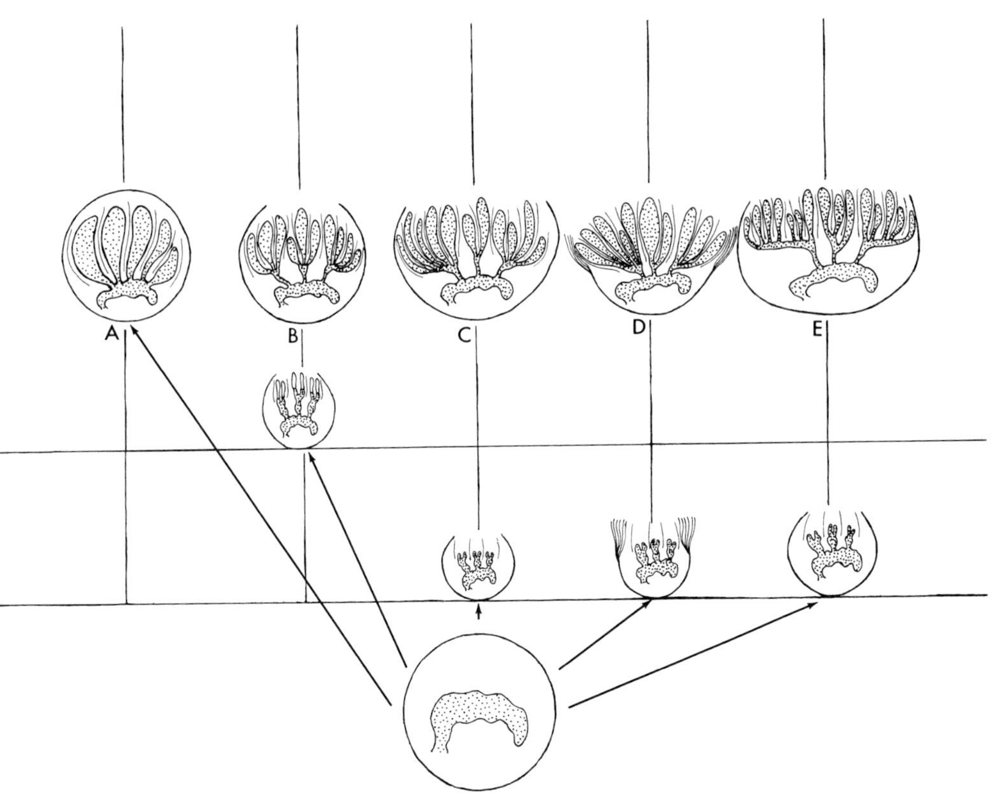

Fig. 7.2A–E. Scheme of ascocarp developmental types in the Ascobolaceae. **A** Closed. **B** Opening in a mesohymenial stage. **C–E** opening in a prohymenial stage. **D** Parahymnohymenial. **E** Eurymnohymenial with an excipulum. (After van Brummelen, 1967.)

Among the inoperculate discomycetes, eugymnohymenial development is noted in *Ascocorticium* Brefield (Oberwinkler et al., 1967) and *Karstenella* Harmaja. There is a real possibility that *Myriogonium* Cain and other reduced genera currently considered among the Endomycetales are in fact inoperculate discomycetes. von Arx (1967) has already established that *Amauroascus* Schroeder, a very reduced Eurotiales, has operculate asci.

In the paragymnohymenial types, the ascogonium is overarched by hyphae of limited growth. In both stipitate and sessile taxa with gymnohymenial development, the paraphyses may grow and branch extensively to form an epithecium around ascal tips. The species of *Helvella* exemplify paragymnohymenial development with an unlimited amount of ascogonial ensheathment. A similar development was also noted in the inoperculate genus *Microglossum*. However, some species develop a thin, membranous epithecium which was confusing to Corner (1930a) in his interpretation of gymnohymenial development. In certain taxa such as *Peziza quelepidotia* and *Thecotheus cinereus*, ascogonia emerge with clusters of conidia. These conidia often persist to form an epithecial layer over the asci (O'Donnell and Hooper, 1974; Conway, 1975; O'Donnell et al., 1976c).

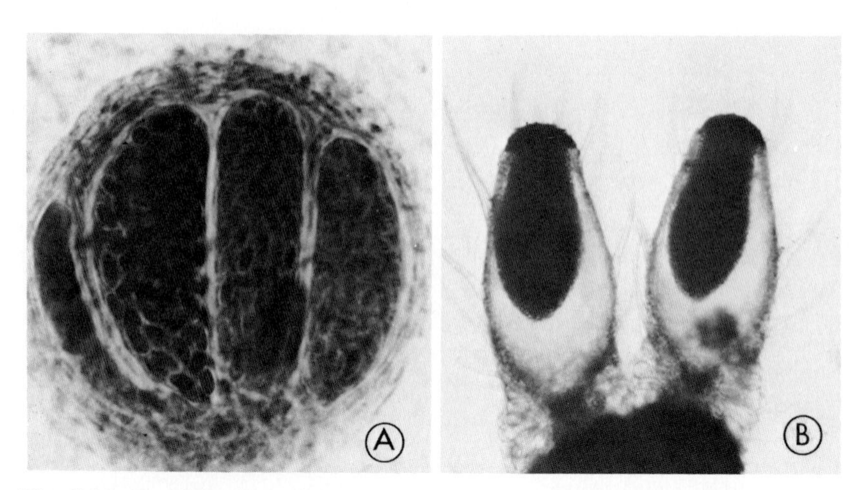

Fig. 7.3A, B. Cleistohymenial ascocarps opening in the telohymenial phase. **A** *Thelebolus crustaceous.* **B** *Lasiobolus monascus.*

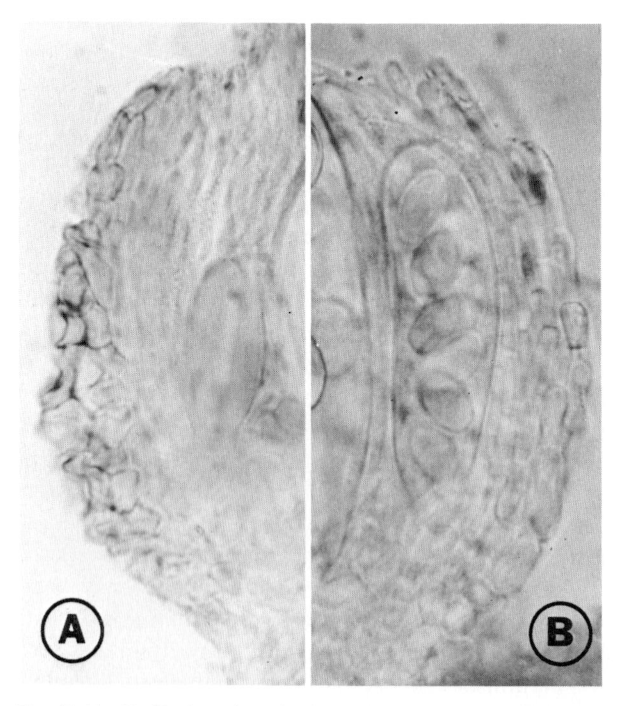

Fig. 7.4A, B. Reduced excipulum. **A** *Pyronema* sp. **B** *Coprotus glacellus.*

From the scanning electron microscope study of *P. quelepidotia* (O'Donnell and Hooper, 1974), it is difficult to determine whether development is cleistohymenial or paragymnohymenial. When serial sections are viewed with the light microscope, however, the development proves to be paragymnohymenial with an epithecium of conidia (O'Donnell et al., 1976c). A unique feature of *P. quelepidotia* is the elevation of the ascogonium within the stalk as it elongates. Similar information about most other stipitate discomycetes is lacking, although Brown (1910) showed that in *Leotia lubrica* (Scopoli) Persoon the ascogonium is basal with the ascogenous hyphae ramifying through the stipe to the hymenial area (Brown, 1910). McCubbins (1910) showed that in *Helvella elastica* Fries the ascogonia were subhymenial.

Paragymnohymenial development is represented in a number of Pezizales and is typified by species of *Saccobolus* Boudier (O'Donnell et al., 1974) and by the stipitate *P. quelepidotia* Korf and O'Donnell (O'Donnell and Hooper, 1974). In order to apply this term to stipitate species, the concept must be modified to include those also with extensive excipular growth. It appears that from the few detailed studies, coupled with numerous superficial observations, the stipitate Pezizaceae, Helvellaceae, and Morchellalaceae (operculates) and the Geoglossaceae and stipitate or capitate Leotiaceae are paragymnohymenial.

With the exception of the Phacidiales, most of the groups ascribed to the discomycetes possess a *Xylaria*-type centrum, or one that may easily be derived from such a centrum. Centrum development appears to be rather uniform in the Helotiales and Ostropales.

Cyttaria Berkeley has not been proved an operculate or inoperculate discomycete. The centrum is basically that of the *Xylaria* type, in which the apical tissues, instead of forming an ostiole, are spread widely apart to expose a wide hymenium of asci. Gamundi (1971) has shown that each locule of this compound ascocarp develops into an open apothecium.

Among those with cleistohymenial development, the ascocarp is usually initiated by a single ascogonial stalk or coil. In the Aleuriaceae–Otidiaceae complex (Kimbrough, 1970), the apothecium is initiated commonly by a stalked ascogonium in which the terminal cell gives rise to ascogenous hyphae—whereas those with ascogonial coils may bear ascogenous hyphae from one to several cells of the coil. The centrum of operculate discomycetes with a cleistohymenial development is essentially that of the *Xylaria*-type centrum. Branches from the stalk cells of the ascogonium, or from neighboring vegetative hyphae, surround the ascogonium and form the apothecium. Hyphal branches with free tips (paraphyses) grow upward and inward from the inner surface of the wall over the base and sides of the apothecium. Pressure exerted by the growth of opposed paraphyses expands the walls to create the cavity. The ascogonium produces ascogenous hyphae that grow up among the paraphyses to form a hymenium of asci. In the *Xylaria* centrum of pyrenomycetes, however, the layer of inwardly growing hyphae continues to expand upward to form a schizogenous ostiole. In the discomycetes, hymenial expansion ruptures the epithecial tissues, resulting in a cupulate or variously shaped apothecium.

It is among the cleistohymenial, inoperculate discomycetes that one finds the greatest modification in centrum development. This is especially true of those with stromatic, hysterothecioid ascocarps. In the Phacidiales, for example, Gordon (1966, 1968) outlined three distinct types of ascocarp development. Type I (Fig. 7.5) consists of pseudoparenchymous tissues in which central cells began to elongate to initially form pseudoparaphyses. These elements soon disintegrate, and the interstitial hyphae become true paraphyses. Anastomosing basal cells give rise to dikaryotic cells from which asci arise. Type II (Fig. 7.5) consists of ascocarps that are initiated by binucleate hyphae. Initial development of pseudoparaphyses is similar to type I. Upper cells of the ascocarp develop into a clypeus, whereas those below elongate. Certain binucleate cells within the pseudoparaphyses enlarge and become asci. These appear to develop at different levels, but the resulting asci position themselves in a true hymenium. Lysis of the interstitial filaments results in a single locule by the time of spore development. In the type III (Fig. 7.5) centrum, ascocarp initials may be subcuticular or subepidermal, developing radially into a parenchymatous sheath. These more or less isodiametric cells develop into columnar hyphae. Columnar cells become deeply pigmented, and those toward the center become somewhat elongate. One or two layers of the central

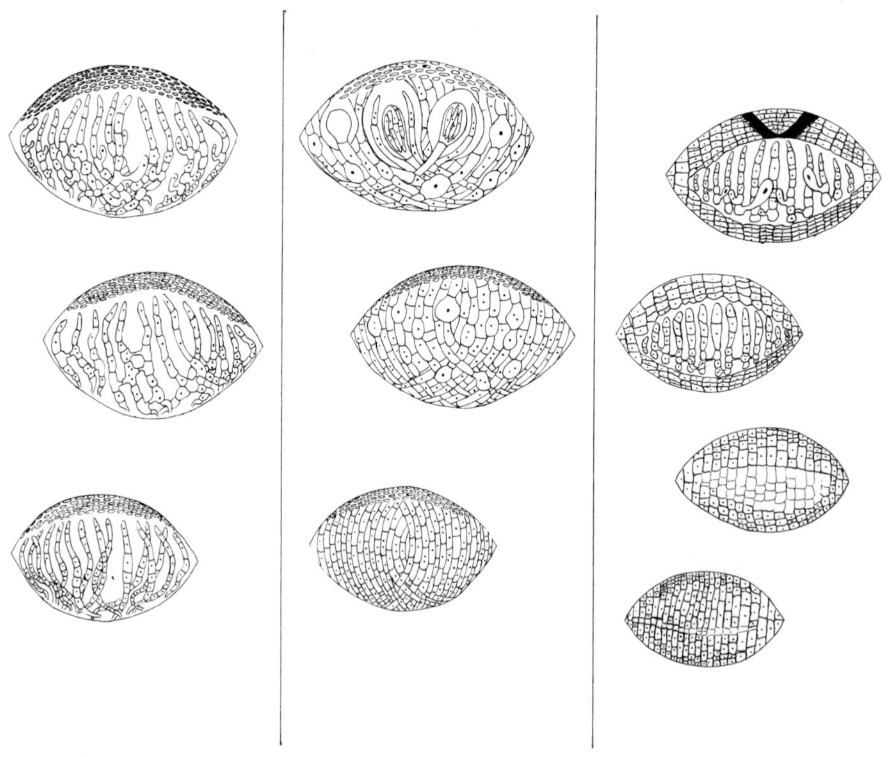

Fig. 7.5. Centrum development in the Hypodermataceae. Schematic drawings of stages in the centrum types I, II, and III. See text for explanation. (After Gordon, 1966, 1968.)

cells become hyaline and undergo divisions, resulting in a hyaline zone within the ascocarp. Some of the hyaline filamentous cells become free apically to form paraphyses. Asci are initiated through the anastomosis of certain basal cells of the pseudoparaphyses. A band of cells within the clypeus become hyaline, initiating the ostiole. Although superficially similar, the Hemiphacidiaceae of the Helotiales do not possess the overlying stromatic tissues, and true paraphyses are present. Most of the cupulate, sessile, and stipitate inoperculates appear to have, at least in the initial stages, a *Xylaria*-type centrum (Bellemère, 1967).

Of the eugymnohymenial discomycetes, both operculates (*Ascodesmis, Pyronema,* and *Coprotus*) and inoperculates (*Ascocorticium, Karstenella,* and perhaps others), there appear to be no comparable centrum types present among the ascomycetes. Most have apically free paraphyses and so may be interpreted as reduced *Xylaria* types in which the vegetative development is essentially depressed. It may be more than coincidental that operculate discomycete taxa with gymnohymenial development are initiated by clusters of paired antheridia and ascogonia. This character is shared with a very large number of plectomycetes.

General Conclusions

Although apothecial morphology is being used to a lesser extent in the contemporary systematics of discomycetes, the cellular, microchemical, and developmental features of these structures can tell us much about the taxonomy and phylogeny of cup fungi. It is becoming increasingly evident that ascocarp morphology alone is unreliable in most instances for showing natural relationships. On the basis of developmental morphology, however, some general conclusions may be reached.

1. The development of vegetative primordia within which ascogonia develop is a central feature of the inoperculate discomycetes. This feature is common among several major taxa of pyrenomycetes and loculoascomycetes but is not present in the Pezizales.
2. The apical pore apparatus in asci of inoperculate discomycetes, although varied at the ultrastructure level, is similar and parallel to those of the pyrenomycetes. The Pezizales have strikingly different asci.
3. Spermatia are common in both the pyrenomycetes and inoperculate discomycetes but are not found in the Pezizales.
4. Inoperculate discomycetes, like the pyrenomycetes and loculoascomycetes, are predominantly plant inhabiting and with many parasitic species. Plant parasitism is an exception in the Pezizales.
5. Naked ascogonia of the Pezizales is a characteristic shared with many plectomycetes.
6. Hysterothecioid ascocarps of the Phacidales, with their variety of centra, bear a close resemblance to the loculoascomycetes.

References

Bellemère, A., 1967. Contribution a l'etude du développement de l'apothecie chez discomycètes inoperculés. Bull. Soc. Mycol. France 83: 395–931.

Blackman, V.H., and H.C.I. Fraser, 1906. On the sexuality and development of the ascocarp in *Humaria granulata*. Proc. Roy. Soc. Bot. London 77: 354–368.

Boedijn, K.B., 1935. Two new Malaysian genera of discomycetes. Bull. Jard. Bot. Buitenzorg (3) 13: 478–483.

Boedijn, K.B., 1936. The genus *Cordierites* in the Netherlands Indies. Bull. Jard. Bot. Buitenzorg (3) 13: 525–529.

Boudier, E., 1905–1910. Icones Mycologicae. 4 vols. Klincksieck, Paris.

Brown, W.H., 1910. The development of the ascocarp in *Leotia*. Bot. Gaz. 50: 443–459.

Carpenter, S.E., 1976. Taxonomy, morphology and ontogeny of *Gelatinodiscus flavidus*. Mycotaxon 3: 209–232.

Conway, K.E., 1975. Ascocarp ontogeny and imperfect state of *Thecotheus* (Pezizales, Ascomycetes). Mycologia 67: 241–252.

Corner, E.J.H., 1929a. Studies in the morphology of discomycetes. I. The marginal growth of apothecia. Trans. Br. Mycol. Soc. 14: 263–274.

Corner, E.J.H., 1929b. Studies in the morphology of discomycetes. II. The structure and development of the ascocarp. Trans. Br. Mycol. Soc. 14: 275–291.

Corner, E.J.H., 1930a. Studies in the morphology of discomycetes. III. The Clavuleae. Trans. Br. Mycol. Soc. 15: 107–120.

Corner, E.J.H., 1930b. Studies in the morphology of discomycetes. IV. The evolution of the ascocarp. Trans. Br. Mycol. Soc. 15: 121–134.

Drayton, F.L., 1934. The sexual mechanism of *Sclerotinia gladioli*. Mycologia 26: 46–72.

Gamundi, I.J., 1971. Las "Cyttariales" Sudamericanas. De Darwiniana 16: 461–510.

Gaumann, E., 1964. Die Pilze. Birkhäuser Verlag, Stuttgart.

Gordon, C.C., 1966. Ascocarp centrum ontogeny of species of the Hypodermataceae of conifers. Amer. J. Bot. 53: 319–327.

Gordon, C.C., 1968. Ascocarpic centrum ontogeny of species of Hypodermataceae on conifers. II. Amer. J. Bot. 55: 45–52.

Harper, R.A., 1900. Sexual reproduction in *Pyronema confluens,* and the morphology of the ascocarp. Ann. Bot. 14: 321–400.

Kimbrough, J.W., 1970. Current trends in the classification of discomycetes. Bot. Rev. 36: 91–161.

McCubbins, W.A., 1910. Development of the Helvellineae I. *Helvella elastica*. Bot. Gaz. 49: 195–206.

Oberwinkler, F., F. Dasagrade, and E. Muller, 1967. Ueber *Ascocorticium anomalum* (Ellis et Harkness) Earle. Nova Hedw. 14: 283–289.

O'Donnell, K.L., W.G. Fields, and G.R. Hooper, 1974. Scanning ultrastructural ontogeny of cleistohymenial apothecia in the operculate discomycete *Ascobolus furfuraceus*. Can. J. Bot. 52: 1653–1656.

O'Donnell, K.L., and G.R. Hooper, 1974. Scanning ultrastructural ontogeny of paragymnohymenial apothecia in the operculate discomycete *Peziza quelepidotia*. Can. J. Bot. 52: 873–876.

O'Donnell, K.L., G.R. Hooper, and W.G. Fields, 1976a. Scanning ultrastructural ontogeny of eugymnohymenial apothecia in the operculate discomycetes *Ascodesmis nigricans* and *A. sphaerospora*. Can. J. Bot. 54: 572–577.

O'Donnell, K.L., G.R. Hooper, and W.G. Fields, 1976b. Apothecial ontogeny in *Saccobolus versicolor* (Pezizales, Ascomycetes). Can. J. Bot. 54: 2055–2060.
O'Donnell, K.L., G.R. Hooper, W.G. Fields, and A.O. Ackerson, 1976c. Apothecial morphogenesis in *Peziza quelepidotia:* scanning electron and light microscopy. Can. J. Bot. 54: 2254–2267.
Paden, J.W., and E.A. Stanlake, 1973. Ascocarp development in *Ascobolus michaudii.* Can. J. Bot. 51: 1271–1273.
van Brummelen, J., 1967. A world-monograph of the genera *Ascobolus and Saccobolus.* Persoonia, Suppl. 1: 1–260.
von Arx, J.A., 1967. Pilzkunde. J. Cramer-Verlag, Lehre, Germany.

Chapter 8

The Ascocarps of Ascohymenial Pyrenomycetes

A. PARGUEY-LEDUC AND M. C. JANEX-FAVRE

Introduction

Pyrenomycete systematics historically are based on the characteristics of the fructification containing the asci, the ascus itself, and the ascospore. In contemporary classifications, ascocarp development and ascus structure are considered most significant.

Ascocarp Ontogeny as Criterion of Classification

Several early authors, in particular von Höhnel (1917), called attention to the "nucleus" of the ascocarp. More recent thought stems from the work of Nannfeldt (1932), who distinguished three fundamental ontogenic types among the euascomycetes. The Plectascales were established for ascomycetes in which the asci occurred at various levels in the fruit body. The Ascoloculares were conceived as ascomycetes in which the asci occurred in stromatic formations within locules that were not surrounded by true perithecial walls. Additionally, a sterile element, the pseudoparaphyses, developing from the apex of the locule to its base, was a component of this type. A third group was called the Ascohymeniales, wherein the asci were produced alongside a sterile element, the paraphyses, which developed from the base of the hymenial chamber to its apex. Additionally, a true perithecial wall surrounds the asci. Munk (1957) regarded the presence of a schizogenous ostiolar pore bordered by periphyses as the essential character of this third group.

Nannfeldt's successors have tried to refine this approach through detailed analysis of ascocarp ontogeny. The main investigators have been F. Moreau and M. Moreau, J. H. Miller, E. S. Luttrell, G. Doguet, M. Chadefaud, R. Hanlin, and A. Parguey-Leduc. In a general study of the pyrenomycetes, Luttrell (1951) defined eight centrum types. Three Ascoloculares (= loculoascomycetes) were exemplified by *Dothidea* Fries, *Pleospora* Rabenhorst, and *Elsinoë* Raciborski. Five ascohymenial types recognized were those representative of *Diaporthe* Nitschke, *Ophiostoma* Sydow, *Xylaria* Hill ex Greville, *Nectria* Fries, and *Phyllactinia* Léveillé. Additional ascohymenial types have been subsequently recognized as the *Eutypa* type (Parguey-Leduc, 1970) and the *Sordaria* type (Huang, 1976).

Nannfeldt's concepts were applied to lichens by Santesson (1952). A majority of the pyrenolichens were interpreted as Ascohymeniales. Only the Arthopyreniaceae and the Mycoporaceae were determined as Ascoloculares. Hale (1961), Poelt (1973), and Henssen and Jahns (1974) adopted the same viewpoint. Janex-Favre (1970) maintained that certain lichenized species in the genus *Arthopyrenia* Massee and Müller-Argonviensis and the order Verrucariales appeared as Ascoloculares, whereas other families studied, including the Dermatocarpaceae, Pyrenulaceae, and Arthopyreniaceae, appear more clearly as Ascohymeniales.

Ascus Structure as Criterion of Classification

Luttrell (1951) distinguished two large ascomycete groups based on ascus wall structure and their mode of dehiscence. The "bitunicate ascus" has two distinct walls or tunicae that undergo dehiscence by the "jack-in-the-box" method. The "unitunicate ascus" has a single wall or two inseparable fused walls that show diverse modes of dehiscence but never the jack-in-the-box type. Both ascus groups are well represented in the pyrenolichens. Dughi (1957) preferred "fissitunicate" to bitunicate and "infissitunicate" to unitunicate so as to characterize dehiscence without prejudicing wall construction. Gaümann (1949) distinguished a third type of ascus, the "protunicate," in which the wall was evanescent. These types of wall structure serve as the basis of the principal divisions of the pyrenomycetes in various classification schemes, notably those of Ainsworth et al. (1973) and Wehmeyer (1975).

The structure of the ascus tip has been studied in detail by M. Chadefaud. Using light microscopy techniques, he distinguished three main apical apparatus types. The nassasceous-type pattern is discerned by the presence of an apical "nasse" and the annellasceous by that of an apical ring. The archaeascus type pattern has both a nasse and an apical ring; it is a synthetic and probably archaic type not known in the pyrenomycetes. However, several pyrenolichens in genera such as *Porina* Müller-Argonviensis appear to have a variation called the "prearcheasceous type" (Janex-Favre, 1970; Letrouit-Galinou, 1973b). This group also possesses the apical ring typical of the annellasceous type. Chadefaud's ascus apical apparatus types are generally applicable to other types of asci that have been defined. Hence, Nannfeldt's Ascoloculares is in principle to be defined as bitunicate and nassasceous and the Ascohymeniales as unitunicate and annellasceous.

Ascocarp Components

The components of the ascocarp that can serve as a basis for demonstrating relationships are the ascocarp wall, the carpocenter, the paraphyses, and the ostiolar pore apparatus. These features are found in nonlichenized as well as lichenized pyrenomycetes. In this review, the portion of the ascocarp enveloping the centrum is considered important and is given equal consideration. Only the ascogonial component of the fertile apparatus is examined because it plays an important part in the initiation of the ascocarp. The ascus has been the object of many other studies.

The Ascothecium

The ascocarp wall, or more properly the ascothecium, is a sheath or envelope that comes to surround the mature fertile apparatus. The ascothecium is defined by its origin rather than its structure and must be examined in primordia and ascocarp initials as well as the mature stages.

The Primordia

The primordium makes up the first stage of the ascocarp; no fertile center or core is present in this structure. In the Ascohymeniales, a fertile apparatus, consisting of one to several ascogenous, often helical, filaments, is formed first, before the primordium. Hyphae are produced at the base of the ascogonium. They are called "covering hyphae" because they progressively envelop the sexual structure. Further development of these enveloping filaments can involve mycelial filaments as well as those arising from stromatic sources. In nonlichenized pyrenomycetes, the ascogonial apparatus and the primordium can be enclosed within a stroma or formed directly on the mycelium. In the pyrenolichens, the primordium is formed in the thallus.

Parguey-Leduc (1967a, b) distinguished several types of primordia. The arbuscular type (Pl. I, Fig. 8.1) is recognized by a fascicle of covering hyphae formed adjacent to the ascogenous hyphae. This type is quite rare, being found in *Epichloë typhina* (Fries) Tulasne (Doguet, 1960a); *Chaetoceratostoma longirostre* Farrow (Doguet, 1955a); diverse sordariaceous species, particularly *Gelasinospora calospora* (Mouton) C. Moreau and M. Moreau (Parguey-Leduc, 1967c); and species of *Valsa* de Norsetier (Chadefaud, unpublished data; Parguey-Leduc, 1967a,b). The fasciculate covering hyphae of the glomerule type (Pl. I, Fig. 8.2) curl around the ascogonium and together they form a small glomerule. This type is widely observed. The arbuscular–glomerule type (Pl. I, Fig. 8.3) combines both types of features in that the covering hyphae are eventually clustered separately from the ascogonium, and then the hyphal tips join together to curl around the ascogonium. This type is known only in *Thielavia terricola* (Gilman and Abbott) Emmons (Lemans, 1962).

Variations of these basic types have been found. In certain members of the Sordariales (Parguey-Leduc, 1967c) (Pl. I, Fig. 8.4) the covering hyphae begin differentiation with the proarchicarp wound in a spiral rather than from an ascogonial filament per se. In *Podospora arizonensis* (Mainwaring and Wilson, 1968), the covering hyphae do not exist, and the primordial glomerule consists only of stromatic filaments. This is equally true in *Chaetomium senegalensis* (Héau, 1968) where the ascothecial envelope can develop in the absence of an ascogonial apparatus. In certain xylariaceous species (Lupo, 1922; Parguey-Leduc, 1972) (Pl. I, Fig. 8.5), the glomerule originates from stromatic cells in advance of the appearance of the ascogonium. This is similar to development in pyrenolichens, where it is in the thallus (Janex-Favre, 1970; Henssen and Jahns, 1974) (Pl. I, Fig. 8.6).

The Ascocarp Bud

A second stage in the differentiation of the ascocarp is characterized by the formation of a fertile center surrounded by an envelope. The envelope becomes that found in a mature ascothecium (Chadefaud, 1944, 1960). It can remain simple or become divided into a dark external wall with pigmented membranes and a clear internal wall of nonpigmented cells. Internally, the bud differentiates plectenchymatous tissue which becomes organized as a carpocenter (sensu Chadefaud) containing the reproductive elements.

The Mature Ascocarp

The elements of the young ascocarps are symmetrically arranged with reference to the longitudinal axis during the course of development. The ascothecial envelope generally remains distinct at maturity. The dark, flattened cells of the envelope are arranged in concentric layers around the ascocarp cavity. Any stromatic

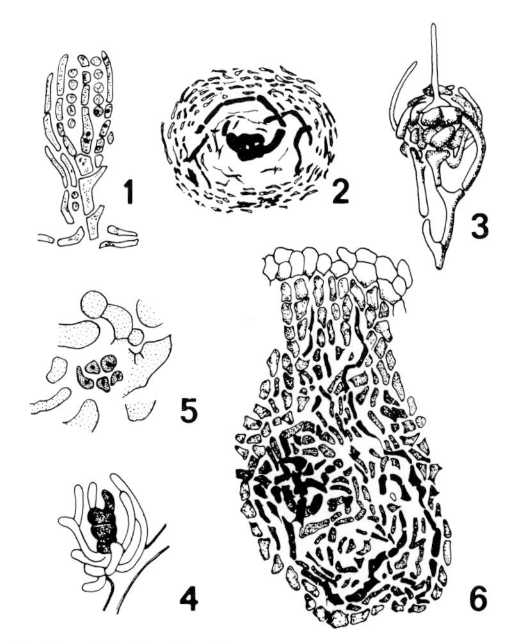

Plate I. The different types of primordia found in the Ascohymenes.
Fig. 8.1 Arbuscular type, *Chaetoceratostoma longirostre*. (After Doguet, 1955.)
Fig. 8.2 Glomerular type, *Hypoxylon coccineum*.
Fig. 8.3 Arbuscular–glomerular, *Thielavia terricola*. (After Lemans, 1962.)
Fig. 8.4 Covering filaments differentiated from the sterile proarchicarp, *Gelasinospora calospora*.
Fig. 8.5 Differentiation of covering filaments arising from basal stromatic cells, *Hypoxylon*.
Fig. 8.6 Plexiform primordium of the pyrenolichen, *Dermatocarpon miniatum*.

cells involved in the ascocarp are most often isodiametric and are positioned external to the envelope. The upper part of the ascothecium generally forms a more or less elongate collar or neck, internally filled with periphyses and opening by an ostiole. Moreau and Moreau (1952, 1953) refer to this bottle shape as "lagynie."

The ascothecium can be difficult to identify in a differentiated ascocarp bud because of its definition by origin rather than by any structural characteristics. In *Lasiosphaeria ovina* (Persoon ex Fries) Cesati and de Notaris in the order Sordariales, the carpocenter is lodged in a stromatic sphere or pyrenosphere. Here the internal part simulates an ascothecium by its structure. Chadefaud (1960) and Parguey-Leduc (1967a) refer to this as the "pseudoascothecium." In the pyrenolichens, which also lack a true ascothecium, the secondary envelope formed on the ascocarp appears to be a homolog. In *Dermatocarpon miniatum* (Linneus) Mann, the secondary envelope forms very early and constitutes the wall of the mature perithecium (Janex-Favre, 1970). The carpocentral envelope of various other pyrenolichens form later from the periphery of the carpocenter and is for the most part comparable to the pseudoascothecium of *Lasiosphaeria ovina.*

The Carpocenter

Different terms have been employed for the central parts of the ascocarp bud: first "nucleus," then "centrum"—created by Wehmeyer (1926)—and now replaced by the term "carpocenter." The carpocenter is formed in general from the central parts of the primordium and in a few pyrenolichens, such as *Dermatocarpon miniatum* (Janex-Favre, 1970), from the entire primordium.

Complete Carpocenter

Chadefaud (1960) described the carpocenter as having a massive amount of nutritive plectenchyma at its center that functions to nourish the development of the ascogenous hyphae and to mechanically assist in the formation of the ascocarp cavity or locule. These functions are not, however, effective in such fungi as *Sordaria* spp. and *Gelasinospora* spp. (Parguey-Leduc, 1967c), which retain the plectenchyma for a prolonged period. In several species in the Diatrypales and Xylariales (Pl. II, Fig. 8.11), the nutritive plectenchyma is replaced by a plexus; in several pyrenolichens the plexus takes on a paraphysoid appearance.

A perilocular layer surrounds the sporophytic apparatus, then asci and the paraphyses. This layer is composed of a subhymenial disc and a suprahymenial bell which consists of a meristematic cone producing the ostiolar apparatus and its periphyses, but never the pseudoparaphyses. The ostiolar canal originates in the splitting of the cone.

Incomplete Carpocenter

The carpocenter is rarely complete in the Ascohymeniales. Exceptions are *Helminthosphaeria clavariarum* (Desmazières) Fuckel apud Munk (Pl. II, Fig. 8.7) (Parguey-Leduc, 1961), *Pleurage minuta* (Fuckel) Kunze (Wicker, 1962), and

Chaetomium spp. (Moreau and Moreau, 1954; Chadefaud and Avellanas, 1967; Héau, 1968). The carpocenter has been found to be incomplete, aberrant, or missing altogether, suggesting a transfer of the nutritive function from sterile tissue to the sporophytic apparatus, and a transfer of the paraphysogenous function from the subhymenial disc to the ascotnecium (Parguey-Leduc, 1972).

Moreau and Moreau (1953) found that the wall of *Aspergillus ruber* Bremer produced centralized pockets of nutritive tissue that were joined into a compact tissue. Parbery (1969a, b) found that the carpocenter of *Amorphotheca resinae* Parbery was formed of radiate filaments forming a network similar to that of the Eurotiales. Moreau and Moreau (1950) (Pl. II, Fig. 8.8) found the carpocenter to be reduced to the perilocular layer in *Triangularia bambusae* (Van Beyma) Boedijn. This layer was interpreted as further reduced to only a subhymenial disc in *Lasiosphaeria ovina* (Parguey-Leduc, 1967c) and *Dermatocarpon miniatum* (Janex-Favre, 1970) (Pl. II, Fig. 8.9). The nutritive plectenchyma seems only transient or even absent in certain of the Chaetomiales. Doguet (1955a, 1960a) found no plectenchyma, plexus, or perilocular layer in *Chaetoceratostoma longirostre* (Pl. II, Fig. 8.10) and *Epichloë typhina.*

The nutritive role seems to be centered in a part of the sporophytic apparatus in *Gnomonia ulmi* (von Schmeinitz) von Thüman (Pomerleau, 1937) and *Erataocystis moniliformis* (Hedgecock) C. Moreau (Moreau and Moreau, 1952). *Gnomonia leptostyla* (Fries) Cesati and de Notaris (Fayret and Parguey-Leduc, 1975) (Pl. II, Fig. 8.12) undergoes an aberrant formation. The subhymenial disc gives rise to files of splitting nutritive cells with a paraphyses-like appearance. These rapidly coalesce laterally, forming pseudoparenchyma and thus a secondary nutritive tissue.

Paraphyses

The presence of paraphyses in the ascothecium is also a fundamental character of the Ascohymeniales. These structures can be clearly recognized in the mature ascocarp. True paraphyses are ascending filaments that arise in the perithecial cavity from the subhymenial disc. Together with the asci, they form the hymenium. Generally, the paraphyses are thin, multicellular, little anastomosed filaments. An exception is the inflated apically tapering cell typical of the Diaporthales.

By contrast, pseudoparaphyses, or apical paraphyses, are descending filaments originating in the suprahymenial bell, and paraphysoids are subvertical filaments formed by the reorganization of the carpocentral plexus that becomes positioned between the asci and the perilocular locule.

True or primary paraphyses generally issue from the surface of the subhymenial disc. Each paraphysogenous cell produces a single paraphyse or many as in *Melogramma spiniferum* (Wallroth) de Notaris (Doguet, 1960b) (Pl. III, Fig. 8.13) and in *Chaetomium elatum* Kunze (Moreau and Moreau, 1954). Primary paraphyses can be found in many species such as *Triangularia bambusae* Boedijn (Moreau and Moreau, 1950) (Pl. II, Fig. 8.8), *Helminthosphaeria clavariarum* (Parguey-Leduc, 1961), *Lasiosphaeria ovina* (Parguey-Leduc, 1967c), *Podos-*

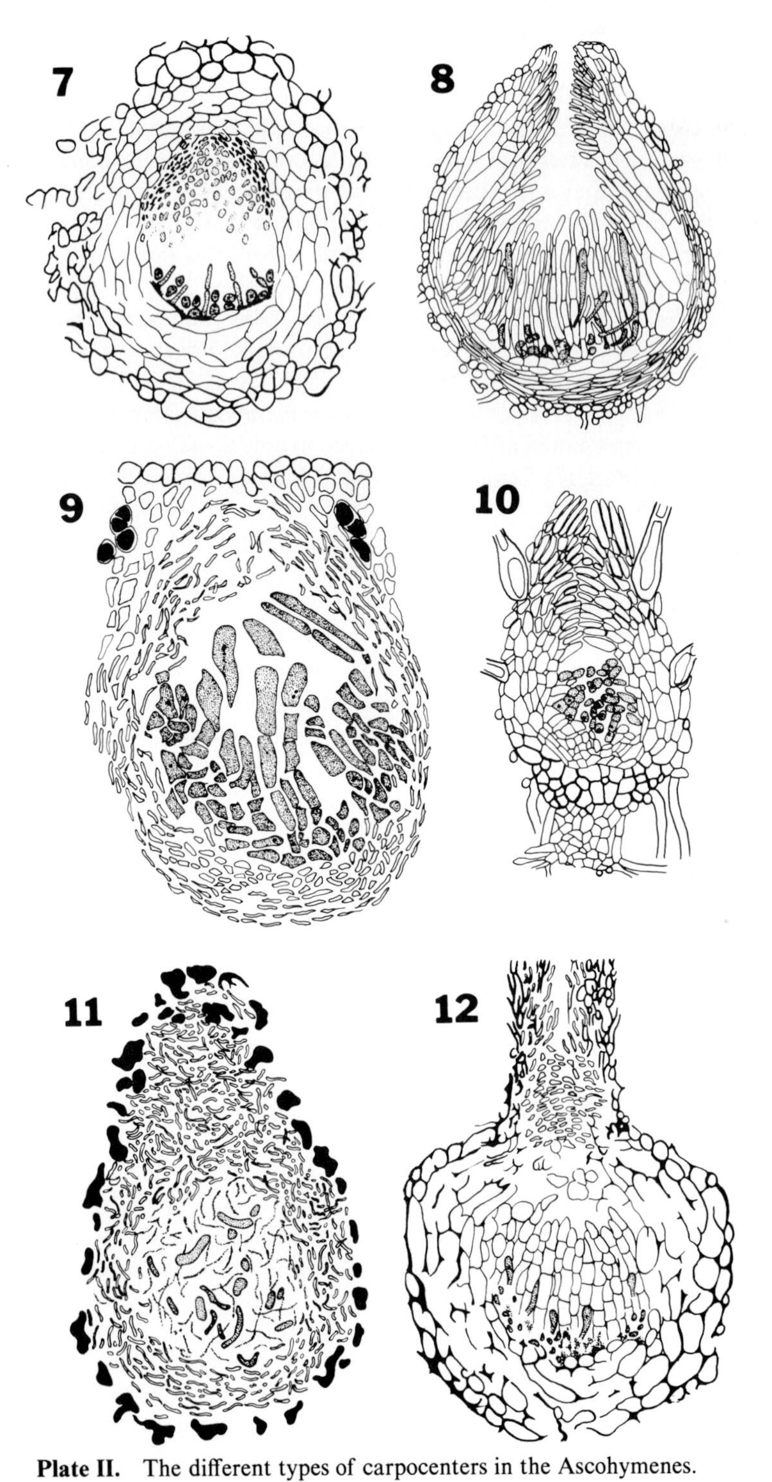

Plate II. The different types of carpocenters in the Ascohymenes.

Fig. 8.7 Complete carpocenter, *Helminthosphaeria clavariarum.*

Fig. 8.8 Carpocenter reduced to perilocular lining, *Triangularia bambusae.* (After Moreau and Moreau, 1950.)

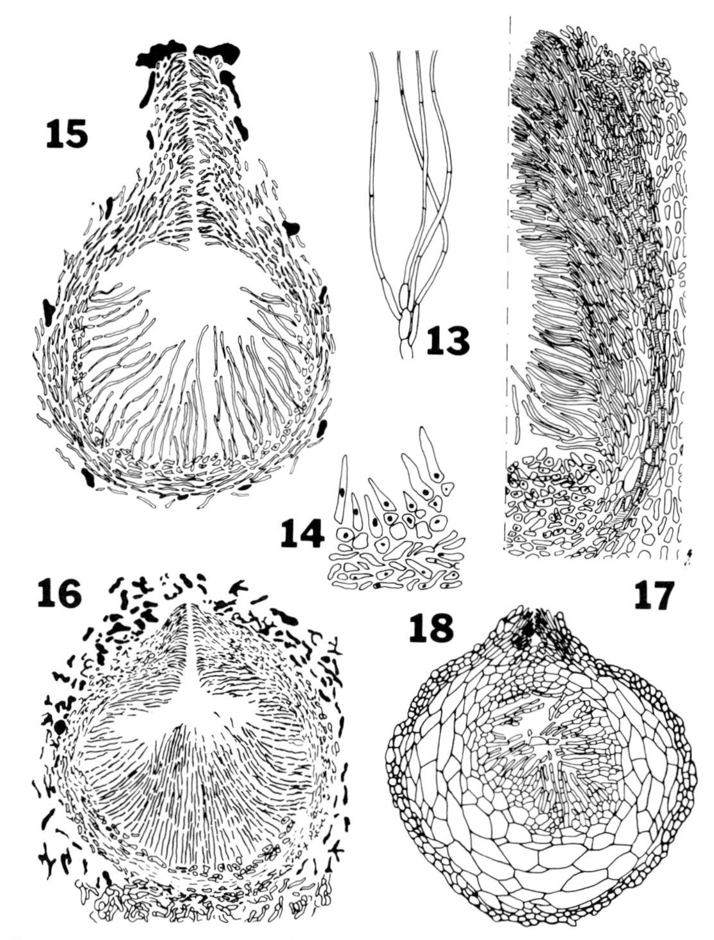

Plate III. The different types of paraphyses in the Ascohymenes.

Fig. 8.13 Formation of many paraphyses from a paraphyogenous cell of the subhymenial meniscus, *Melogramma spiniferum*. (After Doguet, 1960b.)

Fig. 8.14 Formation of paraphyses from the carpocentral network, *Cryptosphaeria eunomia*.

Fig. 8.15 Formation of paraphyses from the ascothecial envelope, *Eutypa lata*.

Fig. 8.16 Primary paraphyses, secondary paraphyses, and periphyses, *Hypoxylon rubiginosum*.

Fig. 8.17 Secondary paraphyses and periphyses, not of primary paraphyses, *Epichloë typhina*. (After Doguet, 1960a.)

Fig. 8.18 Radial filaments issued from the wall, *Microascus stysanophorus*. (After Doguet, 1957b.)

Fig. 8.9 Perilocular lining reduced to a subhymenial meniscus, *Dermatocarpon miniatum*.

Fig. 8.10 Carpocenter absent, *Chaetoceratostoma longirostre*. (After Doguet, 1955a.)

Fig. 8.11 Reticulated carpocenter, *Eutypa lata*.

Fig. 8.12 Secondary nutritive tissue, *Gnomonia leptostyla*.

pora arizonensis (Mainwaring and Wilson, 1968), *Triangularia backusii* (Huang, 1976), and several pyrenolichens, such as *Arthopyrenia conoides* (Nylander) Zopf (Janex-Favre, 1970).

In certain diatrypaceous species, such as *Cryptosphaeria eunomia* Fuckel (Parguey-Leduc, 1970), the paraphyses originate from a carpocentral plexus compressed into the locule base (Pl. III, Fig. 8.14). The paraphyses originate from the internal face of the ascothecial envelope in *Eutypa lata* (Persoon) Tulasne (Parguey-Leduc, 1970) (Pl. III, Fig. 8.15) in the Xylariales (Parguey-Leduc, 1972) and probably the Lasiophaeriaceae (Huang, 1976) where the perilocular layer is not formed.

Doguet (1959a) recognized secondary paraphyses, which differ from the primary type in their location. These paraphyses develop in a periostiolar annular growth zone. Initially, the filaments are oriented toward the axis of the ostiolar canal. The primary and secondary paraphyses become continuous with the diametric expansion of the perithecial cavity as exemplified by *Lasiosphaeria hispida* (Tode ex Fries) Fuckel (Parguey-Leduc, 1967c) and *Hypoxylon rubiginosum* (Persoon ex Fries) Fries (Parguey-Leduc, 1972) (Pl. III, Fig. 8.16). Only the secondary paraphyses differentiate in *Epichloë typhina* (Doguet, 1960a) (Pl. III, Fig. 8.17).

Ostiolar Apparatus

The mature perithecium opens via an ostiole, which may or may not be subtended by a neck and an ostiolar apparatus consisting of a hollow axial canal lined with periphyses permitting spore dispersal from the ascus. According to Miller (1949), the term "ostiole" refers to both the canal and the pore. The schizogenous formation of the ostiole from the perithecial wall is considered by Munk (1957) to be of fundamental significance in true pyrenomycetes (= Ascohymeniales). Cain (1972) and von Arx (1973) regard this event as a variable character at the species level. Chadefaud and Avellanas (1967) and Parguey-Leduc (1967b) suggested three types of differently originating formations in the ostiolar apparatus: the internal one from a carpocentric origin; the intermediate and external ones from the internal and external layers of the perithecial wall.

The internal formation characteristic of the Ascoloculares can be found in *Helminthosphaeria clavariarum* (Parguey-Leduc, 1961), *Sordaria macrospora* Anserwald (Parguey-Leduc, 1967c) (Pl. IV, Fig. 8.19) and probably in *Podospora anserina* (Cesati) Niessl (Mai, 1976.) The internal and intermediate formations coexist in *Nectria flava* Ignotus (Gilles, 1947), *Triangularia bambusae* (Moreau and Moreau, 1950), *Bombaria lunata* Zickler (Zickler, 1952), *Ceratocystis moniliformis* (Moreau and Moreau, 1952), *Gnomonia leptostyla* (Fayret and Parguey-Leduc, 1975) (Pl. IV, Fig. 8.20), and *Triangularia backusii* (Huang, 1976).

The ostiolar apparatus is reduced to the intermediate formation in *Chaetoceratostoma longirostre* (Doguet, 1955) (Pl. IV, Fig. 8.21) and in *Melanospora* spp. (Doguet, 1955b); *Gelasinospora calospora* (Parguey-Leduc, 1967c); certain

Xylariales, i.e., *Hypoxylon* spp. (Parguey-Leduc, 1972); and *Dermatocarpon miniatum* (Janex-Favre, 1970).

The intermediate and the external wall origin formations are found together in *Xylaria hypoxylon* (Linneus) Greville (Parguey-Leduc, 1972). In this species, the lower periphyses of the neck are derived from the inner wall layer, whereas the upper ones are from the external layers. A similar origin occurs in *Chaetomium senegalensis* (Héau, 1968) and *Pyrenula nitida* (Merg.) Acharius (Janex-Favre, 1970), but with reduction of the intermediate wall layers. A crown is formed when only the external and intermediate layers are present from a thickening and reorientation toward the exterior, in *Coniochaeta ligniaria* (Levelle) Massee (Doguet, 1959a) (Pl. IV, Fig. 8.22) and *Porina* spp. (Janex-Favre, 1970). A periostiolar crown also exists in the pyrenolichens *Arthopyrenia conoidea* and *Arthopyrenia sublitoralis* (Leigh) Arnaud formed by hyphae at the wall summit. There is no neck or ostiolar apparatus. The ascocarp opens by a circular pore. There is neither a neck, nor an ostiolar apparatus, nor a pore in the perisporiaceous pyrenomycetes.

The most primitive internal apparatus formation is the only kind found in the Ascoloculares. This formation is found with and replaces the intermediate one,

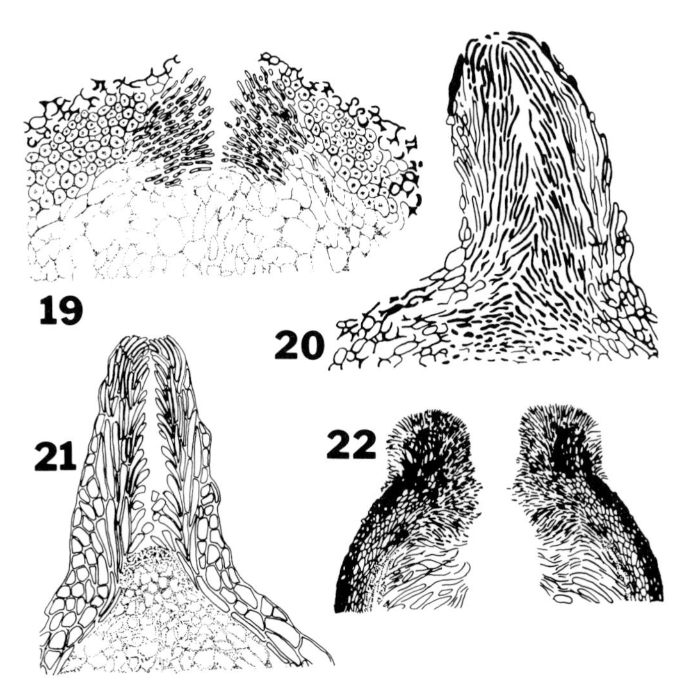

Plate IV. Different types of ostiolar apparatuses in the Ascohymenes.
Fig. 8.19 Internal ostiolar apparatus, *Sordaria macrospora.*
Fig. 8.20 Internal and intermediate ostiolar apparatuses, *Gnomonia leptostyla.*
Fig. 8.21 Intermediate ostiolar apparatus, *Melanospora zamiae* (After Doguet, 1955b.)
Fig. 8.22 Ostiolar apparatuses, intermediate and short external, and forming a crown, *Coniochaeta ligniaria.* (After Doguet, 1959a.)

and finally the external one in more advanced forms of Ascohymeniales. Eventually, the apical apparatus is no longer found in the majority of the Perisporiales.

A transfer can be discerned in the origin of the formations involved in the ostiolar apparatus from the carpocenter to more external parts of the wall. Such a regressive evolution is similar to that of the carpocenter.

Ascocarp Types

Ascothecia and Ascostroma

Ascocarp types may be basically categorized on the basis of association with stromatic tissue or on development types. The ascocarp components can be associated with stromatic tissue in the pyrenomycetes or thalline in the lichens. In some well-developed ascocarps, the ascothecium can be within a more or less voluminous stroma. The stroma is palisade-like in the upper fertile portion and contains many endostromatic ascothecia in the less evolved of xylariaceous and dastrypaceous species. In the case of Trypetheliaceae, the stromatic tissue develops progressively around the differentiating fruit body at the expense of the thallus or perithecial wall. In this case, the term "pseudostroma" is preferred over "stroma" (Johnson, 1940; Letrouit-Galinou, 1957, 1973a) and an analogy can be made to the thalline envelope present in pyrenolichens.

The ascothecium of *Lasiosphaeria ovina* (Parguey-Leduc, 1967c) is semiendostromatic. The ascocarp primordium is initially endostromatic but eventually comes into maturity on the surface of the stroma. *Epichloë typhina* (Munk, 1957; Doguet, 1960a) produces epistromatic ascothecia that are positioned side by side in close proximity.

A reduced ascostroma is recognizable in several types of ascocarp. *Rosellinia* spp. produce an ascocarp composed of side by side ascothecia developed in segments of a large stroma. The pyrenospheres sensu Chadefaud are reduced to stromatic envelopes comprised of a few cell layers on a mycelial surface. *Lasiosphaeria hispida* is an example (Parguey-Leduc, 1967c). Miller (1949) thought that this stromatic sphere existed in all sphaeriaceous species. The pyrenolichen *Pyrenula nitida* (Janex-Favre, 1970) appears to show this type of ascocarp. The wall of the perithecium is comprised of two concentric envelopes—one is an external envelope derived from the primordial periphery, comparable to a pyrenosphere produced by the mycobiont in the lichen thallus. The second is a secondary internal envelope more or less equivalent to an ascothecium.

Helminthosphaeria clavariarum (Parguey-Leduc, 1961) and *Coniochaeta ligniaria* (Doguet, 1959a) produce only rudimentary fragments of the pyrenosphere. The ascothecia of the more evolved Sordariales, Xylariales, and Perisporiales are without stroma. This is the case in the pyrenolichen *Dermatocarpon miniatum*. In other pyrenolichens, the ascothecium is covered with a thalline envelope. This separate envelope is directly formed by encircling thallus. This is not a true stromatic formation.

In *Arthopyrenia conoidea,* a structure analogous to the stromatic conceptacle of certain discomycetes can be seen (Bellemère, 1967).

Developmental Types

The ontogenic study of perithecia has permitted the recognition of various types based on the development and the centrum (Luttrell, 1951), and therefore on their carpocenters. Evolutionary relationships at the generic level can be seen in the modification and eventual suppression of carpocenter characteristics. Five developmental types can be discussed—three in the Ascohymeniales and two basic lichenized types.

The *Diaporthe* Type

In the *Diaporthe* type of development the centrum is composed of pseudoparenchyma and paraphyses are often lacking. The apically positioned ostiole is schizogenously derived. Luttrell (1951) recognized this type in *Diaporthe* Nitschke *Melanconis* Tulasne, *Gnomonia* Cesati and de Notaris, *Melanospora* Corda, *Hypomyces* (Fries) Tulasne, and *Sordaria* Cesati and de Notaris. Confirmation was obtained in the Melanosporaceae by Kowalski (1965a) and in the Sordariales by Parguey-Leduc (1967c).

Variations of the *Diaporthe* type can be found concerning paraphyses and the nutritive plectenchyma. Rapidly evanescent paraphyses may be found in *Melanconis* (Wehmeyer, 1941). They persist longer in some primitive species in the Sordariales, such as *Lasiosphaeria* Cesati and de Notaris (Parguey-Leduc, 1967c), *Helminthosphaeria* Fuckel (Parguey-Leduc, 1961), and *Triangularia bambusae* (Moreau and Moreau, 1950), and in certain Chaetomiaceae—for example, *Chaetomium globosum* Kunze (Chadefaud and Avellanas, 1967). The nutritive plectenchyma may be ephemeral as in *Melanospora moreaui* Doguet (Doguet, 1955b) and certain species of *Chaetomiaceae* Kunze (Héau, 1968; Chadefaud and Avellanas, 1967). The perilocular layer can persist where the nutritive plectenchyma does not form at all, as in *Triangularia bambusae* (Moreau and Moreau, 1950), or not persist, as in *Chaetoceratosoma longirostre* (Doguet, 1955a).

The *Diaporthe* type must be related to the *Sordaria* type, which Huang (1976) recognized in *Triangularia backusii.* This species possesses a centralized pseudoparenchymatous area and paraphyses. Huang considered the development to be intermediate between the *Diaporthe* type and the *Xylaria* type of Luttrell (1951). However, the diaporthacean nature of the carpocenter is significant, but the presence of paraphyses does not justify a distinction of a *Sordaria* type. In fact, the paraphyses of the Sordariales, when present, are unlike those of the Xylariales. They are derived from the carpocenter and not from the perithecial wall.

The *Diaporthe* type is also related to the *Ophiostoma* type, differing only in the irregular disposition of asci in the fruit body locule. This distinction does not justify a separation into two types; the ontogenetic types being based on the sterile components of the centrum. The ascohymenial Eurotiales, with similarly disposed asci, are recognized as having a *Diaporthe* type of centrum.

The *Phyllactinia* type of Luttrell (1951) seems to be the same as the *Diaporthe* type. The centra are comprised of a plectenchyma; the absence of an ostiole is the distinguishing character. Cain (1972) and von Arx (1973) support the view that the ostiolar character is variable even at the specific level. Chadefaud (1960) suggested that this variability is the result of a simplified regressive evolution.

The *Xylaria* Type

In the *Xylaria* type, only nutritive plectenchyma is present. The centrum is comprised of somewhat evanescent paraphyses which have a basal and lateral origin from the internal surface of the ascocarp wall lining the locule. The ostiole is schizogenous.

Xylariales, Clavicipitales, and Chaetomiaceae belong to this type, although the two latter families have not the same disposition of fertile elements in the locule.

The *Eutypa* Type

The *Eutypa* type (Parguey-Leduc, 1970) is a third fundamental type. The reticulate structure of the carpocenter (= plexus) and the development of the paraphyses from both the discoid plexus and the ascothecial envelope characterize this type.

The Developmental Types of the Pyrenolichens

The developmental types found in the pyrenolichens are not directly linked to those of the nonlichenized ascomycetes. Henssen and Jahns (1974) placed all pyrenolichens in the Ascohymeniales with the exception of the Arthopyreniaceae and the Mycoporaceae. However, two developmental types ought to be recognized.

The first of these types is characterized by the presence of paraphysoids in the perithecial bud that persist or more often are replaced by true paraphyses. This type is found in the Porinaceae (unitunicate asci), the Pyrenulaceae, the Microglaenaceae and the Strigulaceae. The second lichen type produces a bud consisting of a closed sphere within a central cavity. Later, radiating hyphae develop, followed by an increase in height and width of the bud. The mature perithecium lacks paraphyses and opens by an ostiolar apparatus. This ascohymenial type is found in the Verrucariales.

Janex-Favre (1970) suggests that the pendant filaments and the carpocentral ostiolar apparatus link the type to the Ascoloculares (it is found in the Verrucariales sensu stricto). *Dermatocarpon miniatum* is an exception in this order because of the reduced carpocenter, a lack of pendant filaments, and the origin of the ostiolar apparatus from a secondary envelope.

Marginal Pyrenomycetes

Some pyrenomycetes do not clearly fall into the Ascohymeniales or the Asco-loculares if both the ascocarp type and the ascus are considered. Problems can be identified concerning species that produce a bitunicate ascus and an ascohymenial type. The reverse is true when a unitunicate ascus is produced in the ascolocular-iaceous fruit body. The ascocarp of the Nectriales, the Clavicipitales, and some pyrenolichens are of special concern.

Annellasceous Bitunicate Ascohymeniales. Melogramma spiniferum (Doguet, 1959b, 1960b) produces a clearly recognizable ascothecium and ascendant primary paraphyses, and asci of the annelasceous type with a chitinoid apical ring and "jack-in-the-box" or bitunicate dehiscence (Parguey-Leduc and Chadefaud, 1963). This is equally the case in *Helminthosphaeria clavariarum* (Parguey-Leduc, 1961).

Unitunicate Ascoloculares. Didymocrea sadasavanii (Ramachandra-Reddy) Kowalski (Kowalski, 1965b) and *Buergenerula spartinae* Kohlmeyer and Gessner (Kohlmeyer and Gessner, 1976) show ascocarp development of the ascolocular-iaceous type. The ascogonium is produced in the bud and pseudoparaphyses are present. However, the asci are unitunicate with a nonamyloid apical ring.

The place of the Coronophorales is always uncertain. Miller (1949) placed them in the Ascoloculares, whereas Nannfeldt (1932, 1975a, b) placed them in the Ascohymeniales as an aberrant group. They appear to be best regarded as unitunicate Ascoloculares. The ascocarp development is of the ascolocular iaceous type and the hymenial locule is directly excavated in the apostromic pyrenosphere. However, the perilocular layer produces periphyses only in the neck canal and not pseudoparaphyses. *Bertia moriformis* (Tode ex Fries) de Notaris (Parguey-Leduc, 1966) produces immature asci which have a bitunicate structure that later becomes unitunicate. *Coronophora gregaria* (Libert) Fuckel (Parguey-Leduc, 1966, 1967d) produces unitunicate asci directly.

Luttrell defined the *Nectria* type of centrum development based on *Thyronectria austro-americana* (Spegazzini) Seeler and *Sphaerostilbe aurantiicola* (Berkeley and Broome) Petch. Their ascocarp is derived from an ascogonium that secondarily surrounds a lump of stromatic hyphae. Therefore a true perithecial wall is formed. A hymenial cavity containing pseudoparaphyses and a schizogenous ostiolar canal bordered by periphyses can be found in mature fruit bodies. In reality, the Nectriales is a very heterogeneous group of species in which the ascocarp is fleshy and brightly colored. Parguey-Leduc (1966) has shown that the development is not uniform and the asci are of both uni- and bitunicate types.

Letendraea padouk Parguey-Leduc (Parguey-Leduc, 1959, 1966, 1967d) is an example of nectioid species that are bitunicate–nassaceous.

A group of nectioid species possesses an ascolocular iaceous ascocarp, and a unitunicate ascus and pseudoparaphyses develop from the suprahymenial bell in or on the stroma. *Gibberella pulicaris* (Fries) Saccarov, *Nectria coccinea* (Persoon ex Fries) Fries and *Nectria cinnabarina* (Strickman and Chadefaud, 1961; Parguey-Leduc, 1966, 1967a–d) differentiate one to several arbuscular primordia that mature as endo- or epistromatic fruit bodies. *Hypomyces aurantius* is initiated as a helical primordium on a mycelial mat (Parguey-Leduc, 1966). Others develop directly on a mycelial filament, as does *Nectria episphaeria* (Tode ex Fries) Fries as a glomerule preceding the differentiation of the archicarp.

The primordium appears in the form of a helical roll in *Hypomyces lactifluorum* (von Schweinitz) Tulasne *Hypomyces trichothecoides* Tubaki and in *Hypocrea schweinitzii* (Fries). Hanlin (1963a, 1964, 1965) considers the roll as the ascogonial initial, which is embedded by the covering filaments to form the ascothecium. This differentiation of the primordium is identical to that observed in *Hypomyces aurantiacus* Peck (Parguey-Leduc, 1966, 1967d), in *Nectria galligena* Bresadola (Parguey-Leduc and De Horter, 1973), in certain of the Sordariales (Parguey-Leduc, 1967c), and in *Thielavia terricola* (Lemans, 1962). In these species the helical roll corresponds to a sterile proarchicarp that organizes into a bud of the pyrenosphere from which the fertile archicarp appears secondarily.

In contrast, a true ascothecium is differentiated in *Nectria gliocladioides* Smalley and Hansen (Hanlin, 1961) and *Nectria haematococca* Berkeley and Broome (Hanlin, 1971). However, the developmental pathway is typically ascoloculariaceous, because of the pseudoparaphyses which unite into a pseudoparenchyma in *Nectria haematococca, N. galligena* (Parguey-Leduc and De Horter, 1973), and *Nectriella rousseliana* (Montagne) Saccardo (Parguey-Leduc, 1966, 1967a–d). The ascothecial bud is associated with covering filaments of a stromatal origin in the pseudoparaphysate species *Thyronectria austro-americana* (Seeler, 1940) and *Sphaerostilbe aurantiicola* (Luttrell, 1944).

The ascothecial formation around the archicarp and the carpocenter is similar to the *Diaporthe* type in *Nectria flava* (Gilles, 1947; Strikmann and Chadefaud, 1961), in *Thyronectria denigrata* (Winter) Seaver (Lieneman, 1938), and in *Neuronectria peziza* (Tode ex Fries) Fries (Hanlin, 1963b). Paraphyses can also be differentiated.

The pyrenospheres are ascoloculariaceous, but the asci are unitunicate with an evanescent wall and without an apical apparatus in *Lulworthia bulgariae* (Parguey-Leduc, 1966, 1967d). *Lulworthia* Sutherland is related to *Ceratocystis* Ellis and Halsted, *Pyxidophora* Brefeld and von Tavel, and *Thielavia* Zopf, which are typically ascohymeniaceous. The comparison can be made based on the resemblances between the ascospores, the presence of phialides of the *Chalara* type, on the perithecia as well as on other characters. An ascoloculariaceous type of development with evanescent asci is recalled in the Eurotiales by the *Phaeotrichum* type of Barr (1956). The asci are deposited without definite order in the locule; the ascothecium produces no ostiole. This same type is found in *Pycnidiophora dispersa* Clum (Kowalski, 1964).

Clavicipitales. The Clavicipitales were placed among the Hypocreales because of their bright coloration and fleshiness. They have also been included in the Sphaeriales (Gäumann, 1949; Miller, 1949) or the Xylariales (Luttrell, 1951). Chadefaud (1960) separated them because of the ascal characteristic of a voluminous apical plug in the apical apparatus. The relationship to the ascohymenial type (Nannfeldt, 1932) is classically recognized; various authors have described the perithecial wall in diverse species (Varitchak, 1927, 1931; Jenkins, 1934; Miller, 1949; Doguet, 1960a). Doguet ascertained that the perithecial wall of *Epichloë typhina* is unique and of a secondary nature. The wall is formed around an archicarp with a spreading arbuscular primordium and is comprised of hyphae originating internally from the archicarp stalk and externally from ordinary stromatic hyphae. The hyphae ramify near the interior and give rise directly to the secondary paraphyses and to the periphyses found on the apical crown. The structure of the ascocarp appears to be somewhat like that of *Cordyceps agariciformia* (Bolton) Seaver (Jenkins, 1934). Chadefaud (1960) believed that the wall of the Clavicipitales represented a nest of coalescent paraphyses encircling a centrum of the *Xylaria* type; it therefore links them to the primitive discomycetes. Parguey-Leduc (1967b) maintained that the ascocarps of the Clavicipitales were pyreniform ascocarps comparable to the apothecia of some discomycetes, which are provided with a perithecioid envelope filled with internal filaments.

Pyrenolichens. The fundamental characteristics of the ascoloculariaceous and ascohymenial types cannot be found in a strict sense in pyrenolichens, which show no strict evolutionary relationships to the ascocarps of the nonlichenized ascomycetes. *Arthopyrenia submicans* (Nylander) Arnaud and *Arthopyrenia fallax* (Nylander) Arnaud are clearly ascoloculariaceous; in other pyrenolichens, the ascoloculariaceous and ascohymeniaceous nature is more ambiguous.

The Verrucariaceae sensu stricto appear to be closely related to the Ascoloculares by the presence of basipetal filaments in the perithecial locule that are comparable to pseudoparaphyses, by the nearly universal absence of true paraphyses and a secondary parietal formation, and by their bitunicate asci. However, the species of this family also possess characteristics of the Ascohymeniales, such as the presence of true paraphyses as seen in *Verrucaria controversa* Massee and *Amphoridium calcisedum* (De Candolle) Servit or a secondary envelope more or less comparable to an ascothecium as seen in *Verrucaria cazzae* Zahlbruckner and *Verrucaria dufourii* De Candolle. *Dermatocarpon miniatum* clearly appears to be of the ascohymenial type seen in certain nonlichenized ascohymenial species including *Chaetoceratostoma longirostre* (Doguet, 1955a) and *Triangularia bambusae* (Moreau and Moreau, 1950). The wall is formed entirely by a secondary envelope, which belongs to the ostiolar apparatus and in which the carpocenter is reduced to a subhymenial disc that is in itself hardly important. The asci are bitunicate.

Pyrenula nitida, Porina spp., and *Arthopyrenia conoidea* represent other variants of the ascohymenial type. Because their ascocarps have an "anteparathecial" type of ontogeny and the asci are bitunicate and occasionally archeasceous, at least in the immature state, these species approach the ascohymenial type of certain discomycetes that are equally bitunicate or archeasceous. Henssen and Jahns (1974) do not agree with this viewpoint in that only the Arthopyreniaceae and the Mycoporaceae are ascoloculariaceous in their concept of the pyrenolichens. The species of the Porinaceae are unitunicate and typically ascohymeniaceous; they must be included in the order Sphaeriales. Other families that are clearly ascohymenial and bitunicate cannot be linked to existing nonlichenized ascohymenial orders.

General Conclusions

The existence of marginal pyrenomycetes gives a certain perspective on the difficult distinction to be made between the Ascoloculares and the Ascohymeniales groups and, in another way, on the difference between the bitunicate and the unitunicate groups. The intermediate characteristics that are present in the pyrenomycetes equally allow for a better understanding of the existing relationships between these groups.

Contrary to the opinion of Nannfeldt (1932), who considered the Ascoloculares and the Ascohymeniales as two fundamentally different groups, the Ascohymeniales are best considered as derived from the Ascoloculares (Strikmann and Chadefaud, 1961; Parguey-Leduc, 1966, 1967d). Several ontogenetic characters are important in this hypothesis.

The primordium of the ascocarp, in particular the arbuscular primordium, is organized in the same manner in both groups, but the ascogonium is produced first in the Ascohymeniales and secondarily in the Ascoloculares. However, in certain of the Xylariales, the ascothecium differentiates before the appearance of the archicarp from the basal stromatic cells. In certain Sordariales, the covering filaments arise from a sterile proarchicarp with the fertile archicarp appearing later in the bud. The ascogonial apparatus differentiates secondarily in the primordium in the pyrenolichens.

The carpocenter of certain Ascohymeniales resembles many found in ascoloculariaceous species because of the presence of a nutritive plectenchyma. This is seen, for example, in the carpocenters of both the *Dothidea* type and the *Diaporthe* type. A more or less complete perilocular layer demonstrates the resemblance. This layer produces pseudoparaphyses in the Ascoloculares and paraphyses in the Ascohymeniales. In several species of the Verrucariales where the perilocular layer is complete the two types of filaments coexist; the pseudoparaphyses, however, remain short and do not make up the interascal filaments. These species therefore show a synthetic type of carpocenter, which is perhaps primitive.

The ascostroma or pyrenosphere type of ascocarps is found in both the Ascoloculares and the Ascohymeniales. In the course of evolution, a supplementary envelope, the ascothecium, appeared between the stromatic layers and the hymenial cavity. Therefore, there is no ascothecium in the Ascoloculares. However, a pseudoascothecium can be found in certain primitive members of the Sordariales, whereas the true ascothecium is present in practically all species of the Ascohymeniales. Correlated with this, the stromatic layers tend to disappear in the course of evolution and the ascocarp consists of only ascothecial tissue in the most highly evolved species.

The perithecial wall evolution in pyrenolichens is noteworthy. It is doubly primitive from being formed of an external part comparable to a pyrenosphere or a stroma and an internal part differentiated around the carpocenter. The external part seems to have been evolutionarily suppressed and, as is in the most evolved Ascohymeniales, the wall is then reduced to a secondary envelope. This latter structure could be compared to the ascothecium of the ascohymeniaceous pyrenomycetes. It recalls the secondary formations of the less evolved ascohymeniaceous discomycetes.

The unitunicates are derived from the bitunicates because in certain asci the bitunicate structure becomes unitunicate in the mature form. It is probably that the two walls tend to be fused together more and more intimately to form only one wall in the course of maturation—or evolution. Likewise, perhaps in regressive evolution, the internal ascal wall has disappeared in certain unitunicates.

References

Ainsworth, G. C., F. K. Sparrow, and A. S. Sussman, 1973. The Fungi, vol. IV. Academic Press, New York.

Barr, M. E., 1956. The development of the ascocarp in *Phaeotrichum hystrianum.* Can. J. Bot. 34: 563–568.

Bellemére, A., 1967. Contribution à l'étude du développement de l'apothécie chez les discomycètes inoperculés. Bull. Soc. Mycol. France 83: 395–931.

Cain, R. F., 1972. Evolution of the Fungi. Mycologia 64: 1–14.

Chadefaud, M., 1942. Etude d'asques II: Structure et anatomie comparée de l'appareil apical des asques chez divers discomycètes et pyrénomycètes. Rev. Mycol. 7: 57–88.

Chadefaud, M., 1944. Biologie des Champignons, vol. 1. Gallimard, Paris.

Chadefaud, M., 1946. Les asques à nasse apicale. Bull. Soc. Bot. France 93: 128–130.

Chadefaud, M., 1953. Sur un *Hypocopra,* sa position systématique, ses spores, et ses asques. C. R. Acad. Sci. Paris 236: 513–514.

Chadefaud, M., 1954a. Sur les asques des Erysiphacées. C. R. Acad. Sci. Paris 238: 1445–1447.

Chadefaud, M., 1954b. Sur les asques de deux Dothidéales. Bull. Soc. Mycol. France 70: 99–108.

Chadefaud, M., 1955. Sur les asques et la position systématique de l'*Ophiobolus graminis* Sacc. Bull. Soc. Mycol. France 71: 325–337.

Chadefaud, M., 1958. Sur les asques des Nectriales, et l'existence de Pléosporales nectrioïdes. C. R. Acad. Sci. Paris 247: 1376–1379.

Chadefaud, M., 1959. Les Pléosporales nectrioïdes et la systématique des pyrénomycètes. C. R. Acad. Sci. Paris 248: 1562–1564.

Chadefaud, M., 1960. Les végétaux non vasculaires (Cryptogamie). *In* M. Chadefaud and L. Emberger (Eds.), Traité de Botanique, Tome I. Masson, Paris.

Chadefaud, M., 1964. Sur l'origine et la structure des asques du type annellascé. C. R. Acad. Sci. Paris 258: 299–301.

Chadefaud, M., 1969a. Remarques sur les parois, l'appareil apical et les rèserves nutritives des asques. Osterreich. Bot. Z. 116: 181–202.

Chadefaud, M., 1969b. Données nouvelles sur la paroi des asques. C. R. Acad. Sci. Paris 268: 1041–1044.

Chadefaud, M., 1972. La coiffe réfringente et la calotte épiplasmique des asques de type annellascé. C. R. Acad. Sci. Paris 275: 2335–2338.

Chadefaud, M., 1973. Les asques et la systématique des Ascomycètes. Bull. Soc. Mycol. France 89: 127–170.

Chadefaud, M., and L. Avellanas, 1967. Remarques sur l'ontogénie et la structure des périthèces des *Chaetomium.* Trav. Biol. vég. dédiés au Professeur P. Dangeard. Le Botaniste, Ser. L.: 59–87.

Chadefaud, M., A. Parguey-Leduc, and M. Boudin, 1966. Sur les périthèces de *Preussia multispora* (Saito et Minoura) Cain et sur la position systematique du g. *Preussia.* Bull. Soc. Mycol. France 82: 93–122.

Doguet, G., 1955a. Etude du développement du *Chaetoceratostoma longirostre* Farrow. Rev. Mycol., Suppl. vol. 20: 132–143.

Doguet, G., 1955b. Le genre *Melanospora.* Le Botaniste 39: 1–313.

Doguet, G., 1956a. Le genre *Thielavia* Zopf. Rev. Mycol., Suppl. vol. 21: 1–21.

Doguet, G., 1956b. Morphologie et organogénie du *Neocosmospora vasinfecta* E. F. Smith et du *Neocosmospora africana* von Arx. Ann. Sci. Nat., Bot. Ser. 11: 353–370.

Doguet, G., 1957a. Organogénie du *Creopus spinulosus* (Fuckel) Moravec. Organogénie comparée de quelques Hypocréales du même type. Bull. Soc. Mycol. France 73: 144–164.

Doguet, G., 1957b. Organogénie du *Microascus stysanophorus* (Matt.) Curzi. Bull. Soc. Mycol. France 73: 165–178.

Doguet, G., 1959a. Organogénie du périthèce du *Coniochaeta ligniaria.* Comparaison avec l'organogénie des *Xylaria* et des discomycètes angiocarpes. Rev. Mycol. 24: 18–38.

Doguet, G., 1959b. Les bases actuelles de la systématique des Pyrénomycètes et le cas du *Melogramma spiniferum.* C. R. Acad. Sci. Paris 249: 2605–2607.

Doguet, G., 1960a. Morphologie, organogénie et évolution nucléaire de l'*Epichloe typhina.* La place des Clavicipitaceae dans la classification. Bull. Soc. Mycol. France 76: 171–203.

Doguet, G., 1960b. Etude du *Melogramma spiniferum* (Wallr.) De Notaris, pyrénomycète ascohyménié, annellascé, bituniqué. Rev. Mycol. 25: 13–37.

Dughi, R., 1957. Membrane ascale et réviviscence chez les champignons lichéniques discocarpes inoperculés. Ann. Fac. Sci. Marseille 26: 3–20.

Fayret, J., and A. Parguey-Leduc, 1975. L'ontogénie des périthèces et les asques du *Gnomonia leptostyla* (Fr.) Ces. et de Not. Ann. Sci. Nat. Ser. 12, 16: 17–41.

Gaümann, E. A., 1949. Die Pilze—Grundzüge ihrer Entwicklungsgeschichte und Morphologie. Birkhäuser Verlag, Basel.

Gilles, A., 1947. Evolution nucléaire et développement du périthèce chez *Nectria flava.* La Cellule 52: 1–32.

Hale, M. E., 1961. Lichen Handbook. Smithsonian Inst. Publ., Washington, D.C.

Hanlin, R. T., 1961. Studies in the genus *Nectria.* II—Morphology of *Nectria gliocladioides.* Amer. J. Bot. 48: 900–908.

Hanlin, R. T., 1963a. Morphology of *Hypomyces lactifluorum.* Bot. Gaz. 124: 395–404.

Hanlin, R. T., 1963b. Morphology of *Neuronectria peziza.* Amer. J. Bot. 50: 56–66.

Hanlin, R. T., 1964. Morphology of *Hypomyces trichothecoides.* Amer. J. Bot. 51: 201–208.

Hanlin, R. T., 1965. Morphology of *Hypocrea schweinitzii.* Amer. J. Bot. 52: 570–579.

Hanlin, R. T., 1971. Morphology of *Nectria haematococca.* Amer. J. Bot. 58: 105–116.

Héau, E., 1968. Recherches sur l'ontogénie et l'organisation des périthèces de deux Chaetomiacées. Diplôme d'Etudes Supérieures de Université Pierre and Maire Curie, Paris (unpublished).

Henssen, A., and M. Jahns, 1974. Lichenes. Georg Thieme Verlag, Stuttgart.

Huang, L. H., 1976. Developmental morphology of *Triangularia backusii* (Sordariaceae). Can. J. Bot. 54: 250–267.

Janex-Favre, M. C., 1970. Recherches sur l'ontogénie, l'organisation et les asques de quelques pyrénolichens. Rev. Bryol. Lichénol. 37: 421–650.

Jenkins, W. A., 1934. The development of *Cordyceps agariciformia.* Mycologia 26: 220–243.

Johnson, G. T., 1940. Contribution to the study of the Trypetheliaceae. Ann. Missouri Bot. Gard., U.S.A. 27: 1–50.
Kohlmeyer, J., and R. V. Gessner, 1976. *Buergenerula spartinae* sp. nov., an ascomycete from salt marsh cordgrass, *Spartina alterniflora.* Can. J. Bot. 54: 1759–1766.
Kowalski, D. T., 1964. The development and cytology of *Pycnidiophora dispersa.* Amer. J. Bot. 51: 1076–1082.
Kowalski, D. T., 1965a. The development and cytology of *Melanospora tiffanii.* Mycologia 57: 279–290.
Kowalski, D. T., 1965b. The development and cytology of *Didymocrea sadasavanii.* Mycologia 57: 404–416.
Lemans, C., 1962. Recherches sur un Pyrénomycète du genre *Thielavia.* Diplôme d'Etudes Supérieures de Université Pierre and Maire Curie, Paris (unpublished).
Letrouit-Galinou, M. A., 1957. Révision monographique du g. *Laurera* (Lichens, Trypéthéliacées). Rev. Bryol. Lichénol. 26: 207–264.
Letrouit-Galinou, M. A., 1973a. Sexual reproduction. *In* V. Ahmadjian and M. E. Hale (Eds.), The Lichens. Academic Press, New York, pp. 59–90.
Letrouit-Galinou, M. A., 1973b. Les asques des lichens et le type archaeascé. The Bryologist 76: 30–47.
Lieneman, C., 1938. Observations on *Thyronectria denigrata.* Mycologia 30: 494–511.
Lupo, P., 1922. Stroma and formation of perithecia in *Hypoxylon.* Bot. Gaz. 73: 486–495.
Luttrell, E. S., 1944. The morphology of *Sphaerostilbe aurantiicola* (B. and Br.) Petch. Bull. Torrey Bot. Club 71: 599–619.
Luttrell, E. S., 1951. Taxonomy of the pyrenomycetes. Univ. Missouri Studies No. 24, pp. 1–120.
Mai, S. H., 1976. Morphological studies in *Podospora anserina.* Amer. J. Bot. 63: 821–825.
Mainwaring, H. R., and I. M. Wilson, 1968. The life cycle and cytology of an apomitic *Podospora.* Trans. Br. Mycol. Soc. 51: 663–677.
Miller, J. H., 1928a. Biologic studies in the Sphaeriales—I. Mycologia 20: 187–213.
Miller, J. H., 1928b. Biologic studies in the Sphaeriales—II. Mycologia 20: 305–339.
Miller, J. H., 1949. A revision of the classification of the ascomycetes with special emphases on the pyrenomycetes. Mycologia 41: 99–121.
Moreau, F., 1914. Sur le développement du périthèce chez une Hypocréale, le *Peckiella lateritia* (Fries) Maire. Bull. Soc. Bot. France 61: 160–164.
Moreau, F., 1952. Les Champignons: Physiologie—Morphologie Développement et Systématique. Lechevalier, Paris.
Moreau, F., and V. Moreau, 1930. Le développement du périthèce chez quelques ascomycetes. Rev. Gén. Bot.: 1–32.
Moreau, F., and V. Moreau, 1950. Etude du développement du *Triangularia bambusae* (van Beyma) Boedijn. Rev. Mycol. 15: 146–158.
Moreau, F., and V. Moreau, 1951. Observations cytologiques sur les ascomycètes du genre *Pleurage* Fr. Rev. Mycol. 16: 198–208.
Moreau, F., and V. Moreau, 1952. Sur le développement du *Ceratocystis moniliformis.* Rev. Mycol. 17: 141–153.
Moreau, F., and V. Moreau, 1953. Etude de développement de quelques Aspergillacées. Rev. Mycol. 18: 165–180.
Moreau, F., and V. Moreau, 1954. Sur le développement des périthèces du *Chaetomium elatum* Kunze. Rev. Mycol. 19: 167–171.
Moreau, F., and C. Moruzi, 1931. Recherches expérimentales sur la formation des périthèces chez les *"Neurospora."* C. R. Acad. Sci. Paris 192: 1476–1478.

Munk, A., 1957. Danish pyrenomycetes. A preliminary flora. Dansk. Bot. Archiv. 17: 1–491.

Nannfeldt, J. A., 1932. Studien über die Morphologie und Systematik der nicht-lichenisierten, inoperculaten Discomyceten. Nova Acta Reg. Soc. Sci. Upsala Ser. IV, 8: 1–368.

Nannfeldt, J. A., 1975a. Stray studies in the Coronophorales (Pyrenomycetes) 1–3. Svensk Bot. Tidskrift 69: 49–66.

Nannfeldt, J. A., 1975b. Stray studies in the Coronophorales (Pyrenomycetes) 4–8. Svensk Bot. Tidskrift 69: 289–335.

Parbery, D. G., 1969a. Isolation of the ascal state of *Amorphotheca resinae* direct from soil. Trans. Br. Mycol. Soc. 53: 482–484.

Parbery, D. G., 1969b. *Amorphotheca resinae* gen. nov., sp. nov.: The perfect stage of *Cladosporium resinae*. Austral. J. Bot. 17: 331–357.

Parguey-Leduc, A., 1959. Le développement de la Pléosporale nectrioïde (?) *Letendrea padouk* n. sp. C. R. Acad. Sci. Paris 248: 1559–1562.

Parguey-Leduc, A., 1961. Etude des asques et du développement de l'*Helminthosphaeria clavariarum* (Desm.) Fuckel ap. Munk. Bull. Soc. Mycol. France 77: 15–33.

Parguey-Leduc, A., 1964. Développement d'une Nectriale: *Gibberella pulicaris*. C. R. Acad. Sci. Paris 258: 2141–2144.

Parguey-Leduc, A., 1965. Sur le développement des périthèces chez les Diatrypacées. C. R. Acad. Sci. Paris 260: 3735–3738.

Parguey–Leduc, A., 1966. Recherches sur l'ontogénie et l'anatomie comparée des ascocarpes des pyrénomycètes ascoloculaires. Ann. Sci. Nat. Bot. Ser. 12, 7: 505–690.

Parguey-Leduc, A., 1967a. Recherches préliminaires sur l'ontogénie et l'anatomie comparée des ascocarpes des pyrénomycètes ascohyméniaux. I. Notions générales. Rev. Mycol. 32(2): 57–68.

Parguey-Leduc, A., 1967b. Recherches préliminaires sur l'ontogénie et l'anatomie comparée des ascocarpes des pyrénomycètes ascohyméniaux. II. Structure et développement des ascothécies. Rev. Mycol. 32(4): 259–277.

Parguey-Leduc, A., 1967c. Recherches préliminaires sur l'ontogénie et l'anatomie comparée des ascocarpes des pyrénomycètes ascohyméniaux. III. Les asques des Sordariales et leurs ascothécies, du type "Diaporthe." Rev. Mycol. 32(5): 369–407.

Parguey-Leduc, A., 1967d. Recherches sur l'ontogénie et l'anatomie comparée des ascocarpes des pyrénomycètes ascoloculaires. Ann. Sci. Nat. Bot. Ser. 12, 8: 1–110.

Parguey-Leduc, A., 1970. Recherches préliminaires sur l'ontogénie et l'anatomie comparée des ascocarpes des pyrénomycètes ascohyméniaux. IV—Les asques des Diatrypacées et leurs ascothécies du type "Eutypa." Rev. Mycol. 35: 90–130.

Parguey-Leduc, A., 1972. Recherches préliminaires sur l'ontogénie et l'anatomie comparée des ascocarpes des pyrénomycètes ascohyméniaux. V. Les asques des Xylariales et leurs ascothécies du type "Xylaria." Rev. Mycol. 36: 194–237.

Parguey-Leduc, A., 1973. Recherches préliminaires sur l'ontogénie et l'anatomie comparée des ascocarpes des pyrénomycètes ascohyméniaux. VI. Conclusions générales. Rev. Mycol. 37: 60–82.

Parguey-Leduc, A., and M. Chadefaud, 1963. Les asques du *Cainia incarcerata* (Desm.) von Arx et Muller et la position systématique du genre *Cainia*. Rev. Mycol. 28(3/4): 200–234.

Parguey-Leduc, A., and B. Dehorter, 1973. Les ascocarpes et les asques du *Nectria galligena* Bres. Le Botaniste 66: 157–175.

Poelt, J., 1973. Systematic evaluation of morphological characters; Classification. *In* V. Ahmadjian and M. E. Hale (Eds.), The Lichens. Academic Press, New York and London. pp. 91–116, 599–632.

Pomerleau, R., 1937. Recherches sur le *Gnomonia ulmea* (Schw.) Thüm. Natural. Can. 1–139.

Santesson, R., 1952. Foliicolous lichens I. A revision of the taxonomy of the obligately foliicolous, lichenized fungi. Symb. Bot. Upsala 12: 1–577.

Seeler, E. V., 1940. A monographic study of the genus *Thyronectria*. J. Arnold Arb. 21: 429–460.

Strikmann, E., and M. Chadefaud, 1961. Recherches sur les asques et les périthèces des *Nectria*, et réflexions sur l'évolution des Ascomycètes. Rev. Gen. Bot. 68: 725–770.

Varitchak, B., 1927. Sur le développement des périthèces chez le *Cordyceps militaris* (Linn.) Link. C. R. Acad. Sci. Paris 184: 622–624.

Varitchak, B., 1931. Contribution à l'étude du développement des Ascomycètes. 6 Botaniste, serie 23, Paris.

Vincens, F., 1917. Recherches organogéniques sur quelques Hypocréales. Thèse de Doctorat Paris (unpublished).

von Arx, J. A., 1973. Ostiolate and nonostiolate pyrenomyctes. Proceedings Koninkluke Nederl. Akademie van Vetenschappen C, 76: 289–296.

von Höhnel, F., 1917. Mykologische Fragmente. Ann. Mycol. 15: 293–363.

Wehmeyer, L. E., 1926. A biologic and phylogenetic study of the stromatic Sphaeriales. Amer. J. Bot. 13: 575–645.

Wehmeyer, L. E., 1941. A revision of *Melanconis, Pseudovalsa, Prosthecium* and *Titania*. Univ. Michigan Studies No. 132, pp. 1–161.

Wehmeyer, L. E., 1975. The Pyrenomycetous Fungi. J. Cramer, Lehre, Germany.

Wicker, M., 1962. Le mycélium et les périthèces dans une souche de la Sordariale *Pleurage minuta* (Fuckel) Ktze. Bull. Soc. Mycol. France 78: 291–326.

Zickler, H., 1952. Zur Entwicklungsgeschichte der Ascomyceten *Bombardia lunata* Zickl. Arch. Protistenk. 98: 1–70.

Chapter 9

The Pyrenomycete Centrum—Loculoascomycetes

E. S. Luttrell

Introduction

Ascoloculares (Nannfeldt, 1932) and loculoascomycetes (Luttrell, 1955) are different names for the same taxon. They refer to essentially the same series of ascomycetes, delimited in essentially the same way. "Loculoascomycetes" was proposed as a variant of Ascoloculares, primarily to make the ending conform to that for fungus taxa, -mycetes. Syllables, wisely or not, were transposed—from ascoloculo to loculoasco—to follow the pattern set by euascomycetes and hemiascomycetes. This may strain the definition of orthographic variant, but the intent was clearly stated. If the rank of subclass is maintained, the name must be modified to Loculoascomycetidae.

The purpose of establishing the subclass Loculoascomycetidae was to give the ascolocular series formal standing in the taxonomic hierarchy. There were two reasons for doing so. First, there are formal categories for expressing taxonomic ideas; "groups" or "series" of undesignated status have uncertain status in the system. Second, a taxonomic hypothesis should be boldly and precisely stated in a form that invites challenge. Use of the vague words "group" or "series" avoids the issue. The use of *Anhang* is rejected for the same reason.

The outlines of the Loculoascomycetidae were sketched by von Höhnel (1909) when he expanded the Dothideales to include the families Myriangiaceae, Dothideaceae, and Pseudosphaeriaceae. Theissen and Sydow (1918) established the superordinal taxon Dothideneae for the Myriangiales, Dothideales, and Pseudosphaeriales. The basic contribution to this taxonomic concept was Nannfeldt's (1932) division of the ascomycetes into three major series: Plectascales, Ascohymeniales, and Ascoloculares. Subsequent contributions to the classification of Loculoascomycetidae have been made by Miller (1949), Luttrell (1951), von Arx and Müller (1975), and Barr (1976). These treatments are neither uniform nor in complete agreement. Chadefaud (1960, 1973) and Parguey-Leduc (1966, 1967) have considered the separation of distinct series represented by the Euascomycetidae and Loculoascomycetidae to be a simplistic interpretation of phylogeny in the ascomycetes and have postulated lines leading from primitive discomycetes through loculoascomycetes to euascomycetes (Parguey-Leduc, 1967, diagram on p. 4). von Arx and Müller (1975) included all loculoascomycetes in the single order Dothideales, as did von Höhnel (1909). Their division of this order into the suborders Dothideneae and Pseudosphaerineae reflects their earlier (Müller and von Arx, 1962) recognition of the orders Dothiorales and Pseudosphaeriales and Theissen's (1918) recognition of the orders Myriangiales, for loculoascomycetes with apothecioid ascocarps, and Dothideales, for those with perithecioid ascocarps. In contrast, Barr (1976) recognized the class Loculoascomycetes with four subclasses and eight orders—a treatment more nearly in accord with that of Luttrell (1973). Although these differences in opinion on phylogeny and the organization of higher taxa must be noted, they should not obscure the fact of a developing consensus on the delimitation of families in the Loculoascomycetidae.

Having arrived at the Loculoascomycetidae intuitively, it is necessary in order to communicate the revelation to search for specific characters that can be used to delimit the taxon. Such characters may be found in the centrum, the ascogenous system and asci, and the associated delicate sterile tissues in which they develop.

Ascocarp Origins

One basic characteristic of the Loculoascomycetidae is the origin of the ascocarp and, consequently, of the centrum tissues directly from a stroma. The ascocarp is ascolocular or ascostromatic. It is a pseudothecium. The distinction between perithecium and pseudothecium is most clearly apparent in a comparison between a fungus with multiloculate pseudothecia, such as *Dothidea* Fries, and one with perithecia immersed in a stroma, such as *Xylaria* Hill. In *Xylaria* (*Xylaria carpophila* Persoon and Fries, Pl. I, Fig. 9.1), the ascocarps obviously are perithecia surrounded by a definite peridium distinct from the stromal tissue in which they are embedded. The individual perithecia can be dissected out of the stroma intact and are no different from individual perithecia developing separately on a mycelium. In *Dothidea* (*Dothidea puccinioides* Fries, Pl. I, Fig. 9.2), clusters of asci occur in cavities (locules) in the stroma. No peridium separates the cluster of asci from the surrounding stromal tissue. There is no perithecium. This is a multiloculate ascostroma.

If large ascostromata with many locules exist, then smaller ascostromata with fewer locules are to be expected. These occur, as for example, in *Mycosphaerella killianii* Petrak (Pl. I, Fig. 9.3). Furthermore, as the stroma becomes smaller and the number of locules fewer, a very small pseudothecium containing a single locule can be predicted. This transition, in fact, may occur in a single species. *Mycosphaerella killianii*, for example, produces uniloculate (Pl. I, Fig. 9.4) as well as multiloculate (Pl. I, Fig. 9.3) pseudothecia. Such transitional species provide a link to the numerous genera, such as *Guignardia* [*Guignardia bidwellii* (Ellis) Viala and Ravaz; Pl. I, Fig. 9.5], that are characterized by uniloculate pseudothecia. We know that such ascocarps exist. The problem is how to distinguish the wall of stromal tissue in a small, globoid pseudothecium from the wall formed by the peridium in a simple perithecium.

Miller (1928, 1949) attempted to distinguish between pseudothecia and peri-

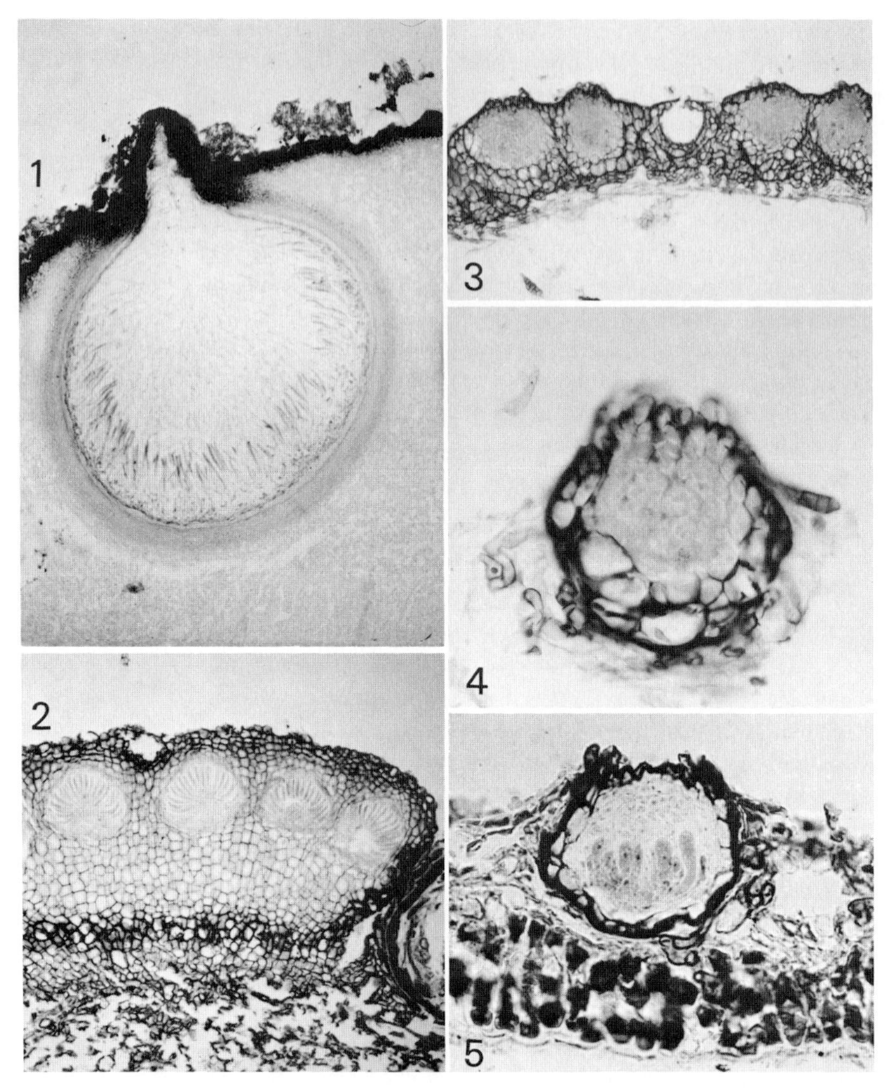

Plate I

Fig. 9.1 *Xylaria carpophila.* Perithecium with young asci immersed in stroma. ×130.

Fig. 9.2 *Dothidea puccinioides* Fries. Compound pseudothecium with clusters of young asci in locules in a pulvinate stroma erumpent through bark. An empty spermogonial locule is present in the stroma at top center. ×130.

Figs. 9.3 and **9.4** *Mycospaerella killianii* in dead leaf of *Trifolium incarnatum.* **Fig. 9.3** Multiloculate pseudothecium with locules prior to formation of asci in thin stroma. An empty spermogonial locule is present in the center of the stroma. ×130. **Fig. 9.4** Uniloculate pseudothecium with pseudoparenchymatous centrum. ×420.

Fig. 9.5 *Guignardia bidwellii* in dead leaf of *Vitis rotundifolia.* Simple perithecioid pseudothecium containing cluster of young asci in single locule. A single layer of dark cells surrounds the centrum. ×420.

thecia on the basis of ontogeny. The pseudothecium develops from a stroma in which the ascogonia later appear. In the ontogeny of a perithecium, the ascogonium appears first, and the peridium develops from hyphae that grow up around the ascogonium from the stalk cells of the ascogonium (Parguey-Leduc, 1973) or from adjacent hyphae as a result of a sexual stimulus. In pseudothecial forms, such as *Botryosphaeria berengeriana* de Notaris, the stromatic initials in which the ascogonia appear are massive (Pl. II, Fig. 9.6), and the course of development postulated by Miller may be easily demonstrated. In many pseudothecial forms, however, the "stromata" in which the ascogonia develop may be so small—a knot of cells or a weft of hyphae—that they are almost nonexistent. In many perithecial forms, the tissue that develops into the peridium may appear before the ascogonia are readily identifiable. In *Mycosphaerella* Johansen, for example, the ascogonium prior to plasmogamy (Pl. II, Fig. 9.7) appears as an enlarged, globose cell with a single conspicuous nucleus embedded in a small stromatic initial. At an earlier stage (Pl. II, Fig. 9.8), this initial consists of a tiny cluster of cells. A single cell in the center with a conspicuous nucleus might be interpreted as an ascogonium. As the cells of the cluster are brought individually into sharp focus, all cells appear similar. The interpretation then is that this cluster of cells represents a stromatic initial in which none of the cells has yet become differentiated as an ascogonium. This, however, is a matter of interpretation.

The stromatic nature of the pseudothecium can often be inferred at later stages of development. In the uniloculate pseudothecium of *Pleospora herbarum* (Persoon ex Fries) Rabenhorst (Pl. II, Fig. 9.9), the young centrum composed of vertically arranged pseudoparaphyses and ascogenous cells occupies a small area near the apex of a massive sclerotioid structure. This structure is obviously stromatic, and no peridium surrounds the centrum. However, compare the ascocarps of species in *Pleospora* Rabenhorst (Pl. II, Fig. 9.9) and *Dothidea* (Pl. I, Fig. 9.2) with an ascocarp of a species in *Phyllachora* Nitchke (Pl. II, Fig. 9.10). The centrum of *Phyllachora lespedezae* (von Schweinitz) Saccardo is typically xylariaceous in that it is composed of paraphyses; yet it would be exceedingly difficult to distinguish a peridium separating the centrum from the surrounding stroma. It is understandable why *Phyllachora* has often been included in the Dothideales, although it unquestionably is a "perithecial" form. We consider the overall characters of *Phyllachora* give intuitive recognition of a position in the Euascomycetidae. The ascocarp is deduced a perithecium.

The concept of ascostroma or pseudothecium is not being questioned; however, characters are not always sharply defined, and single characters at times may fail.

Developmental Patterns

Assume that, although there may be difficulty in defining and recognizing them consistently, pseudothecia exist. Centrum tissues in the Loculoascomycetidae, along with other ascocarpic tissues, therefore, have a common origin: from stroma. The second question is, "Do these tissues have a common pattern of development?" Theissen and Sydow (1918) considered the essential character

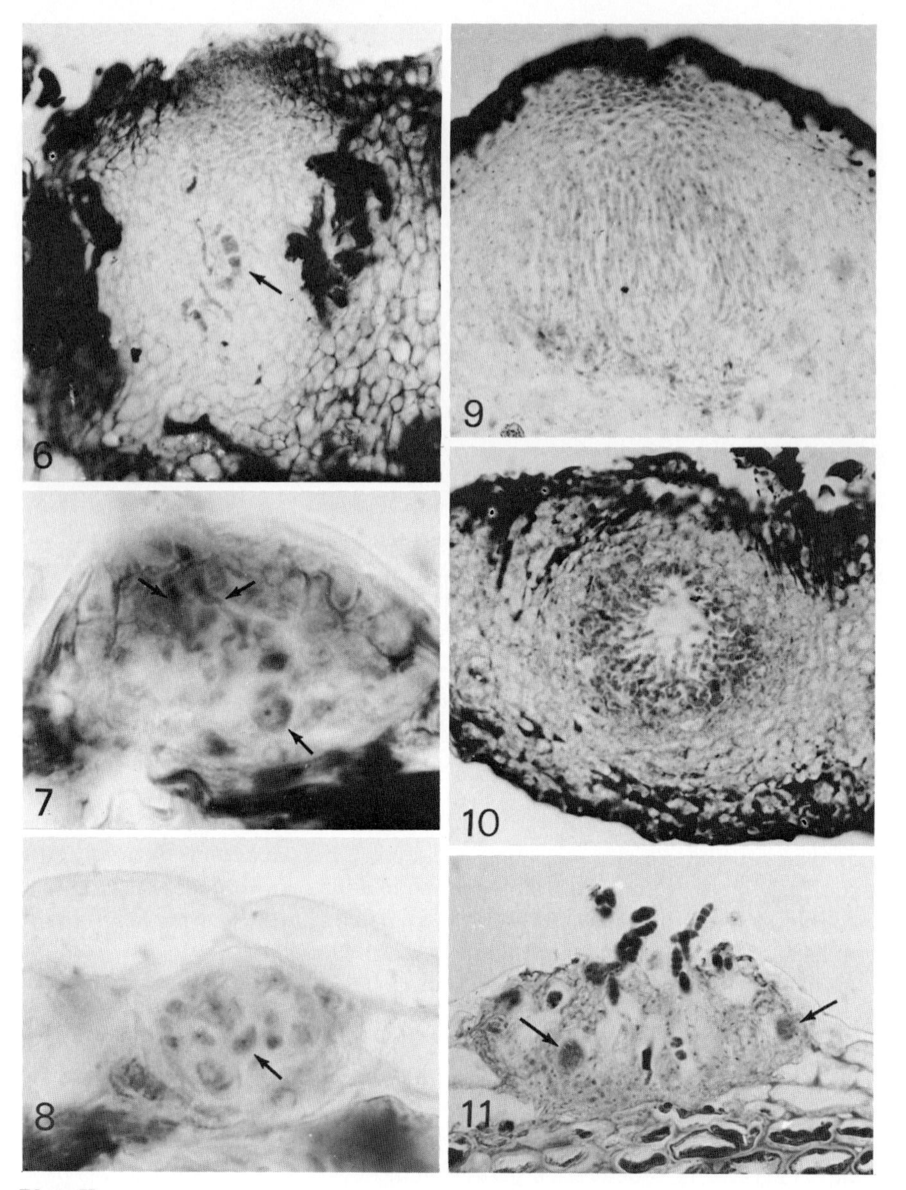

Plate II

Fig. 9.6 *Botryosphaeria berengeriana* (= *B. ribis* Grossenbacher and Duggar). Multicellular ascogonium (arrow) embedded in massive stromatic pseudothecial initial. ×420.
Figs. 9.7 and **9.8** *Mycosphaerella* sp. in necrotic leaf spot on *Lespedeza sericea*. **Fig. 9.7** Small stromatic pseudothecial initial raising outer walls of host cells, showing globose uninucleate ascogonium (arrow) in base and segments of two trichogynes (arrows) near apex. ×1600. **Fig. 9.8** Younger initial composed of globose mass of pseudoparenchymatous cells within a single host mesophyll cell beneath epidermis. A single cell (arrow) in the center of the stroma is more conspicuous, but the surrounding stromal cells appear similar when brought into focus. ×1600.
Fig. 9.9 *Pleospora herbarum*. Tip of young sclerotioid pseudothecium containing a single locule filled with vertical pseudoparaphyses prior to the development of asci. ×420.

uniting members of the Loculoascomycetidae to be formation of asci in monascus locules in a stromatic ascocarp. In the Myriangiales, the stromal tissue separating the irregularly distributed asci persists at maturity and each mature ascus occupies a single locule. In the Dothideales, the asci become grouped in clusters. Stromal tissue separating the asci disintegrates, and at maturity, each locule contains a cluster of asci. In the Pseudosphaeriales, the asci likewise become closely grouped in a single layer, but the stromal tissue separating the asci persists and becomes compressed between the expanding asci to paraphysoidal filaments that retain their attachment to the stroma at both top and bottom of the locule. Will this scheme work? As a model vaguely suggesting possible phylogenetic relationships among Loculoascomycetidae, the answer is "Perhaps." As an outline of a common ontogenetic pattern, the answer is, regrettably, "No."

The concept of the monascus locule may tie together a small and fluctuating group of fungi placed in the Myriangiales. The basic type is usually assumed to be represented by *Elsinoë* Raciborski. The ascocarp of *Elsinoë veneta* (Burkhalter) Jenkins is a small, irregular, erumpent stroma (Pl. II, Fig. 9.11). The globose asci are in monascous locules scattered randomly at various levels throughout the stroma. The unaltered stromal tissue persists between the asci at maturity.

The flat, superficial ascocarps of *Schizothyrium* Desmazières [*Schizothyrium pomi* (Montagne ex Fries) von Arx; Pl. III, Fig. 9.12] have a similar structure. In the stroma beneath the dimidiate covering shield, the asci may be irregularly arranged, as in *Elsinoë*, but commonly they are in a single layer. The interascicular stromal tissue may persist as compressed remnants or may be disintegrated so that the asci form a more or less continuous layer. There are numbers of such fungi that may be related to *Elsinoë* on this basis. The ascocarps are irregular, pulvinate, discoid, flattened, or perithecioid. Asci are ovoid to short cylindric. They apparently arise in monascous locules. The interascicular tissue persists unaltered, is compressed to paraphysoidal strands, or is more or less disintegrated at maturity. In large part, these are small, insignificant fungi on the surfaces of leaves, stems, or fruits. The point is that no complete developmental studies have been made of any of them. Even in *Elsinoë*, the course of development is largely inferred. In extenuation, it must be said that often there is very little to study. In *Aphanopeltis* Sydow, mature ascocarps on host leaves are black flecks that can hardly be seen without a hand lens. The circular ascocarp consists of a single layer of dark cells arranged in a radiate pattern. Beneath the flat shield is a hyaline layer one or two cells thick. Most of these cells become globoid asci. Some remain as packing cells among the asci. No great leap of the imagination is required to

---

Fig. 9.10 *Phyllachora lespedezae*. Young perithecium with developing paraphyses. No distinct peridium delimits the perithecium from the stroma in which it is embedded. ×420.

Fig. 9.11 *Elsinoë veneta*. Mature pseudothecium erumpent through epidermis of *Rubus* stem. Some asci are discharging ascospores, which appear as dark columns in expanded endoasci above the center of the ascocarp; the positions they occupied now appear as cavities in the stroma. Younger asci (arrows) fill monascous locules in the stroma. ×420.

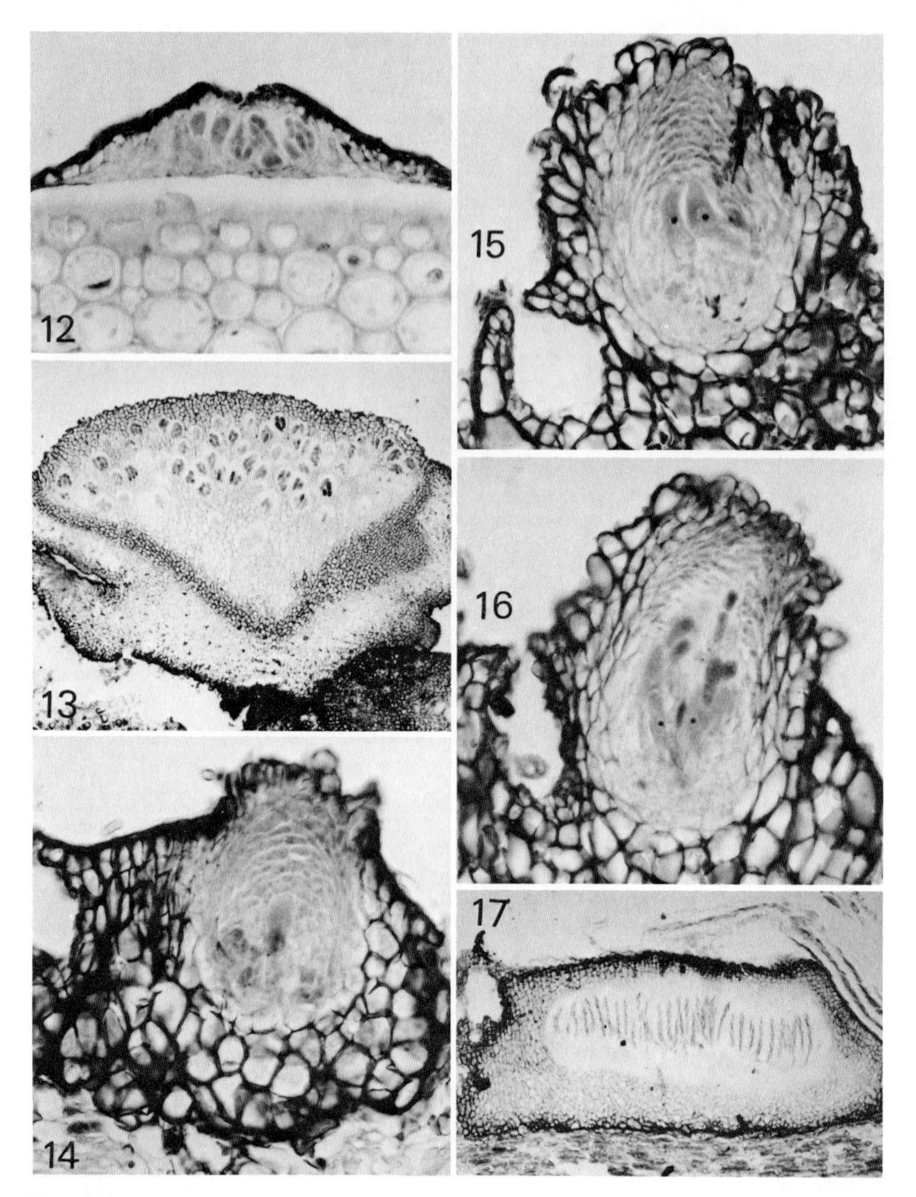

Plate III

Fig. 9.12 *Schizothyrium pomi* on surface of cuticle of *Smilax* stem. Mature pseudothecium with asci distributed individually in thin stroma beneath dark covering shield. ×420.

Fig. 9.13 *Myriangium duriaei*. Apothecioid conceptacle with asci in monascous locules at various levels in pseudoparenchymatous tissue. ×420.

Figs. 9.14–9.16 *Mycosphaerella killianii*. **Fig. 9.14** Ascogenous cells and first uninucleate ascus in pseudoparenchymatous centrum. An empty spermatogonial locule appears in the stroma at the left. ×420. **Fig. 9.15** Group of uninucleate asci pushing up into centrum. ×500. **Fig. 9.16** Nearly mature pseudothecium with small locule containing a few relatively large asci in disintegrating centrum pseudoparenchyma. ×420.

arrive at the conclusion that *Aphanopeltis* may represent a reduced *Elsinoë*. However, it may even more easily be considered a reduced *Microthyrium*.

A number of bothersome little genera may be conveniently disposed of by placing them in the Myriangiales on the basis of inferred development of monascous locules in an ascostroma. These may, however, be a miscellany of odd forms, most of them reduced and similar in appearance only because of extreme reduction. The type genus *Myriangium* Montagne and Berkeley if the results of Miller's (1938) study are accepted, fits nowhere in this series. The ascocarps of *Myriangium duriaei* Montagne and Berkeley apud Berkeley (Pl. III, Fig. 9.13) are among the more massive of the Myriangiales. The stroma may be 5 mm in diameter, and the apothecioid conceptacles clustered on the stroma may be 1.5 mm in diameter. According to Miller (1938), the sterile tissue of the conceptacle in which the globoid asci are individually and irregularly distributed is composed entirely of ascogenous cells. This is unique and surely open to question. Yet, Miller's observations have never been challenged. His paper is cited, these unusual results are summarized, and the discussion proceeds as if Miller's observations were in accord with the pattern of inferred development. They are not.

More information is available on the Dothideales. These are characterized by the production of aparaphysate asci in fascicles in locules. The centrum tissue is a pseudoparenchyma (Pl. I, Fig. 9.4, and Pl. III, Fig. 9.14). The asci originate in clusters and push up as a group (Pl. III, Fig. 9.15) into the centrum pseudoparenchyma, which distintegrates as the asci mature (Pl. III, Fig. 9.16). There is no interascicular sterile tissue at any stage. The locules and uniloculate pseudothecia are small. Compound pseudothecia containing many small locules may be conspicuous (Pl. I, Figs. 9.2 and 9.3). This pattern of development is uniform and diagnostic. It distinguishes a common group of Loculoascomycetidae. The number of genera is relatively low, but the single genus *Mycosphaerella* comprises many of the most frequent species of Loculoascomycetidae, and many of these are of economic importance. The definition of the Dothideales is stretched, on the one hand, by such fungi as *Euryachora* Fuckel and *Omphalospora* Theissen and Sydow (Müller and von Arx, 1962), in which the stroma is so thin and the locules so small that they intergrade with some Myriangiales, and on the other hand by such fungi as *Dothiora sorbi* (Wahl) Fuckel (Pl. III, Fig. 9.17), in which the locules are so broad and contain such an extended layer of asci that the ascocarps appear apothecioid.

The Pleosporales (Pseudosphaeriales) contains the bulk of the Loculoascomycetidae. If we find some common pattern of development that characterizes this order, we have found a means of recognizing the great majority of Loculoascomycetidae. The Pleosporaceae is the core family of the order and of the subclass. It includes the greatest number of the most conspicuous, most diverse, and most commonly encountered members of the Loculoascomycetidae. The centrum is

---

Fig. 9.17 *Dothiora sorbi*. Mature pseudothecium with continuous layer of aparaphysate asci occupying single broad locule in pulvinate stroma. When the stroma above the asci ruptures, the ascocarp appears apothecioid. ×130.

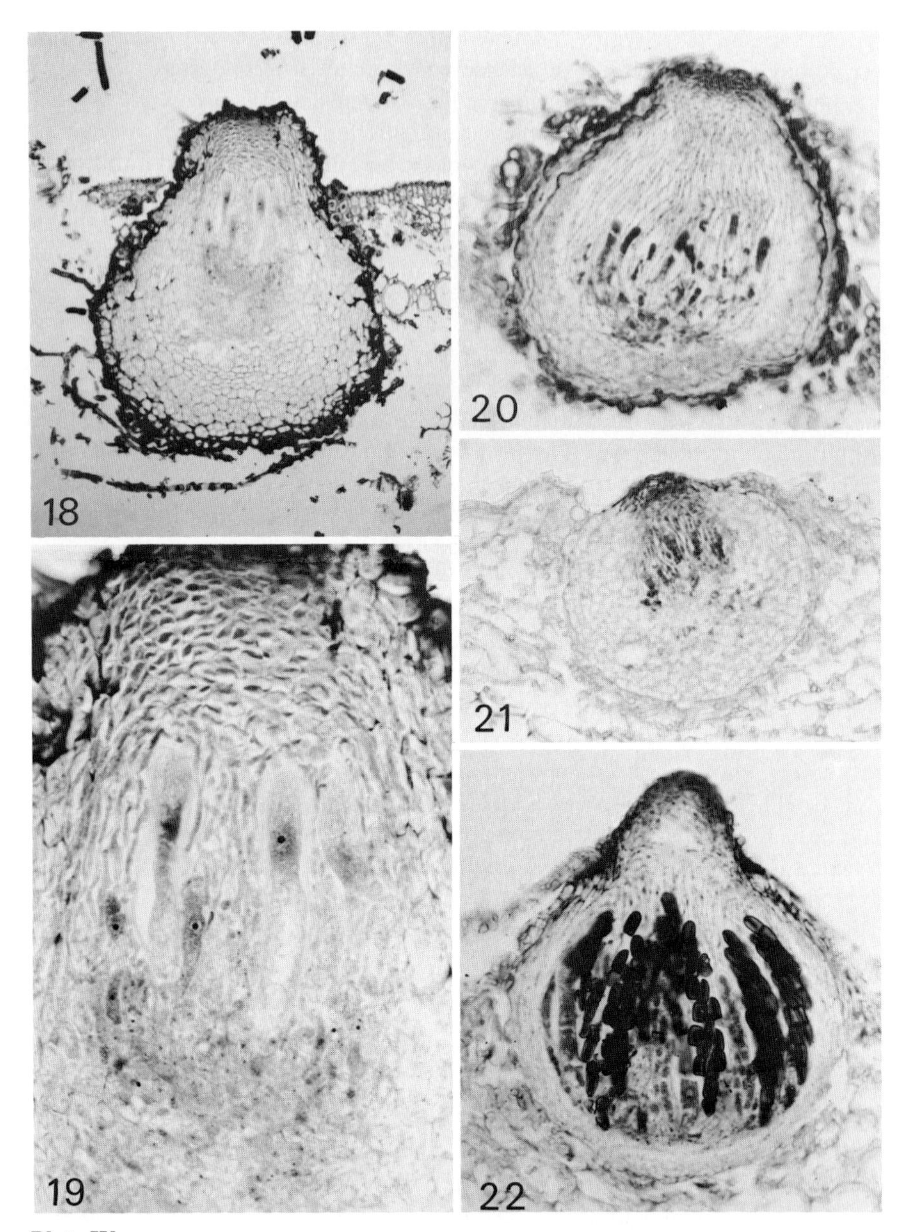

Plate IV

Figs. 9.18 and **9.19** *Pyrenophora avenae*. **Fig. 9.18** Young sclerotioid pseudothecium containing a single locule in the apical region. ×130. **Fig. 9.19** Enlargement of locule showing uninucleate asci among pseudoparaphyses. ×420.

Figs. 9.20–9.22 *Sporormiella* sp. **Fig. 9.20** Uniloculate perithecioid pseudothecium with uninucleate asci growing upward among pseudoparaphyses. ×420. **Fig. 9.21** Young pseudothecium with uninucleate asci among sparse pseudoparaphyses. ×420. **Fig. 9.22** Mature pseudothecium with asci containing brown, phragmosporous ascospores. Pseudoparaphyses are evident above the tips of the mature asci but are very sparse. ×420

composed of vertically arranged hyphae that develop before the asci appear (Pl. II, Fig. 9.9). Consequently, they cannot represent strands of stromal tissue compressed between expanding asci and are not homologous with the interascicular tissues in members of the Myriangiales. Pseudoparaphyses usually originate in the upper part of the ascocarp and grow downward, but their tips almost immediately become enmeshed with the subascogonial tissues, and their subsequent growth is largely intercalary (Corlett, 1973). They elongate as the ascocarp expands, and the asci grow up among them (Pl. IV, Figs. 9.18–9.22; Pl. V, Figs. 9.23 and 9.24). The pseudoparaphyses usually persist to maturity of the ascocarp, and their attachment at the apex of the locule remains discernible (Pl. V, Fig. 9.24), although in minute forms they may be sparse (Pl. IV, Figs. 9.21 and 9.22). So, we have a developmental pattern, a structure of centrum tissues recognizable in mature specimens, that characterizes the Pleosporaceae (Corlett, 1973).

The pseudoparaphysate centrum permits assignment to the Pleosporales of the Pleosporaceae, Lophiostomataceae, Sporormiaceae, Botryosphaeriaceae, Stigmateaceae, and Dimeriaceae. It would permit inclusion also of the dimidiate–scutate forms in the Microthyriaceae and Micropeltidaceae and the apothecioid forms in the Hysteriaceae. The pseudoparaphyses in both *Botryosphaeria berengeriana* (Pl. V, Fig. 9.23) and *Pyrenophora avenae* Ito and Kuribayashi in Ito (Pl. IV, Figs. 9.18 and 9.19) are so broad and densely packed that they often give the impression of a pseudoparenchymatous tissue. Although the origin of the pseudoparaphyses in these genera has not been adequately documented, they are present as recognizable filaments before the asci develop and seem to be acceptable as pseudoparaphyses.

In addition to such apparently acceptable variations, two major deviations occur. First, in the Metacapnodiaceae (Corlett et al., 1973), Chaetothyriaceae, and Herpotrichiellaceae (Barr, 1976), and in the genus *Capnodium* Montague (Reynolds, 1978), the asci develop in aparaphysate clusters. Pendent hyphae extend downward and inward from the roof of the locule. These hyphae may be short or may fill the upper part of the locule. My inclination is to consider these pendent hyphae homologous with periphyses and to ally the families in which they occur with the families of the Dothideales. Corlett et al. (1973) considered the pendent hyphae in *Metacapnodium* to be modified pseudoparaphyses and placed the Metacapnodiaceae in the Pleosporales. Barr (1976) placed all families with this structure in a separate order, the Chaetothyriales.

The second major deviation occurs in the Patellariaceae (Bellemère, 1971; Pirozynski and Reid, 1964) and the Parmulariaceae (Luttrell and Muthappa, 1974), both families of apothecioid forms. In *Rhytidhysterium* Spegazzini of the Patellariaceae, paraphysioidal hyphae originate in much the same fashion as pseudoparaphyses [*Rhytidhysterium rufulum* (Sprengera) Spegazzini; Pl. V, Figs. 9.25 and 9.26]. They are attached to the surrounding stroma at top and bottom, and there is never any indication of free tips at either end. These hyphae elongate greatly as the long cylindrical asci develop. As the overlying stroma crumbles and the ascocarp becomes apothecioid, the paraphysoidal hyphae break loose at the top of the locule. The freed tips regenerate and branch to form an epithecium. In the elongated hysterioid ascocarps of *Aulacostroma parvispora*

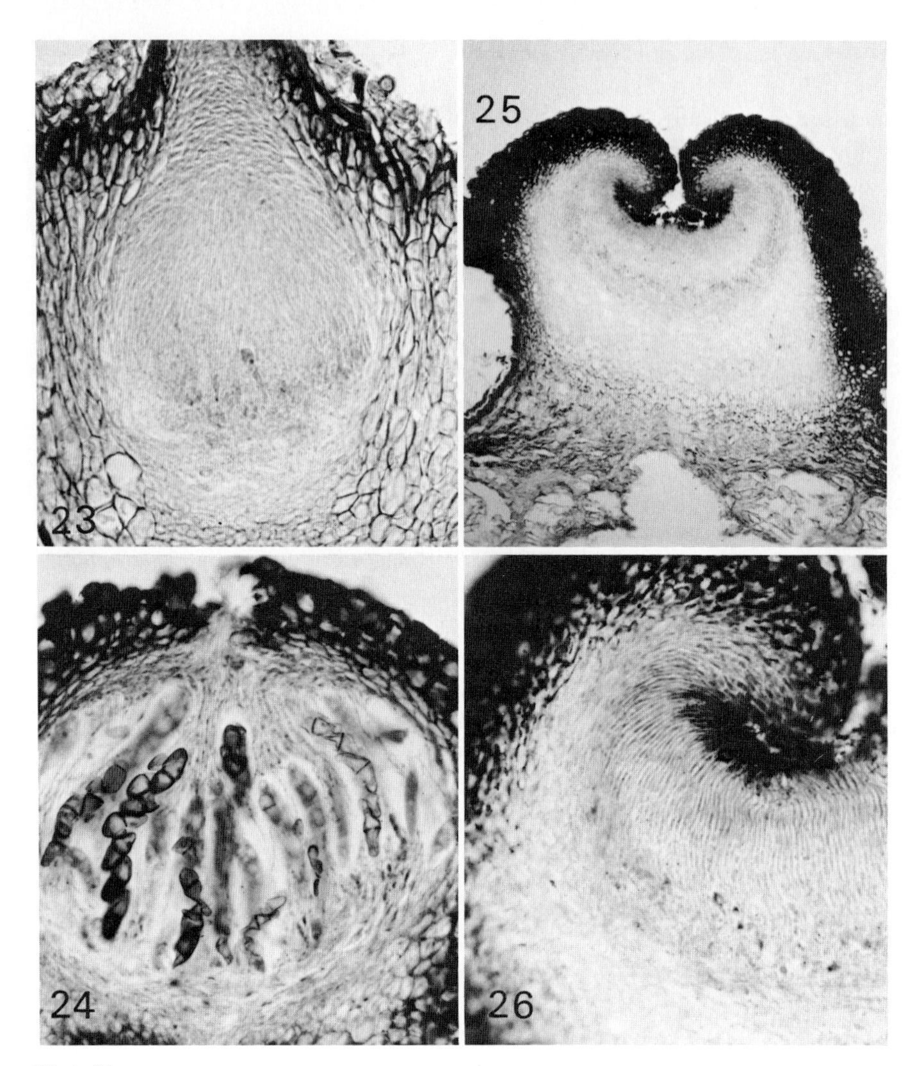

Plate V

Fig. 9.23 *Botryosphaeria berengeriana.* Centrum at stage in which first uninucleate asci appear composed of densely packed pseudoparaphyses. ×420.

Fig. 9.24 *Parodiella hedysari.* Mature asci with uniseptate, brown ascospores among persistent pseudoparaphyses, the attachments of which at the top of the locule are still evident. ×420.

Figs. 9.25 and **9.26** *Rhytidhysterium rufulum.* **Fig. 9.25** Young pseudothecium with ascogenous cells forming a cup-shaped layer below a zone of vertically arranged paraphysoidal hyphae. The depression in the center of the pseudothecium results from the bulging outward and upward of the periphery of an originally pulvinate stroma. At maturity, the incurved lips will break loose from the center and unfurl into an apothecioid structure. ×130. **Fig. 9.26** Enlargement of the zone of paraphysoidal hyphae attached at top and bottom of locule. These hyphae break away from the stroma at the top of the locule. The freed tips regenerate, produce short branches with swollen tips that form an epithecium, and become indistinguishable from paraphyses in the mature pseudothecium. ×420.

Luttrell and Muthappa (Luttrell and Muthappa, 1974) in the Parmulariaceae, the paraphysoidal hyphae seem to originate by elongation of stromal cells. Attachment of these hyphae to the roof of the locule, however, is transitory, and most of their growth occurs as hyphae with free tips that are indistinguishable from paraphyses.

It would be convenient to append the Patellariaceae and Parmulariaceae to the Pleosporales as an *Anhang* without making a commitment as to their proper position. In view of their origin, the simplest interpretation of these paraphysoidal hyphae is that they are pseudoparaphyses extremely modified as a consequence of the apothecioid nature of the ascocarp. The Patellariaceae and Parmulariaceae then may be related to the Pleosporaceae.

General Conclusions

The problem with such relatively simple organisms as fungi is the lack of readily discernable morphological characters. In the ascomycetes, there are too few such characters to permit us to ignore any of them. Yet, we have a remarkably profligate attitude toward characters. As soon as the slightest difficulty is encountered, there is a throwing up of the hands: the character has failed, the classification has collapsed. This is the "sky is falling" syndrome.

As the multilocular, ascostromatic ascocarps of *Dothidea* and the series in *Mycospaerella* leading from multi- to uniloculate ascostromata demonstrate, pseudothecia exist; they characterize the Loculoascomycetidae. Monascous locules characterize the Myriangiales; fascicles of aparaphysate asci characterize the Dothideales. The pseudoparaphysate centrum characterizes the Pleosporales. These are generalizations; there are exceptions. There are difficulties in defining and determining characters. There are difficulties in interpretations of homologies between structures and of relationships between organisms.

As indicated in *Sporormiella* (Pl. IV, Figs. 9.21 and 9.22), there are species of Pleosporales in which the ascocarps are small and the pseudoparaphyses are difficult to demonstrate. There are species of *Didymella* Saccardo in which ascocarps are small and pseudoparaphyses are sparse or evanescent. At this point, it is difficult to draw a sharp line between *Didymella* in the Pleosporales and *Mycosphaerella* in the Dothideales. These are difficulties in discerning characters.

Because of the lack of pseudoparaphyses in *Guignardia* (Pl. I, Fig. 9.5), this genus belongs in the Dothideaceae. The ovoid, hyaline, nonseptate ascospores, however, are similar to those of *Botryosphaeria* species. This type of ascospore is so unusual among Loculoascomycetidae that it is impossible to ignore the hypothesis (von Arx and Müller, 1954, 1975) that places all species with such ascospores in the Botryosphaeriaceae. Pseudoparaphyses are lacking also in *Wettsteinina* von Höhnel (*Pseudosphaeria* von Höhnel) and *Leptosphaerulina*. On this account, these genera belong in the Dothideales. In a general classification, considering overall characters and especially ascus structure, which strikingly resembles ascus structure in *Pyrenophora* Fries, *Wettsteinina*, and *Leptosphaerulina* McAlpine represent pleosporaceous fungi in which pseudoparaphyses have been lost by reduction or, with greater reluctance, that they represent primitive intermediates. These are difficulties of interpretation.

In a hierarchial system, there is no escape from such questions. If they are suppressed at one point, they emerge at another. If the Patellariaceae are placed in a separate order, Patellariales, the question of homology between their paraphysoidal hyphae and pseudoparaphyses and of the relationship between the Patellariaceae and the Pleosporaceae is avoided. The question then becomes one of the relationship between the Patellariales and the Pleosporales, or the Helotiales.

An uncertainty principle surely applies to all biologic systems. We deal in probabilities, odds, percentages. Any character that can be discriminated in 90% of the fungi we examine is a good and useful character. Any classification that enables us to satisfactorily place 90% of the species is a good and useful classification. Difficulties in determining characters are merely a challenge to be more precise in our observations, and occurrences of apparent intermediates between taxa are a challenge to be more penetrating in our interpretations.

References

Barr, M. E., 1976. Perspectives in the Ascomycotina. Mem. New York Bot. Gard. 28: 1–8.

Bellemère, A., 1971. Les asques et les apotheciés des discomycètes bituniqués. Ann. Sci. Nat. Bot. Paris, Ser. 12, 7: 429–464.

Chadefaud, M., 1960. Les végétaux non vasculaires. *In* M. Chadefaud and L. Emberger (Eds.), Traité de Botanique Systématique, Vol. 1. Masson, Paris.

Chadefuad, M., 1973 Les asques et la systématique des ascomycètes. Bull. Soc. Mycol. France 89: 127–170.

Corlett, M., 1973. Observations and comments on the *Pleospora* centrum type. Nova Hedw. 24: 347–366.

Corlett, M., S. J. Hughes, and M. Kaufert, 1973. New Zealand fungi 19. Centrum organization in some Euantennariaceae and Metacapnodiaceae. New Zealand J. Bot. 11: 213–230.

Luttrell, E. S., 1951. Taxonomy of the pyrenomycetes. Univ. Missouri Studies No. 24, pp. 1–120.

Luttrell, E. S., 1955. The ascostromatic ascomycetes. Mycologia 47: 511–532.

Luttrell, E. S., 1973. Loculoascomycetes. *In* G. C. Ainsworth, F. K. Sparrow, and A. S. Sussman (Eds.), The Fungi, Vol. IV, Part A. Academic Press, New York.

Luttrell, E. S., and B. N. Muthappa, 1974. Morphology of a new species of *Aulacostroma*. Mycologia 66: 563–579.

Miller, J. H., 1928. Biologic studies in the Sphaeriales. Mycologia 20: 187–213; 305–339.

Miller, J. H., 1938. Studies in the development of two *Myriangium* species and the systematic position of the order Myriangiales. Mycologia 30: 158–181.

Miller, J. H., 1949. A revision of the classification of the ascomycetes with special emphasis on the pyrenomycetes. Mycologia 41: 99–127.

Müller, E., and J. A. von Arx, 1962. Die Gattungen der didymosporen Pyrenomyceten. Beitr. Kryptogamenfl. Schweiz 11: 1–922.

Nannfeldt, J. A., 1932. Studien über die Morphologie und Sytematik der nicht-lichenisierten inoperculaten Discomyceten. Nova Acta Reg. Soc. Sci. Upsala, Ser. 4, 8(2): 1–368.

Parguey-Leduc, A., 1966. Recherches sur l'ontogénie et l'anatomie comparée des ascocarps des pyrénomycètes ascoloculaires. Ann. Sci. Nat. Bot. Paris, Ser. 12, 7: 505–690.

Parguey-Leduc, A., 1967. Recherches sur l'ontogénie et l'anatomie comparée des ascocarps des pyrénomycètes ascoloculaires. Ann. Sci. Nat. Bot. Paris, Ser. 12, 8: 1–110.

Parguey-Leduc, A., 1973. Recherches préliminaires sur l'ontogénie et l'anatomie comparée des ascocarps des pyrénomycètes ascohymeniaux. VI. Conclusions générales. Rev. Mycol. 37: 60–82.

Pirozynski, K. A., and J. Reid, 1964. Studies in the Patellariaceae I. *Eutrylidiella sabina*. Can. J. Bot. 44: 655–662.

Reynolds, D. R., 1978. Foliicolous ascomycetes 2: *Capnodium salicinum*. Mycotaxon 7: 501–507.

Theissen, F., 1918. Neuen original Untersuchungen über Ascomyceten. Verh. Zool.-Bot. Gesell. Wien 68: 1–24.

Theissen, F., and H. Sydow, 1918. Vorentwurfe zu den Pseudosphaeriales. Ann. Mycol. 16: 1–34.

von Arx, J. A., and E. Müller, 1954. Die Gattungen der amerosporen Pyrenomyceten. Beitr. Kryptogamenfl. Schweiz 11: 1–434.

von Arx, J. A., and E. Müller, 1975. A re-evaluation of the bitunicate ascomycetes with keys to the families and genera. Stud. Mycol. 9: 1–159.

von Höhnel, F., 1909. Fragmente zur Mykolgie VI. No. 244. Revision der Myriangiaceen und der Gattung *Saccardia*. Sitzungsber. Acad. Wiss. Wien, Math-Naturwiss. Kl. Abt. I. 118: 349–376.

Chapter 10
The Lecanoralean Centrum

A. HENSSEN, IN COOPERATION WITH G. KEUCK, B. RENNER, AND G. VOBIS

Introduction

The Lecanorales are considered to be the typical lichenized discomycetes, including most of the lichen families with fruit bodies in the form of apothecia. All the large foliose lichens and the majority of the large fruticose lichens belong to this group. The order is characterized by an ascohymenial development, inoperculate asci, and the consistency and duration of the hymenium. The fruit bodies are long lived; their hymenium is strongly gelatinized with the paraphyses cemented together by mucilage. Earlier, a unitunicate structure and amyloid reaction of the ascus were mentioned as further important features of the order. However, the restriction of the amyloid reaction to the hymenial gelatin or its complete absence has since become known for many members of the order, and recent electron microscopy studies (Honegger, 1978) have given the final proof for the existence of bitunicate asci in some genera. The last mentioned are bitunicate in the functional sense of Luttrell (1951), with a gliding site between the ascus outer and its expansive inner layer. For the varying use of the term "bitunicate" in the literature, see Honegger (1978).

In a traditional concept, the order Lecanorales is a huge and heterogeneous one, including some 10,000 species. Attempts have recently been made to subdivide the order, either mainly on the basis of ontogenetic characters (Henssen and Jahns, 1974) or on the basis of the ascus structure (Poelt, 1974); these classifications are rather preliminary, however. Some of the suggested suborders are well defined; others still need consideration on the basis of more investigations, especially with regard to developmental morphology.

This treatise surveys some of the main lines of ascocarp development in the order Lecanorales with genera of 13 exemplary families. The sequence of the families treated implies no strict arrangement according to taxonomic relationships but has been chosen for practical purposes to demonstrate the different evolutionary lines. The presumed affiliation of the families to certain suborders is considered. Members of the Pertusariineae, a suborder previously included under the Lecanorales (Henssen and Jahns, 1974), are not treated. Their particular ontogeny has been described elsewhere (Henssen, 1976). The pattern of development deviates to such an extent from the rest of the Lecanorales that a separation as distinct order seems justified.

In the development of the apothecium, the formation of the margin and stipe is of interest equal to that of the differentiation of the associated centrum. The Lecanorales with a well-developed thallus are of special interest because somatic tissue is often involved in the formation of the fruit bodies—leading to lecanorine, superlecideoid, and zeorine apothecia, or to the differentiation of the thallinocarps (the terminology used here follows Henssen and Jahns, 1974; Henssen, 1976). In these ascocarp types, the somatic thallus functions more or less as a substitute for the excipulum proprium.

In general, the ascogonia originate in the upper part of the thallus. In heteromerous lichens, they are produced in the medulla or on the border of medulla and cortex. A few cases are known (*Lichinodium* Nylander and *Moelleropsis* Gyelnik, for instance) in which the ascogonia lie free between the thallus lobes; a development corresponding to that in unlichenized ascomycetes. The ascogonia usually develop in groups. They are rarely produced singly. They may be formed directly by the thallus hyphae or by hyphae of the so-called generative tissue. The hyphae of the generative tissue can be distinguished from those of the rest of the thallus by their thin walls, their rich protoplasmatic contents, and their different cell forms. In the course of further development, they form the monokaryotic tissue of the primordium and give rise to the paraphyses and the excipulum. In most cases in which the ascogonia are produced directly from normal thallus hyphae, they become enclosed at a later stage by the developing hyphae of the generative tissue. In general, the generative tissue originates from the somatic hyphae adjacent to the ascogonium. In the case of the Collemataceae, however, this tissue arises mainly from the stalk cells of the ascogonium. In most types of development in the Lecanorales, the primordial paraphyses are richly branched and anastomosing paraphysoids. They are often replaced by true paraphyses arising from the subhymenial layer which have free tips from the beginning.

Main Lines of Ascocarp Development

The families chosen to demonstrate the main evolutionary lines are grouped into three sections according to the morphology of the ascocarps. The first group is characterized by zeorine apothecia. The ascocarps of the second section are lecanorine, lecideine or superlecideoid. The last group is distinguished by hemiangiocarpous apothecia.

Families with Zeorine Apothecia

Collemataceae

The seven genera recognized in the Collemataceae sensu stricto (Henssen, 1965) form a natural group that resembles nonlichenized discomycetes in its mode of fruit body development. The monokaryotic mycelium of the primordium enclosing the ascogenous hyphae originates from the stalk cells of the ascogonium as in the classical ontogeny of the apothecium outlined by Corner (1929 a,b). The Collemataceae differ in this respect from other families of the Lecanorales and remain

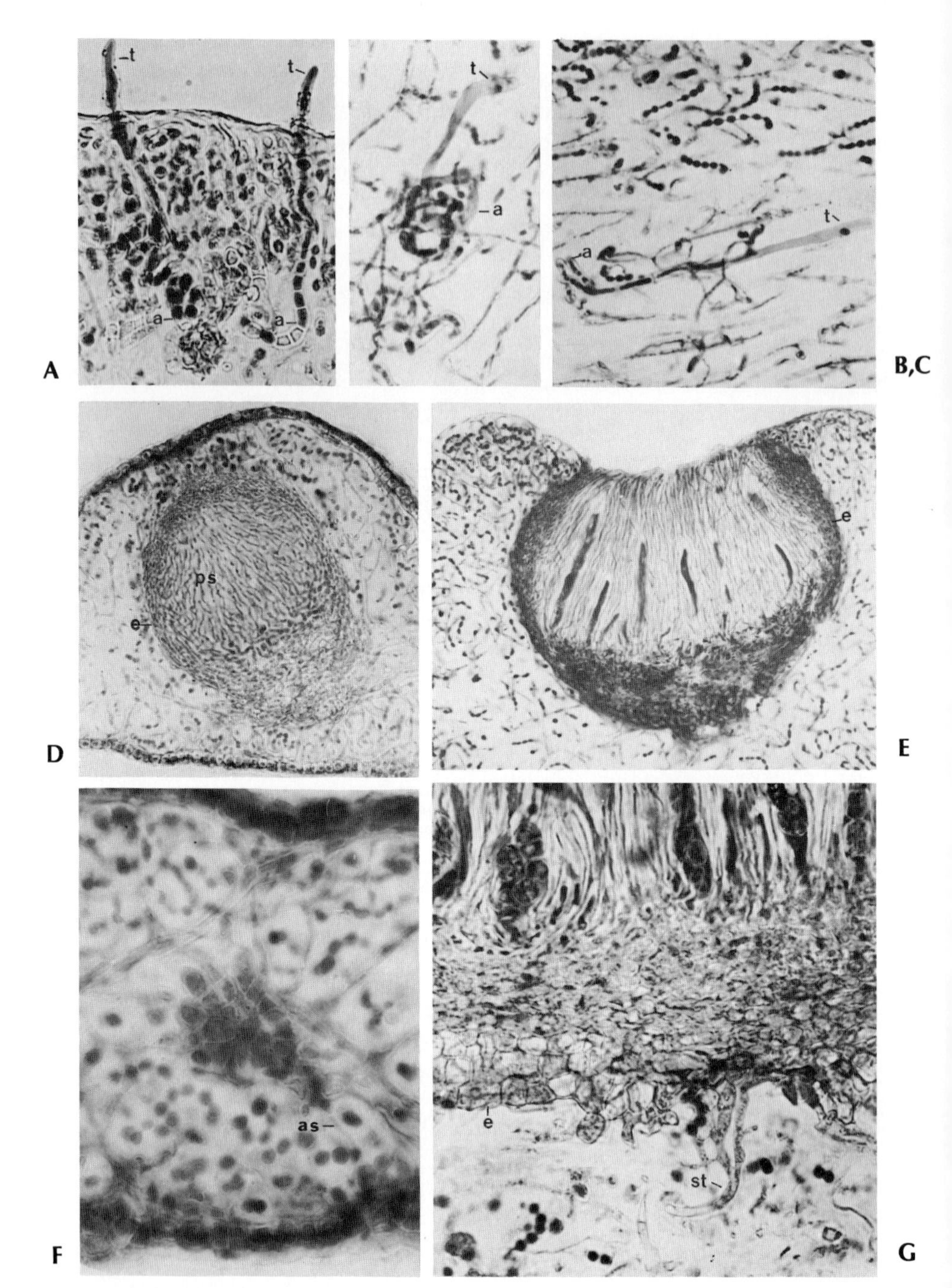

Fig. 10.1 A–G. Development in the Collemataceae. **A** *Collema subnigrescens,* ascogonia with protruding trichogyne. X540. **B** and **C** *Collema bachmanianum,* horizontally extending end-cells of trichogynes in the thallus center. **B** X430; **C** X360. **D–F** *Leptogium sinuatum.* **D** Apothecial primordium containing paraphysoids. Inclined sections. X250. **E**

rather isolated within the suborder Lecanorineae. The asci have an amyloid apical ring or amyloid caps in *Collema* Wiggers and *Leptogium* (Acharius) S. Gray (Dughi, 1956). All species of *Homothecium* Massalongo have an amyloid ring in the apical thickening. Only the mucilage covering the ascus wall stains blue in Lugol's solution in *Ramalodium* Nylander in Crombie (Henssen, 1979).

The development of the ascocarp commences with the differentiation of a multicellular, spirally coiled ascogonium by ordinary thallus hyphae within or below the main algal zone (Stahl, 1877a; Baur, 1898). They are uniformly distributed in the thallus (Fig. 10.1A) or are agglomerated in groups (Fig. 10.2A) interspersed by parts of the thallus free from ascogonia. The ascogonia bear vertically growing trichogynes that protrude through the thallus surface (Fig. 10.1A). In some species, the trichogynes grow more or less laterally (Fig. 10.1, B, C) and the end cell may be considerably extended (Bachmann, 1912, 1913; Degelius, 1954). In the next stage of development, short-celled hyphae sprout from the ascogonial and stalk cells on all sides (Figs. 10.1F and 10.5A). A dense web of generative tissue is formed in which remnants of the ascogonium and its stalk may still be recognized (Fig. 10.2D). Where ascogonia lie together, several may be included in the primordium. The first paraphyses form directly from the generative tissue. They are richly branched and anastomosing paraphysoids, marginally surrounded by the first row of the excipulum hyphae and connected in the upper part with a covering layer (Figs. 10.1D and 10.9A). The covering layer bursts open when the primordium enlarges, and the paraphysoids at a later stage become secondarily free tips (see Fig. 10.9B). A limited hemiangiocarpy may be observed in some species, e.g., *Ramalodium austro–americanum* Henssen (Henssen, 1979) and *Leciophysma finmarkicum*. Later on, true paraphyses grow from the formative layer, especially in the marginal part of the hymenium where they develop as side branches from anticlinal hyphae (see Fig. 10.8F). This development corresponds to unlichenized discomycetes (Bellèmere, 1967). The excipulum develops gradually to form a close cup-shaped layer, or it remains an annular structure (Fig. 10.3).

The ascogonia of *Collema* became known by the studies of Stahl (1877a). He and Baur (1898) observed pycnoconidia adhering to and fusing with the trichogyne, and they inferred a fertilization. Stahl (1877a) did not observe the sprouting stage of the ascogonium and its stalk, and he was not able to decide on the origin of the hyphal web surrounding the enlarged ascogonium cells in young primordia. He assumed the participation of the neighboring thallus hyphae, but he stressed that hyphae adjacent to the ascogonium extended in a direction other than the hyphae of the surrounding sheath. Such hyphae can clearly be seen, especially in young primordia of *Collema subnigrescens* Degelius (Henssen, 1976) and can be

---

Young apothecium. X250. **F** Short-celled hyphae developing from the ascogonium and its stalk. X675. **G** *Collema occultatum* var. *populinum*, lower part of apothecium with partly pseudoparenchymatic excipulum upon the supporting hyphae. X400. (Microtome sections.) *a* ascogonium, *as* ascogonial stalk, *e* excipulum, *pa* paraphysoids, *st* supporting tissue, *t* trichogyne. (A–G are microtome sections. A after Henssen and Jahns, 1974.)

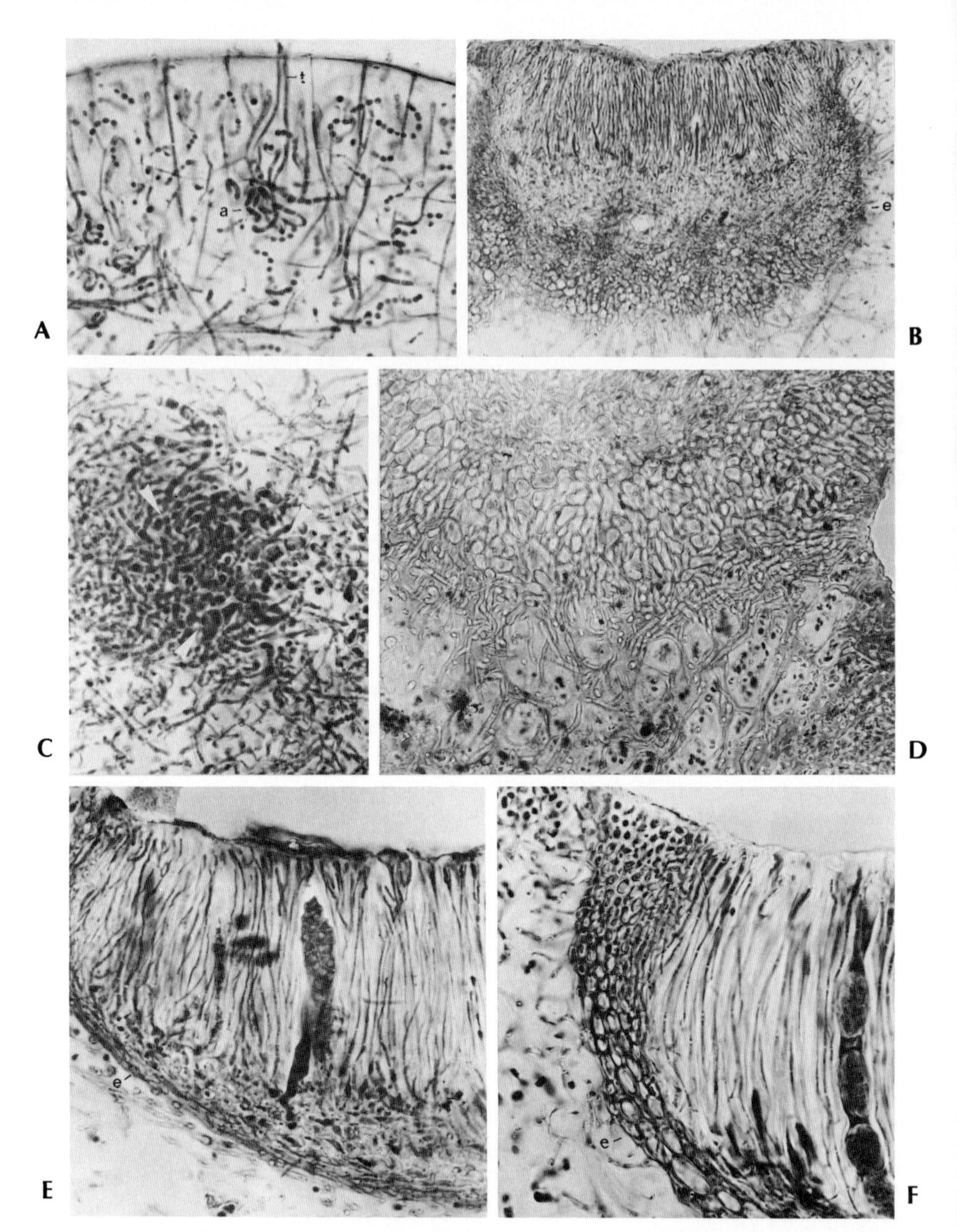

Fig. 10.2 A–F. Development in the Collemataceae. **A–C** *Collema multipartitiens*. **A** Group of ascogonia with protruding trichogynes. X300. **B** Young apothecium, the cupular excipulum being formed by anticlinally arranged hyphae. X150. **C** Lower part of excipulum consisting of interwoven hyphae with enlarged cells. X240. **D** and **E** *Collema bachmanianum*. **D** Primordium of hyphae densely interwoven with the rests of the ascogonium and its stalk (arrows). X330. **E** Part of mature apothecium, the excipulum formed by undifferentiated hyphae. X300. **F** *Collema ceraniscum,* marginal part of apothecium, the excipulum consisting of periclinally oriented hyphae with enlarged cells. X300. *a* ascogonium, *e* excipulum, *t* trichogyne. (Microtome sections.)

called "anchoring hyphae." Stahl's hypothesis of the origin of the auxiliary tissue from the vegetative thallus hyphae was later on taken to be fact (Sturgis, 1890; Bachmann, 1913; Moreau and Moreau, 1929; Degelius, 1954). In reality, the Collemataceae is the only family of the lichenized ascomycetes known in which the generative tissue is derived from the ascogonial stalk. *Collema flaccidum* (Acharius) Acharius is an exception in which the thallus hyphae adjacent to the ascogonium divide and take part in the formation of the surrounding hyphal sheath. This species therefore follows the developmental pattern of other lichen families.

Bachmann (1912, 1913) reported in a *Collema* species trichogynes with horizontally growing tips (Fig. 10.1, B and C). The lichen was later named in her honour *Collema bachmanianum* (Fink) Degelius. She described a fusion of the trichogyne with small hyphal branches produced in clusters from the tip of a medullary hypha. The small branches were termed "internal conidia." Degelius (1954) found internal conidia also in *Collema multipunctatum* Degelius. Both

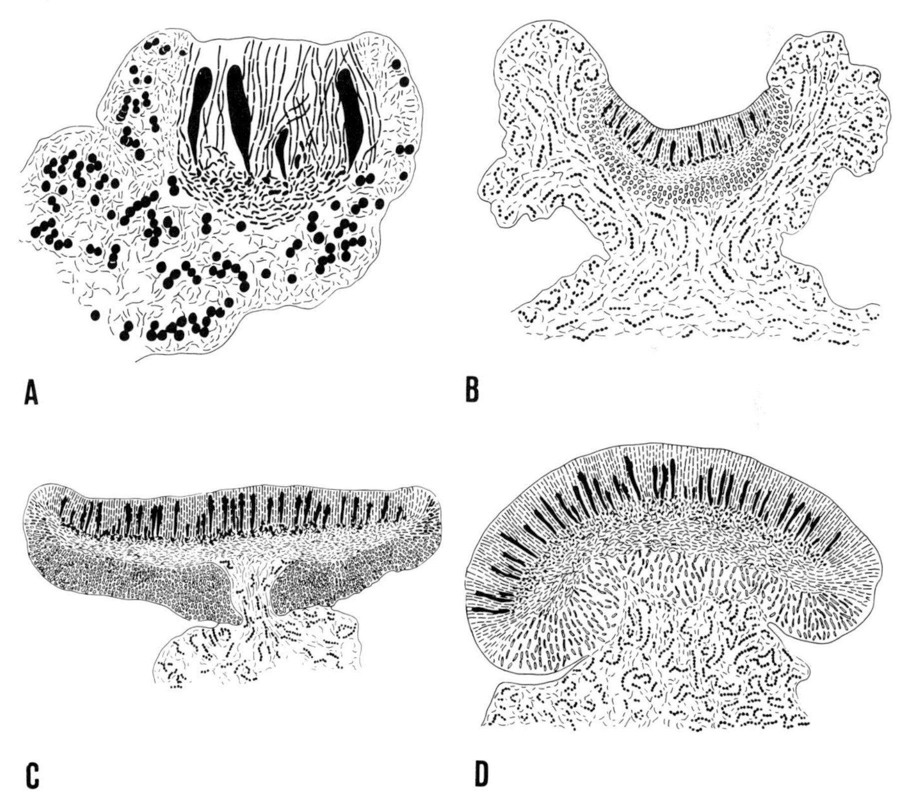

Fig. 10.3 A–D. Types of fruit body in the Collemataceae. **A** *Collema occultatum,* lecanorine apothecium with completely reduced excipulum. **B** *Physma boryanum,* zeorine apothecium with cupulate excipulum and corrugated thalline margin. **C** *Homothecium opulentum,* lecideine apothecium with annular excipulum. **D** *Ramalodium succulentum,* lecideine apothecium with cupulate excipulum. (A–D after Henssen and Jahns, 1974.)

authors claimed that a species with internal conidia does not produce pycnidia. This could, however, not be confirmed. Small but otherwise normal pycnidia were observed in *C. bachmanianum* and *Collema callibotrys* Tuckerman. According to Degelius (1954, 1974), *C. callibotrys* is probably conspecific with *C. multipunctatum.* In *C. bachmanianum,* trichogynes were observed fusing with ordinary thallus hyphae or with hyphae developing from the excipulum. the so-called anchoring hyphae. Obviously, a deuterogamy between the ascogonium with hyphae of different size and origin can be assumed. Trichogynes with adhering conidia have more recently been seen in Collemataceae and many other lichen families (Henssen and Jahns, 1974; Jahns, 1970a), but nobody has yet succeeded in proving that nuclear migration occurs. The known difficulties in culturing lichens or mycobionts make this information gap understandable.

The first stages of development described above can be found in all genera of the family. The ascogonia and primordia are always situated within the lichen thallus. Later on, the ascocarp either becomes elevated above the thallus surface to form lecideine apothecia in *Homothecium, Leciophysma* Th. Fries, and *Ramalodium,* or it remains immersed, giving rise to zeorine, exceptionally lecanorine, apothecia in *Collema, Leptogium, Leightoniella* Henssen, and *Physma* Massalongo (Figs. 10.3 and 10.4). Because the auxiliary tissue typically arises solely from the ascogonial stalk, the primordia and frequently also the mature apothecia give the impression of foreign bodies in the gelatinous matrix of thallus, balanced on the hyphae originally bearing the ascogonium (Fig. 10.1G). Special methods have been developed to attach and lend stability to the ascocarp, giving rise to a fruit body morphology characteristic for this family (Henssen and Jahns, 1974).

Anchoring hyphae grow on all sides from the excipulum into the thallus (in species of the genera with zeorine apothecia, such as *Collema* and *Leptogium*), and on reaching the boundary layer of the latter they may form either peripherally along the margo thallinus or beneath the apothecium supporting tissues, composed of rounded cells that appear to have mechanical strength. Produced in greater amount, these cells are united in a pseudoparenchyma. A basal pseudoparenchyma is especially well developed in *Collema coilocarpum* Müller-Argoviensis (Figs. 10.4C and 10.6A) as has already been pointed out by Degelius (1954). This structure had been misinterpreted by earlier authors as the excipulum (Degelius, 1954). Supporting tissues are principally formed from the cortex by division and elongation of the cells in species with a corticated thallus and may become reinforced at a later stage by cells of the anchoring hyphae (Figs. 10.4D and 10.6, C, D). In *Leptogium diaphanum* (Swartz) Montagne, supporting tissues are developed marginally, and below the apothecium (Fig. 10.5) in *Leptogium corticola* (Taylor) Tuckerman, they are restricted to the thalline margin (Fig. 10.6, B, E).

The development of hyphae growing from the excipulum into the thallus has been discussed by Dughi (1954) in connection with the formation of the margo thallinus. The thalline margin ought to be a new formation produced by the anchoring hyphae. In the Collemataceae, as in most families with lecanorine or zeorine apothecia (Henssen and Jahns, 1974), the surrounding thallus enlarges together with the primordium and forms the margo thallinus. The different steps

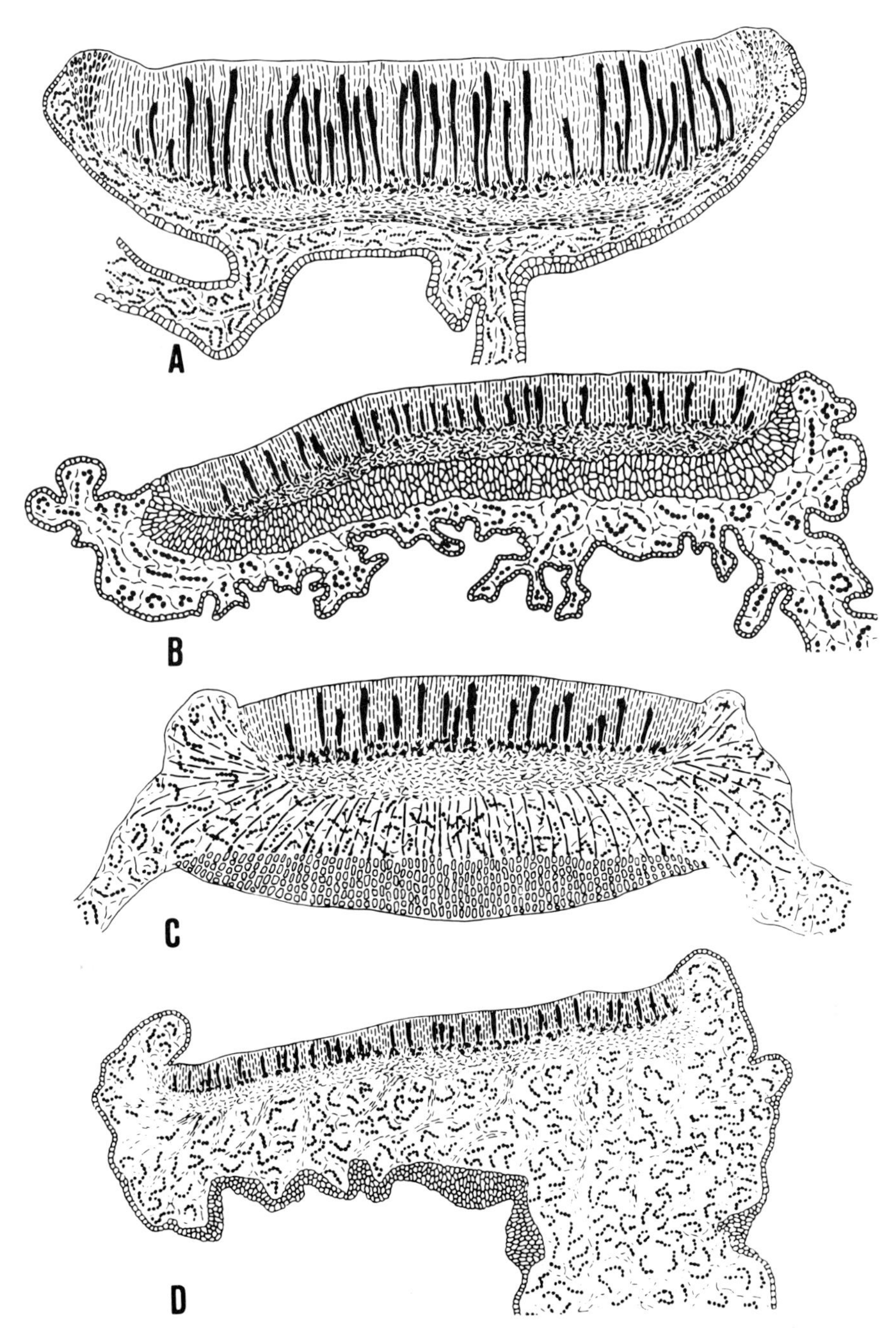

Fig. 10.4 A–D. Types of fruit body in the Collemataceae. **A** *Leptogium lichenoides,* zeorine apothecium with cupulate excipulum composed of periclinal hyphae. **B** *Leptogium milligranum,* zeorine apothecium with pseudoparenchymatous excipulum composed by anticlinal hyphae. **C** *Collema coilocarpum,* apothecium with basal supporting tissue formed from the anchoring hyphae. **D** *Leightoniella zeylanensis,* zeorine apothecium with basal supporting tissue formed by the thalline cortex. (A–D after Henssen and Jahns, 1974.)

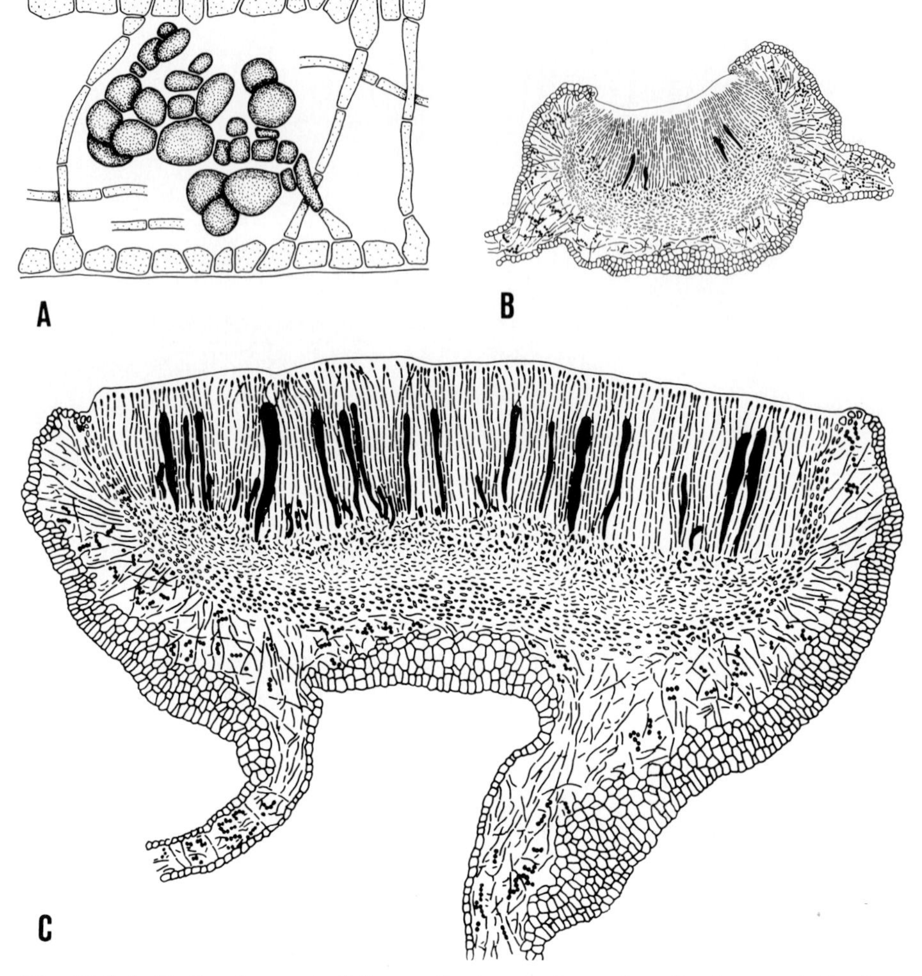

Fig. 10.5 A–C. Development of the apothecium in *Leptogium diaphanum.* **A** Section through thallus with an ascogonium which, together with its supporting hyphae, proliferates on all sides into short-celled hyphae. **B** Section through a young apothecium with basal supporting tissue composed of thalline cortex. **C** Section through an old apothecium with basal and marginal supporting tissue. (A–C after Henssen and Jahns, 1974.)

→

Fig. 10.6 A–G. Anatomy of the fruit bodies in the Collemataceae. **A** *Collema coilocarpum,* marginal part of apothecium with basal supporting tissue produced from the anchoring hyphae. X170. **B–E** *Leptogium corticola.* **B** Part of apothecium with marginal supporting tissue. X170. **C** Part of a young apothecium with anchoring hyphae. X420. **D** Anchoring hyphae forming cells of the supporting tissue. X720. **E** Central part of apothecium with pseudoparenchymatous excipulum. X170. **F** *Leptogium trichophorum,* pseudoparenchymatous excipulum. X170. **G** *Leptogium hirsutum,* lower part of excipulum and basal supporting tissue. X170. *an* anchoring hyphae, *e* excipulum, *st* supporting tissue. (A–G are microtome sections. F after Henssen and Jahns, 1974.)

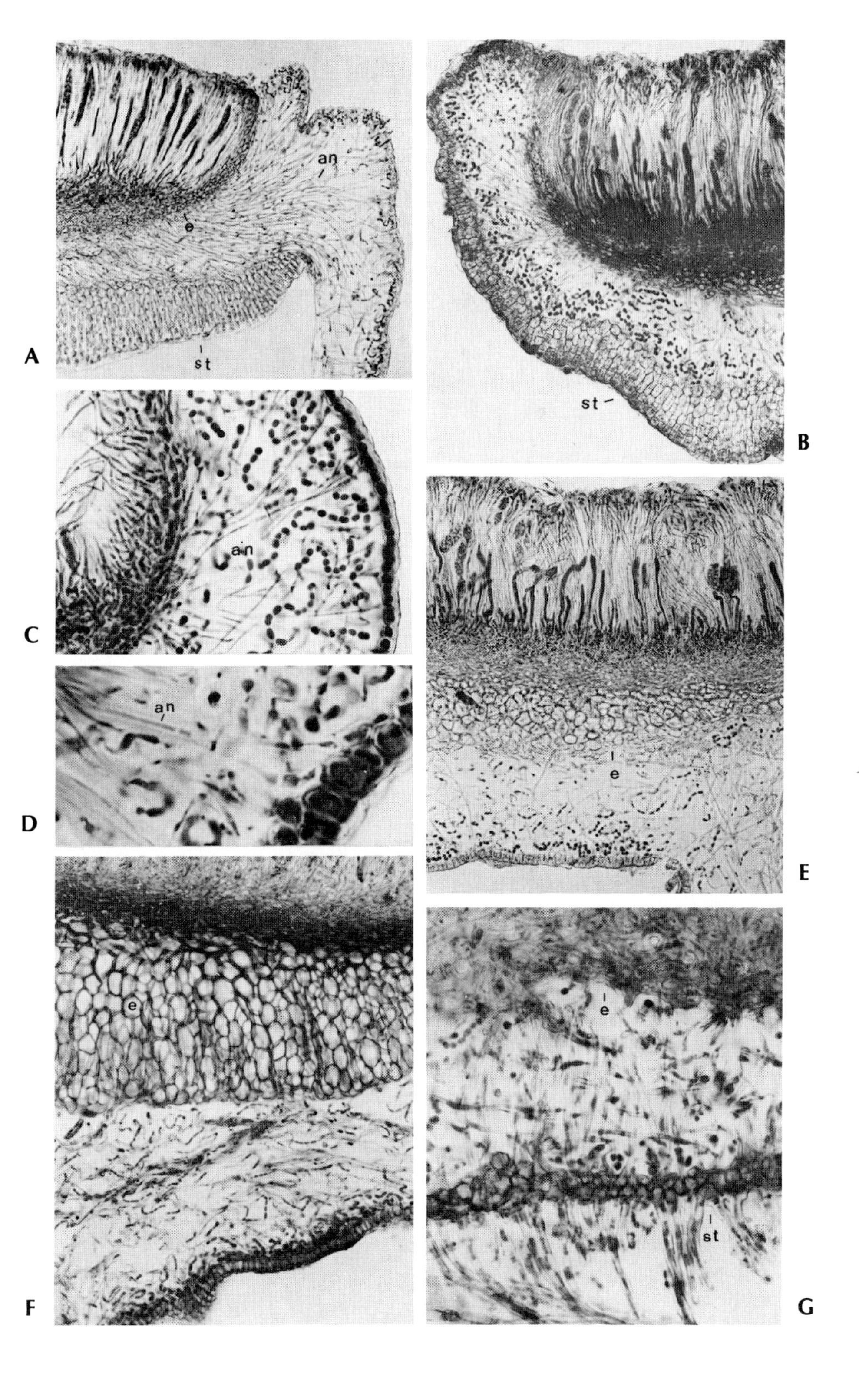
A
an
e
st
B
st
C
an
D
an
E
e
F
e
G
e
st

of such development have been demonstrated for *Collema subnigrescens* (Henssen, 1976). The margo thallinus is developed massively in *Physma* (Fig. 10.3B) and a number of *Leptogium* species, and it forms annular or anticlinal folds around the apothecial disc or lobuliform appendages as in *Leptogium burgessii* (Linneus) Montagne. Occasionally, the apothecia are elevated by the formation of thalline stalks, which may be swollen and inflated as in *Leptogium javanicum* Montagne.

Further methods of acquiring stability can be achieved by the production of a massively developed excipulum. In *Collema* and *Leptogium,* the structure of the cupular excipulum shows considerable variation. The relatively thin-walled hyphae may run anticlinally or periclinally, or they may be irregularly interwoven. The cells are often enlarged to form pseudoparenchyma (Figs. 10.2, 10.4, and 10.6). In rare instances, an excipulum may be lacking entirely, the hymenium being surrounded only by a thalline margin as in *Collema occultatum* Baglietto (Fig. 10.3A). In the genus *Leptogium,* the differentiation of the excipulum and the formation of the supporting tissues are often mutually exclusive. Either the excipulum consists of well-developed pseudoparenchyma and supporting tissues are lacking (Figs. 10.4B and 10.6F), or the latter are present and the excipulum is then only formed of undifferentiated hyphae (Fig. 10.6B).

Additional stability is acquired by the gelatinization of the hyphae walls in *Physma* and the lecideine genera and by the reticulate branching of the excipulum hyphae in *Homothecium* and *Ramalodium* (Figs. 10.3B; 10.8C–F; and 10.9C–F). The excipulum of the mature apothecium is very thick in all of these genera. It is cupulate in the genus *Physma,* annular in *Leciophysma* and *Homothecium,* and first annular but later cupulate in *Ramalodium* (Henssen, 1979). The initial stages are immersed in the thallus (Fig. 10.9A). In *Homothecium,* the young primordium grows vertically, producing an excipulum in the marginal part, and gradually rises above the thallus surface (Figs. 10.7, A, B, and 10.8A). The maturing apothecium, develops horizontally, the excipulum remaining annular. Anchoring hyphae grow from the subhymenial layer into the thallus, passing the inner edges of the excipulum (Figs 10.3C and 10.7C). The outer cells of the excipulum develop hyphae, especially in the central part where the apothecium is near to the thallus surface. These hyphae may form a firm connection with the adjacent thallus (Fig. 10.8E). In *Homothecium sorediosum* Henssen, such hyphae come into contact with the soredia produced by the thallus and enclose and eventually incorporate them into the excipulum (Fig. 10.8, A, C, D).

In *Ramalodium* the formation of the excipulum starts marginally, as in *Homothecium.* Gradually, however, a plectenchyma is formed by the anchoring hyphae connecting the inner edges of the excipulum (Fig. 10.9C). In maturing apothecia this texture takes on the structure of the marginal primary excipulum (Fig. 10.9, D, E). Like in *Homothecium,* hyphae developing from the outer cells of the excipulum establish contact with the adjacent thallus. Plectenchyma may be formed between the excipulum and thallus in *Ramalodium austro–americanum* and the hyphal tips invading the thallus may create a primitive cortical structure along the surface (Fig. 10.9F). A different successive development of a cupulate excipulum is known from *Umbilicaria* Hoffmann species.

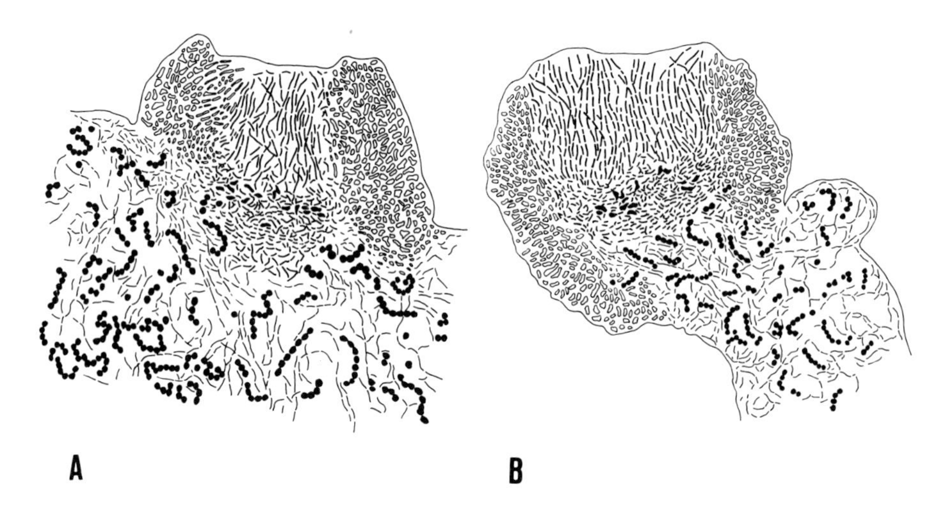

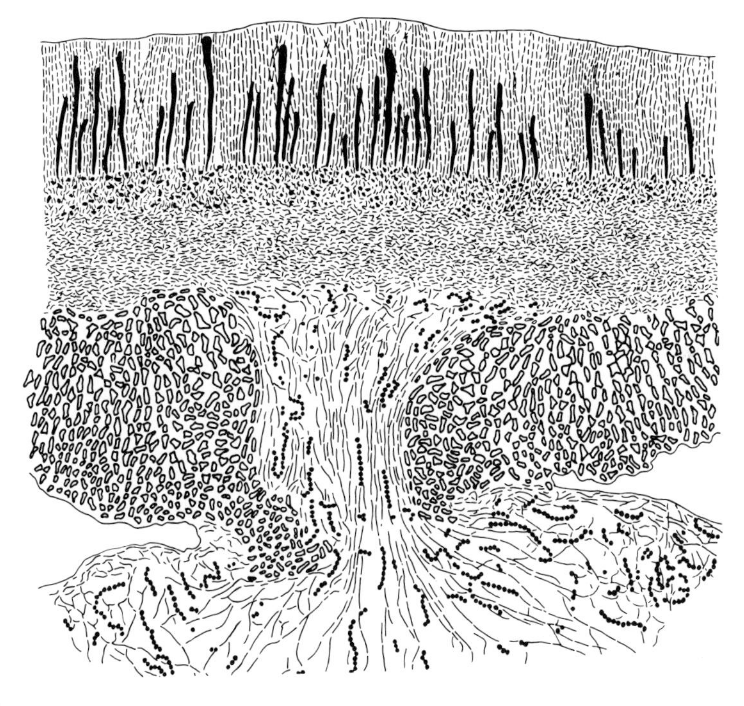

Fig. 10.7 A–C. Development of apothecium in *Homothecium opulentum.* **A** Primordium with ascogenous hyphae and paraphysoids and the commencement of differentiation of the annular excipulum. **B** Somewhat later stage. **C** Portion of a mature apothecium in section. (A–C after Henssen and Jahns, 1974.)

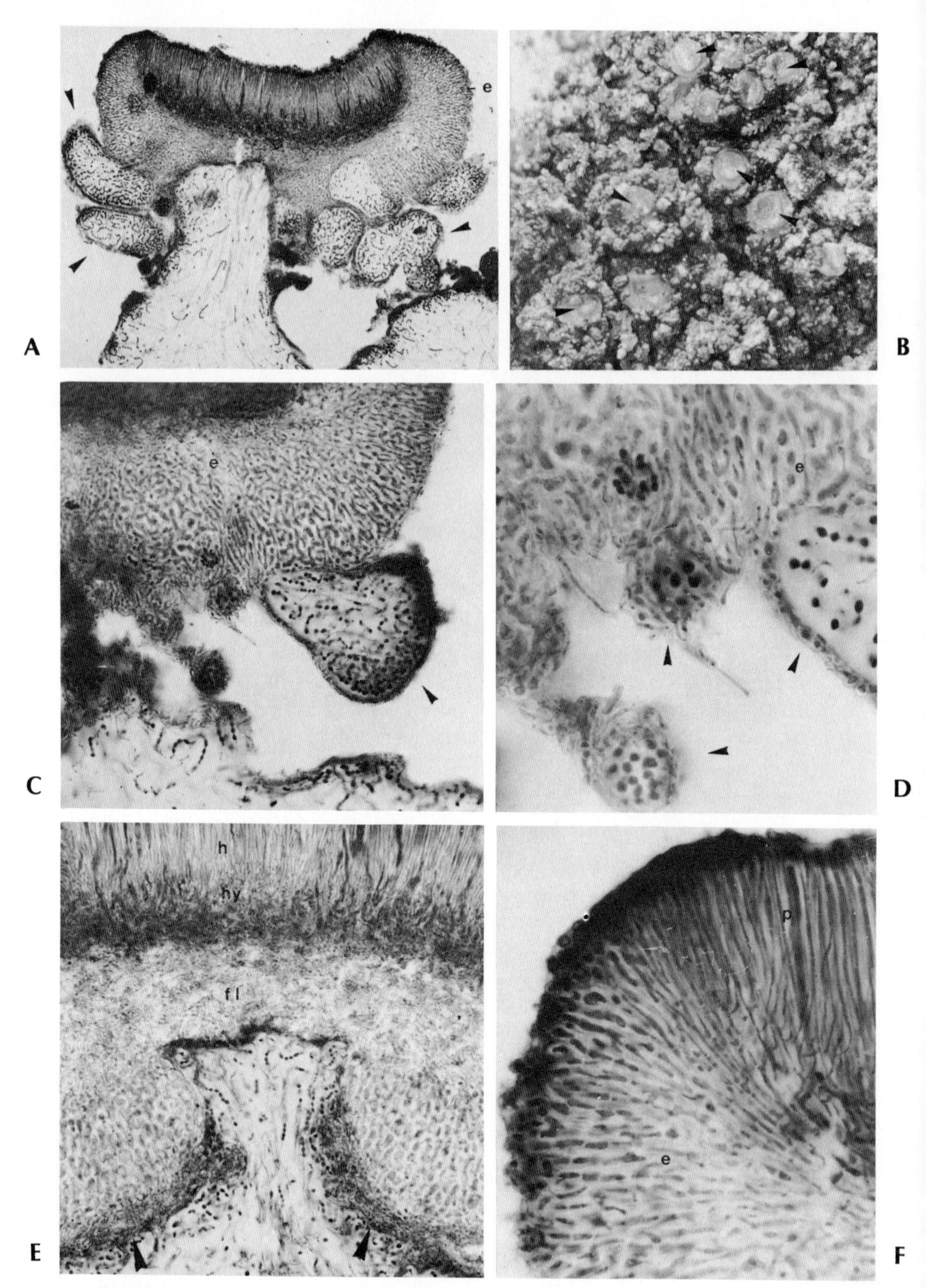

Fig. 10.8 A–F. *Homothecium sorediosum.* **A** Apothecium with soredia (arrows) enclosed in the excipulum. X65. **B** Sorediose thallus with apothecia (arrow). X7. **C** Lower part of excipulum incorporating soredia. X170. **D** Section from **C** in higher magnification of soredia (arrows). X400. **E** Central lower part of an old apothecium, arrow indicating connection of the thallus surface and hyphae developing from the excipulum. X170. **F** Apothecial margin with formation of paraphyses and excipulum hyphae. X400. *e* excipulum, *fl* formative layer, *h* hymenium, *hy* hypothecium, *p* paraphyses. (With exception of **B**, all are microtome sections. E–F after Henssen, 1980.)

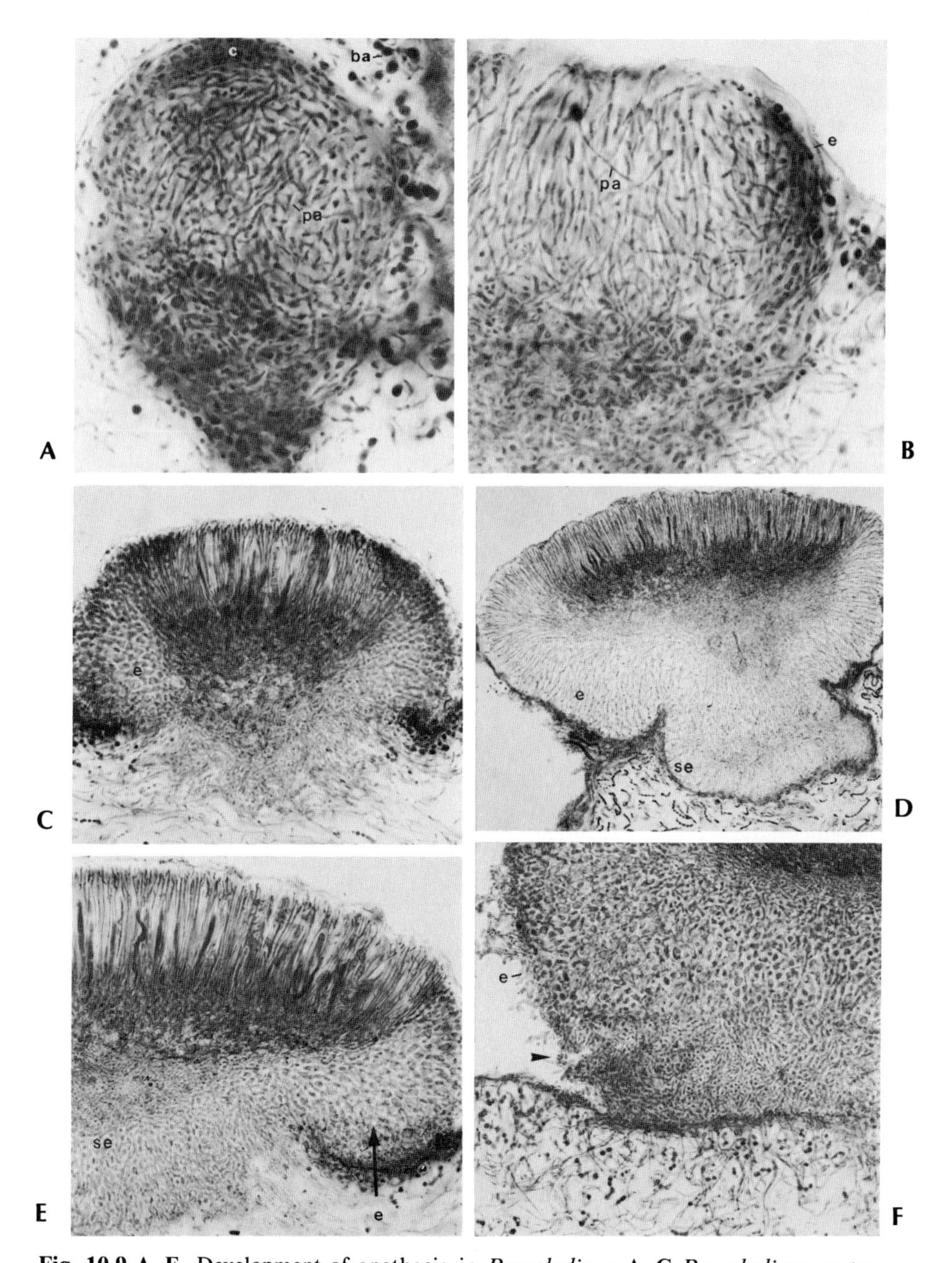

Fig. 10.9 A–F. Development of apothecia in *Ramalodium*. **A–C** *Ramalodium austro-americanum*. **A** Primordium with covering layer and paraphysoids transversely inserted in the thallus; thallus upper surface on the right side of the photograph. X400. **B** Somewhat later stage, the first rows of excipulum being formed. X400. **C** Young apothecium with annular excipulum and a connecting plectenchyma formed by the anchoring hyphae. X170. **D** *Ramalodium succulentum*, mature apothecium with cupulate excipulum. X85. **E** and **F** *R. austro–americanum*. **E** Marginal portion of old apothecium with first and secondarily (arrow), developed parts of the cupulate excipulum. X170. **F** Plectenchyma formed between excipulum and thallus indicated by an arrowhead. X170. *ba* blue-green algae, *c* covering layer, *e* excipulum, *pa* paraphysoids, *se* secondarily formed excipulum. (A–F are microtome sections. B, C, E and F after Henssen, 1980.)

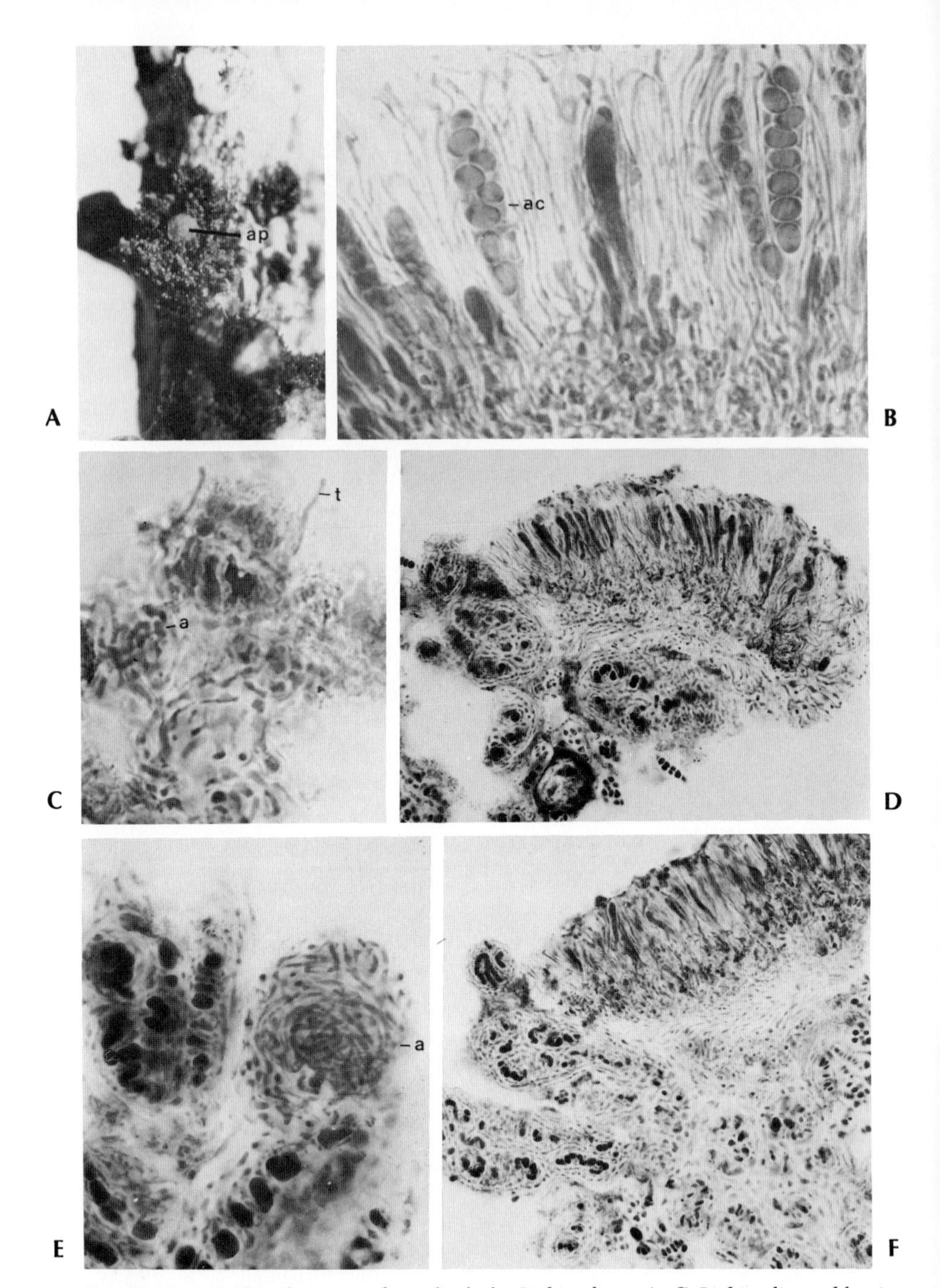

Fig. 10.10 A–F. Development of apothecia in *Lichinodium.* **A–C** *Lichinodium ahlneri.* **A** Thallus with apothecium. X25. **B** Hymenium with asci and paraphyses. X540. **C** Primordium with ascogonia and trichogynes. X675. **D–F** *Lichinodium sirosiphoideum.* **D** Young apothecium between thallus lobes. X425. **E** Generative tissue with ascogonia. X425. **F** Marginal portion of apothecium with thalline margin. X180. *a* ascogonium, *ac* ascus, *ap* apothecium, *t* trichogyne. (With exception of A, all are microtome sections. A after Henssen, 1968; C after Henssen, 1963a.)

Lichinaceae

The family Lichinaceae in the emended description (Henssen, 1963a, 1980) includes a majority of the genera placed in the Ephebaceae and the Pyrenopsidaceae by Zahlbruckner (1926). The Lichinaceae are well characterized by their asci and the development of the ascocarp. They have been located in the separate suborder Lichinineae (Henssen and Jahns, 1974). The asci are prototunicate, have mostly thin walls, and contain more then eight spores in many species (Figs. 10.10B and 10.12I). The ontogeny of the fruit body shows manifold variations;

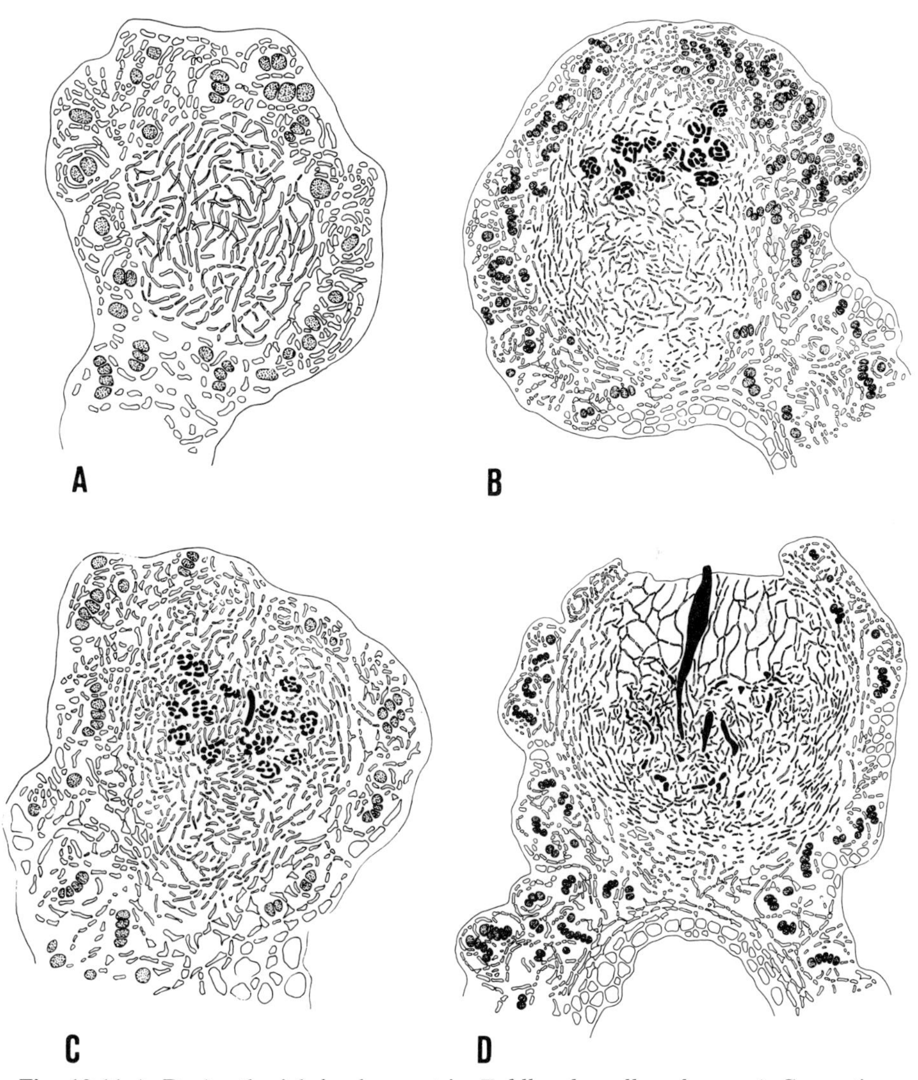

Fig. 10.11 A–D. Apothecial development in *Zahlbrucknerella calcarea.* **A** Generative tissue in a swelling of the thallus. **B** Spirally contorted ascogonia in a generative tissue. **C** Ascogonia and the first ascus. **D** Young apothecium with paraphysoids and margo thallinus. (A–D after Henssen and Jahns, 1974.)

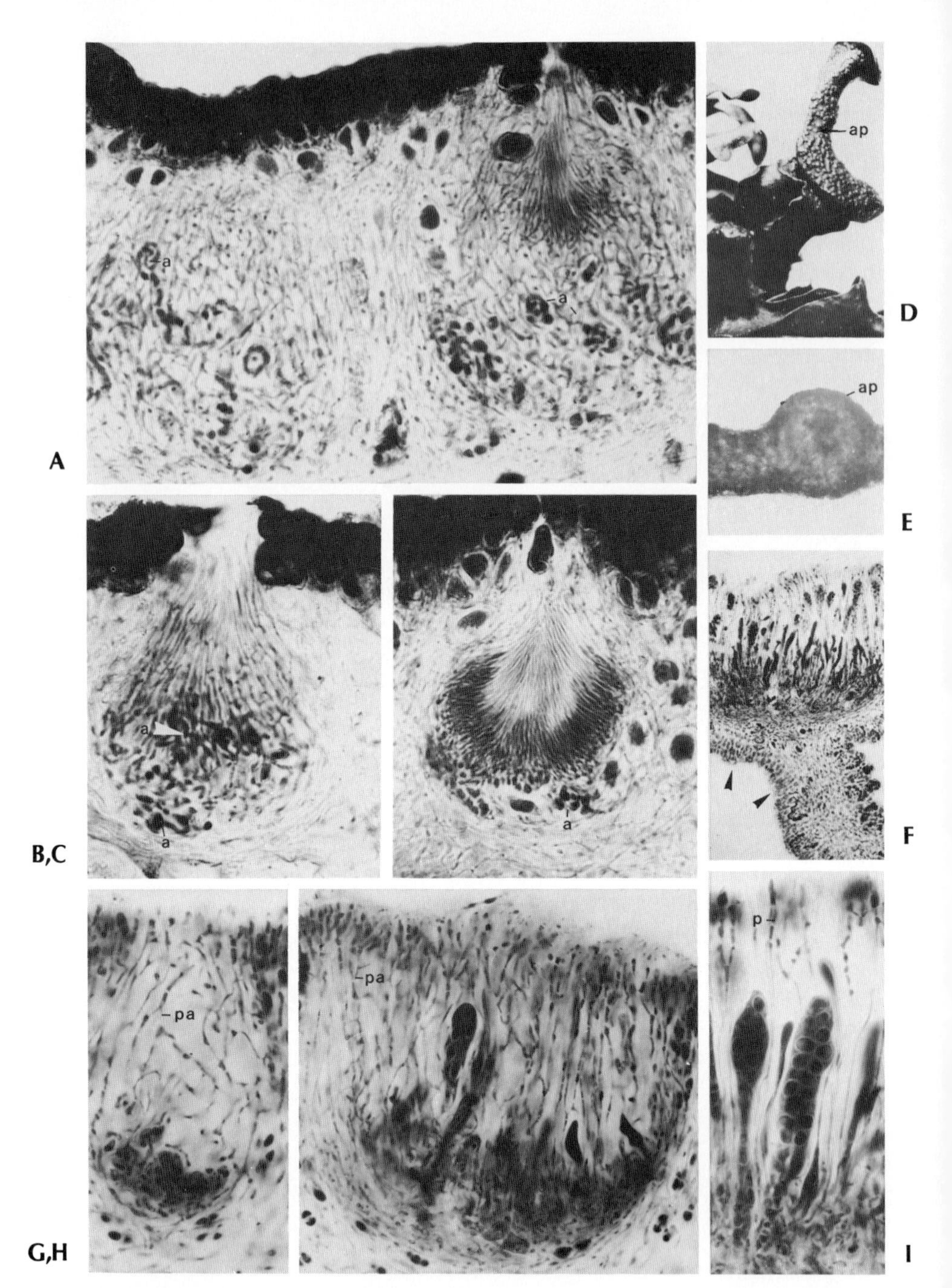

Fig. 10.12 A–I. Ascocarp development in the *Lichinaceae*. **A–C** *Phylliscum demangeonii*. **A** Generative tissue with ascogonia (left) and transition stage between generative tissue and pycnidium (right). X400. **B** Transition stage with stretched hyphae. X480. **C** Pycnidium with ascogonia. X440. **D** *Jenmania osorioi*, reflexed tip of fertile lobe. X7. **E** *Zahlbrucknerella maritima*, apothecium (X55). **F** *Zahlbrucknerella patagonica*, longitudinal section of apothecium basal supporting tissue below the hymenium extending into the adjacent thallus (arrowheads). X50. **G** and **H** *Jenmania goebelii*. **G** Generative tissue

normal apothecia, pycnoascocarps, or thallinocarps occur; interthecial filaments may be paraphysoids (Figs. 10.11D and 10.12, G, H); true paraphyses with frequent characteristic swellings (Figs. 10.12F and 10.13G) elongated conidiophores (Fig. 10.13, C, F) or vegetative hyphae of the thallus (Fig. 10.15, B, E). The immersed or sessile fruit bodies are zeorine, lecanorine, or rarely lecideine (e.g., *Thermutis* Fries). The cupular excipulum is frequently rudimentary and in some instances massively developed in the upper part (Henssen, 1963a, 1980). The apothecia often have a punctiform disc that gives them a spurious angiocarpic appearance (Fig. 10.12, D, E). The width of the opening depends on the quantity of paraphyses present.

Two fundamental types of ascocarp ontogeny may be distinguished; one, the development of the apothecia from a generative tissue, and two, the transformation of pycnidia into apothecia, the so-called pycnoascocarp.

The normal type of development from generative tissue is found in *Zahlbrucknerella calcarea* (Herre) Herre (Henssen, 1963a, 1977). The first stage consists of a hyphal weft in a swelling of the filamentous thallus (Fig. 10.11A). The formation of the ascogonium is quickly followed by the development of the first asci (Fig. 10.11, B, C), between which the hyphae of the generative tissue elongate and intrude as paraphysoids. The hymenium is very gelatinous. Because of the increasing pressure exerted by the growing initial, the thallus bursts open at its upper end (Fig. 10.11D). Young apothecia contain only paraphysoids but true paraphyses may be formed later on (Fig. 10.12I). In *Zahlbrucknerella patagonica* Henssen, a basal supporting tissue develops below the apothecium and extends into the adjacent thallus (Fig. 10.12F). In *Zahlbrucknerella maritima* Henssen, the ascogonia may occur first and then be enclosed by the generative tissue.

This type of development has been observed in many genera of the Lichinaceae. In general, the primordium arises within the thallus. An exception is the genus *Lichinodium* Nylander (Henssen, 1963a, 1968a) where the ascogonia lie in a gelatinous matrix between the thallus lobes (Fig. 10.10E). The strongly gelatinous apothecia only have a rudimentary excipulum and, therefore, a convex shape (Fig. 10.10, A, B). The developing margin may secondarily fuse with adjacent parts of the thallus. Such apothecia are spurious lecanorine in sections (Fig. 10.10, D, F).

The development of pycnoascocarp was first studied in detail in species of *Ephebe* (Henssen, 1963a). The pycnidia develop from wefts in thallus swellings similar to the apothecia of the first development type (Fig. 10.13A), indicating the close connection between these two organs. The formation of the pycnoascocarps commences with the formation of ascogonia below a pycnidium (Fig. 10.13E). From this point onward, two types of development are possible. In the first type, the conidiophores first elongate to form paraphyses (Fig. 10.13F) and

---

with spirally shaped ascogonia. X650. **H** Apothecium with paraphysoids. X520. **I** . *Zahlbrucknerella patagonica,* part of the hymenium with polysporous asci and true (secondary) paraphyses. X520. *a* ascogonium, *ap* apothecium, *p* paraphyses, *pa* paraphysoids. (A–I are microtome sections. A–C and E after Henssen and Jahns, 1974; H after Henssen, 1968; F and I after Henssen, 1973.)

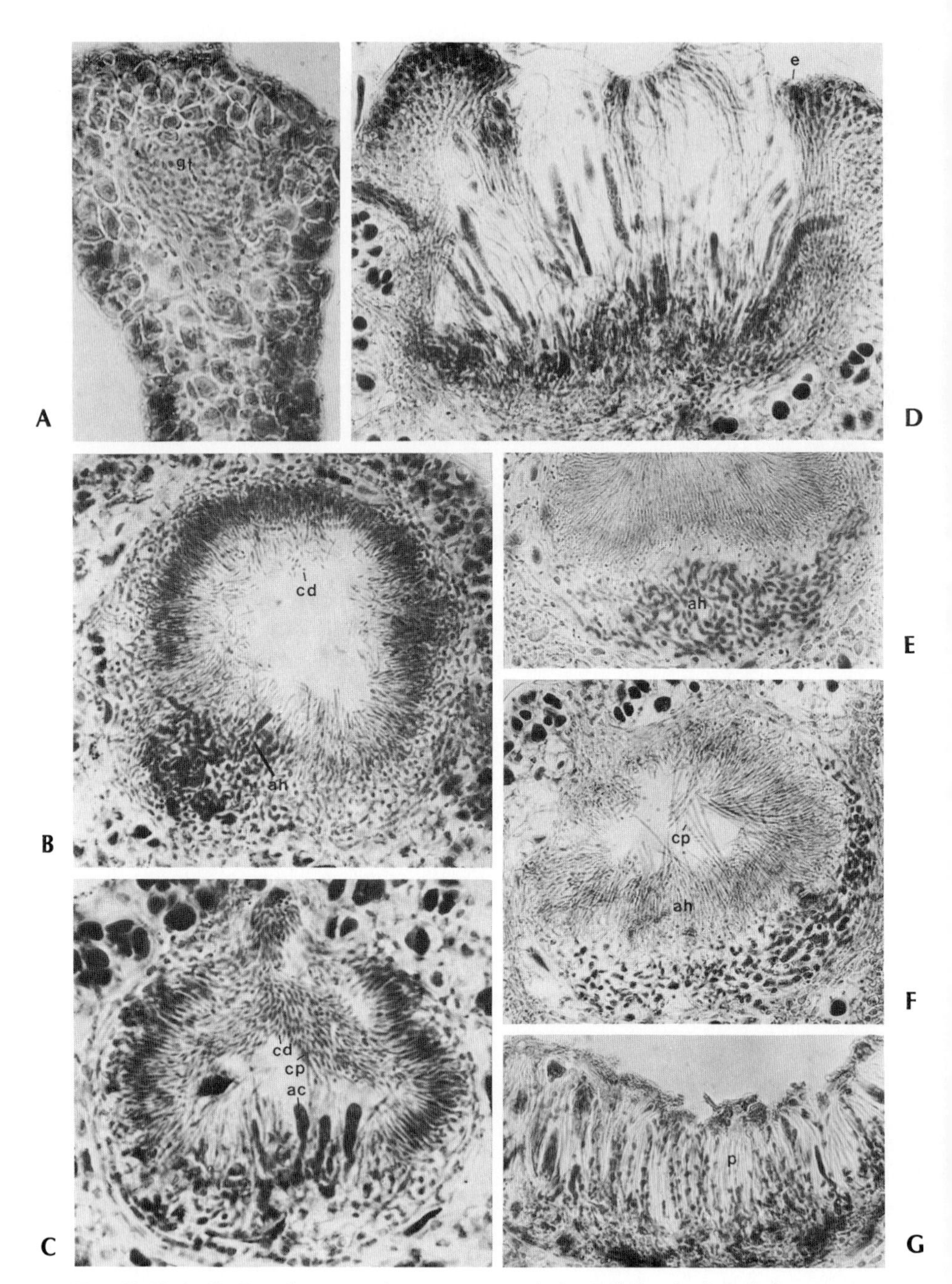

Fig. 10.13 A–G. Development of pycnoascocarpia in *Ephebe*. **A–C** *Ephebe americana*. **A** Generative tissue in a hyphal swelling. X240. **B** Asci developing between conidiophores still producing conidia. X240. **C** Somewhat later stage with elongating conidiophores at the pycnidium base. X460. **D** *Ephebe hispidula*, mature pycnoascocarp with true paraphyses and enlarged upper edge of the excipulum. X280. **E** and **F** *Ephebe brasiliensis*,

the asci grow among the latter. In the second type, the first asci intrude between the conidiophores while conidia are still abstricted (Fig. 10.13, B, C), and the transformation of the conidiophores into paraphyses takes place only later. In both types of development true paraphyses with irregular thickenings at the septa (Fig. 10.13G) are formed secondarily. Young apothecia may contain both types of paraphyses as well as the remains of nonelongated conidiophores, which resemble lateral paraphyses, in the upper part (see Fig. 10.16K). The pycnidial wall gives rise to the excipulum. In mature apothecia, the upper edges of the excipulum may be considerably thickened by the enlargement of the cells (Fig. 10.13D).

Pycnoascocarps are as common in this family as normally developed apothecia. In some genera, one type of development has been observed exclusively. For instance, in *Zahlbrucknerella* Herre only the first developmental type occurs, and in *Ephebe* only pycnoascocarps are formed. In others, such as *Lichina* Argardh and *Jenmania* Wächter (Henssen, 1973), both types occur. In *Phylliscum demangeonii* (Mougeot and Montagne) Nylander the apothecia arise either from pycnidia or from the generative tissue (Henssen, 1963a). In addition, transition forms between these two types have been observed (Fig. 10.12, A–C). In this species, the paraphyses gradually deliquesce into mucilage and the mature apothecia simulate pycnidia (see Fig. 10.16J).

The thallinocarp represents a strongly modified form of the first type of development. The ascogonia, ascogenous hyphae, and asci lie directly among the thallus hyphae in a swelling of the thallus and the hymenium is covered by groups of algal cells. A thallinocarp, considering its outward morphology, resembles a lichen gall (Fig. 10.15A). Thallinocarps are the characteristic fruit bodies in the genera *Lichinella* Nylander (Henssen, 1968b) and *Gonohymenia* Steiner [including *Thallinocarpon* Dahl and *Rechingeria* Servit (Henssen, 1980)]. The sequence of development morphology in *Gonohymenia cribellifera* (Nylander) Henssen and *Gonohymenia inflata* Henssen (Henssen, 1980), is illustrated in Figures 10.14 and 10.15. In Figure 10.14A, a spirally coiled ascogonium is seen beside a more developed one with swollen cells. The ascogonia become surrounded by a weft of generative tissue, which remains rather limited (Fig. 10.14B). Obviously stimulated by the growth of the primordium, the adjacent thallus hyphae start to multiply and function as paraphyses (Fig. 10.14, C, D). The asci grow up among the thallus hyphae. This is most clearly seen in the marginal part of the horizontally spreading ascocarp (Fig. 10.15, B–E). Early stages have a continuous hymenium and are uniformly covered by the algal layer. Subsequently, the hymenium segregates into small parts that become more or less surrounded by excipulary structures, and a few paraphyses may be formed (Figs. 10.14, F, G and 10.15, F, G). At the end of development (Fig. 10.14, G, H), sterile parts are seen between the diverging fertile parts of the hymenium and the overlaying algal cells are mainly

---

ascogenous hyphae beneath elongated conidiophores. **E** Inclined section; **F** Almost median section; both X240. **G** *E. americana,* hymenium of old pycnoascocarp with true paraphyses that show thickening at the septa. X320. *ac* ascus, *cd* conidia, *cp* conidiophore, *e* excipulum, *gt* generative tissue, *p* paraphyses. (A–G are microtome sections. A–G after Henssen, 1963a.)

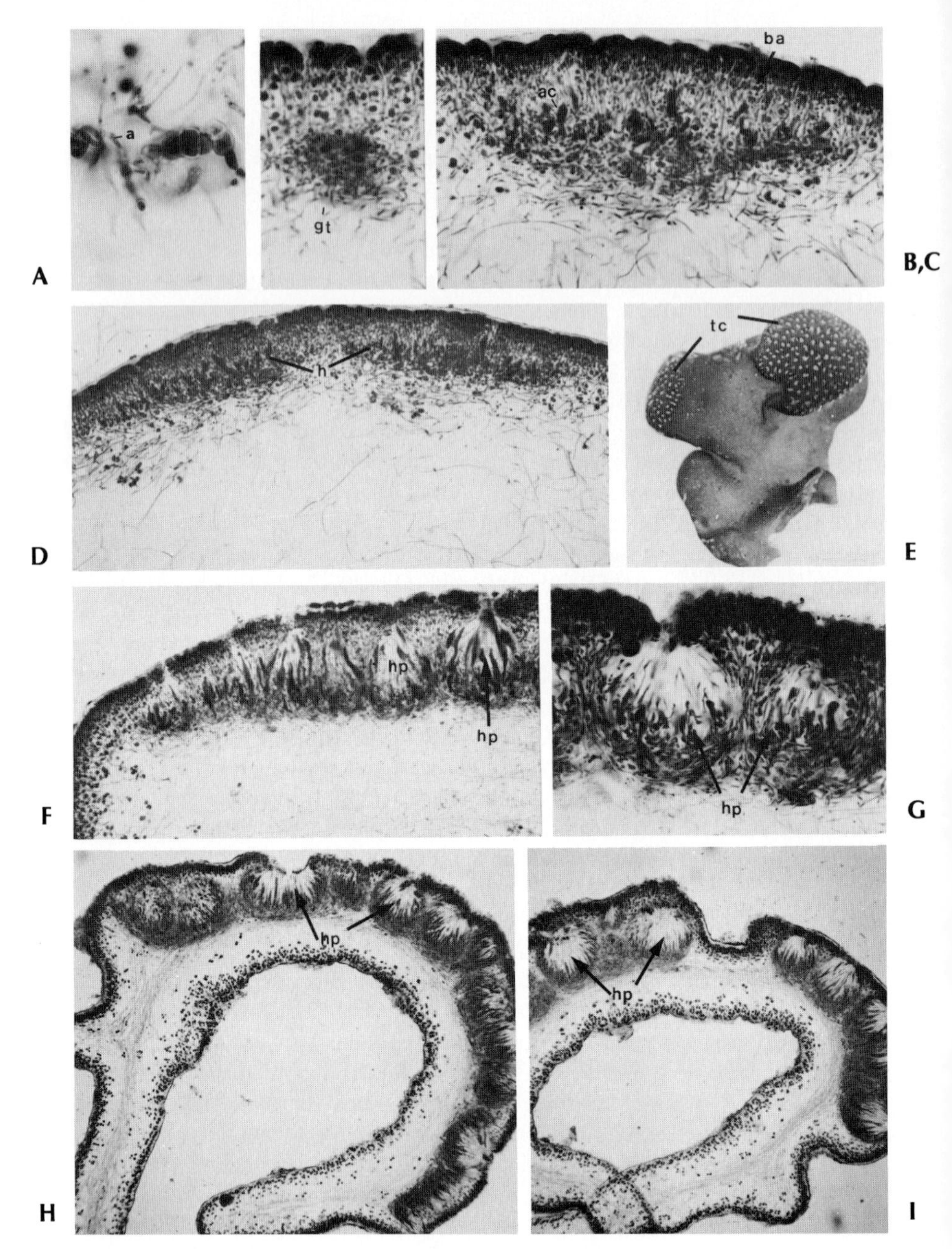

Fig. 10.14 A–I. Development of thallinocarp in *Gonohymenia cribellifera.* **A** Spirally contorted ascogonium (left) beside ascogonium with enlarged cells. X520. **B** Generative tissue surrounding ascogenous hyphae. X250. **C** Young hymenium with asci developing between thallus hyphae. X200. **D** Discontinuous young hymenium. X130. **E** Habit photograph of lobe tip bearing two thallinocarps. X10. **F** Hymenium segregating into small parts. X100. **G** Portions of hymenium with true paraphyses and a shell of pericline hyphae. X200. **H** and **I** Two sections of the same thallinocarp. X50. *a* ascogonium, *as* ascogonial stalk, *ba* blue-green algae, *gt* generative tissue, *h* hymenium, *hp* hymenial portion, *tc* thallinocarp. (With exception of E, all are microtome sections.)

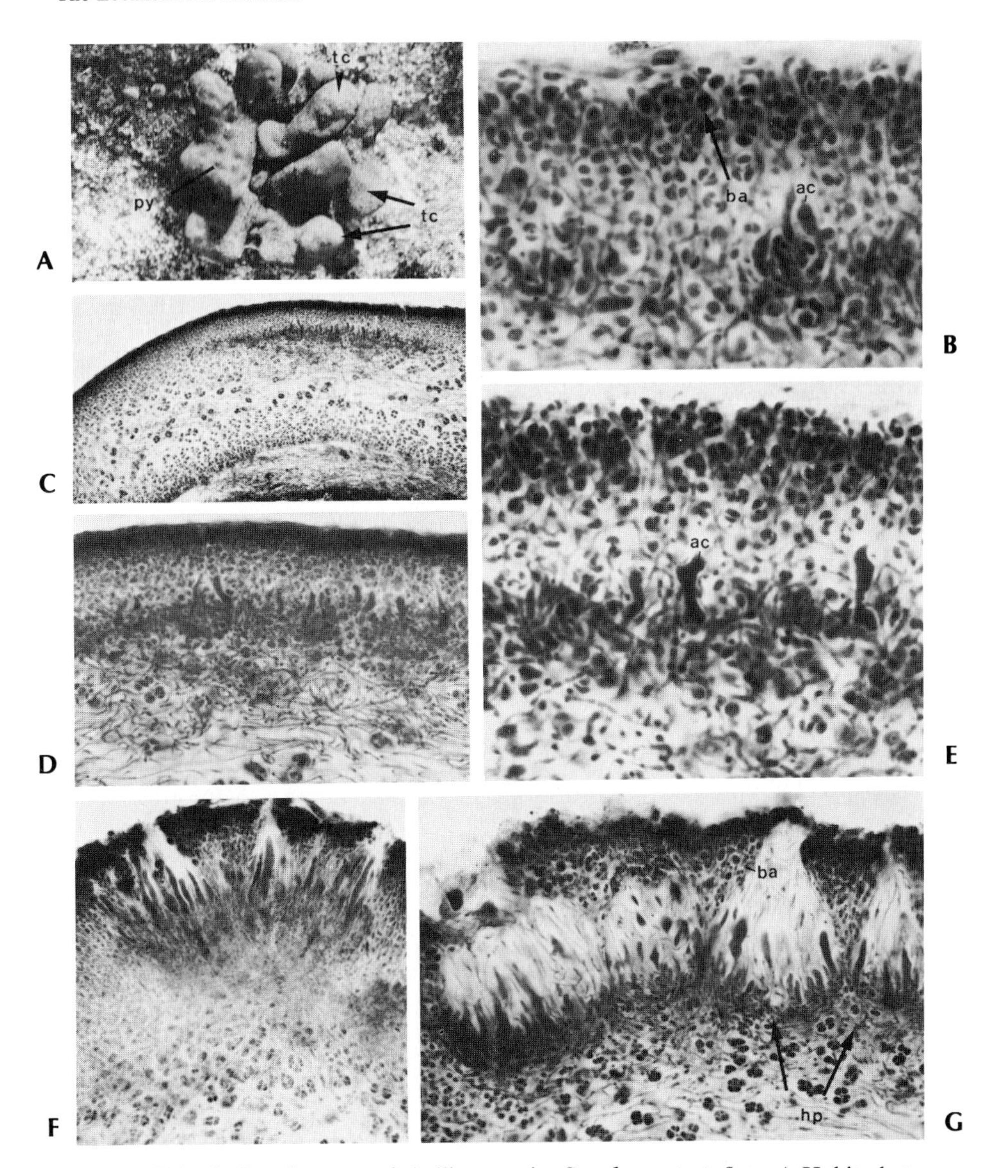

Fig. 10.15 A–G. Development of thallinocarp in *Gonohymenia inflata*. **A** Habit photograph of thallus bearing thallinocarps (arrows) and pycnidia. X8. **B** Margin of developing hymenium with asci growing between thallus hyphae. X520. **C** and **D** young hymenium immersed in the thallus tissue. X80 and X200. **E** Portion of hymenium with young asci. X520. **F** Hymenium with segregating parts. X380. **G** Portions of hymenium with true paraphyses. X520. *ac* ascus, *ba* blue-green algae, *hp* hymenial portion, *py* pycnidium, *tc* thallinocarp. (B–G are microtome sections. A after Henssen, 1980.)

restricted to the sterile parts. In this stage, the thallinocarp has the spurious appearance of a thallus with grouped individual fruit bodies. Seen from above, especially in moist condition, the hymenium has a sievelike perforation (Fig. 10.14E).

The genus *Rechingeria* was delimited on the basis of the alleged aggregation of fruit bodies. However, the pycnidia are aggregated only in the type species. Exceptionally small apothecia may arise within a group of pycnidia. The normal development, however, is that of a thallinocarp and not distinguishable from many other *Gonohymenia* species. The genus *Rechingeria* therefore has to be considered as a synonym of *Gonohymenia* (Henssen, 1980).

Regarding the formation of fruit bodies, a thallinocarp may be interpreted as an ascocarp in which not only the excipulum but also the paraphyses are replaced by elements of the thallus.

The main possibilities of ascocarp development in the Lichinaceae are schematically illustrated in Figure 10.16. The first type of development is summarized on the left side of the drawing, and the development of a pycnoascocarp on the right. The nature and occurrence of paraphyses in mature apothecia is demonstrated in Figure 10.16, G–K; Figure 10.16G typifies certain species of *Lichinella* and *Gonohymenia* in which the hymenium remains continuous and no true paraphyses are formed.

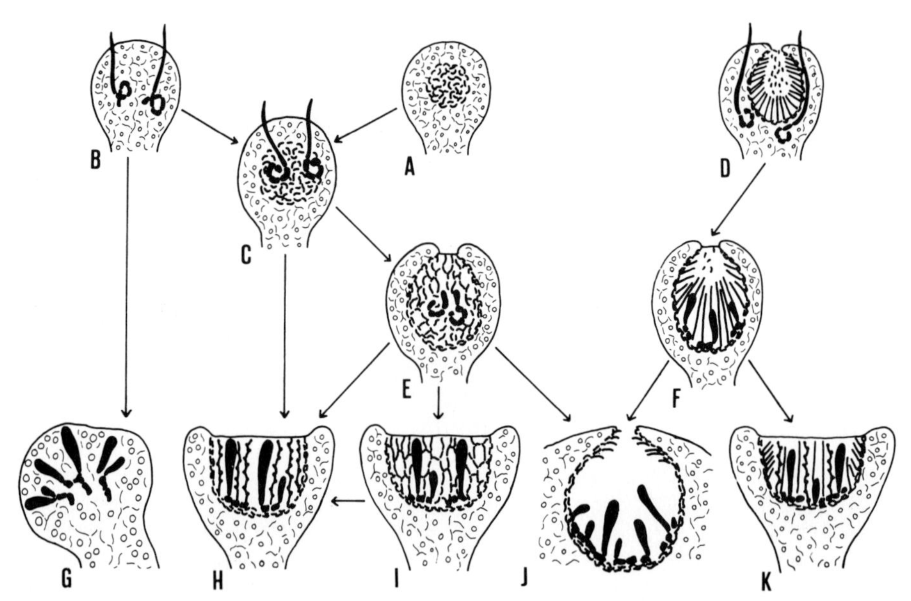

Fig. 10.16 A–K. Development of the fruit bodies in the Lichinaceae (schematic). **A–C** Initial stages. **D** Ascogonia beneath pycnidium. **E** Young ascocarp with paraphysoids. **F** Young pycnoascocarp with elongated conidiophores. **G** Thallinocarp. **H** Apothecium with true paraphyses. **I** Apothecium with paraphysoids. **J** Apothecium lacking paraphyses. **K** Mature pycnoascocarp with true paraphyses and rests of conidiophores in the upper part. (A–K after Henssen and Jahns, 1974.)

Parmeliaceae

The large family Parmeliaceae, including the Usneaceae (Henssen and Jahns, 1974), is separated from the other families of the suborder Lecanorineae by the ontogeny of the ascocarp, involving the formation of a meristematic cupular excipulum and a complicated differentiation of the centrum. The structure of the amyloid ascus corresponds to that of *Lecanora* type (Letrouit and Lallemant, 1970a). Letrouit and Lallemant (1970a) described the development of *Parmelia conspersa* (Acharius) Acharius. They discuss the previous investigations and correct older misinterpretations. The ontogeny of *Parmelia exasperata* de Notaris, outlined briefly in Henssen and Jahns (1974), follows the same pattern as that recognized by Letrouit and Lallemant (1970a) for *Parmelia conspersa*. The ascocarp development of species in usneoid genera has been studied by J. Hähndel; the descriptions of *Letharia* (Th. Fries) Zahlbruckner and *Protousnea* (Motyka) Krog given below as well as the notes on other usneoid genera are taken from his thesis, submitted to the University of Marburg in 1977.

The ascocarp development in *Parmelia exasperata* commences with a compact generative tissue originating between the cortex and the algal layer and includes a group of ascogonia. The long ascogonia are irregularly bent and twisted, and their trichogynes project above the surface of the thallus (Baur, 1904) (Figs. 10.17A and 10.18B). In the next stage illustrated (Figs. 10.17B and 10.18C), the generative tissue differentiates into an outer layer of radiately aggregated, plasma-rich cells embedded in mucilage—the cupular meristematic excipulum—and into a strongly gelatinized central plexus interlaced with hyphae connecting the outer shell. The ascogenous hyphae are difficult to recognize in this stage of development. They appear with the first growth of paraphyses in the upper part of the central plexus (Fig. 10.18, D, E). The paraphyses are branched. The tips of some paraphyses are partly free (true paraphyses) and some are connected with a covering layer (paraphysoids). The covering layer is derived from the generative tissue in the upper part of the primordium in varying degree and a hemiangiocarpy is accordingly more or less well developed (Figs. 10.17, B, C and 10.18, C–F). A hyaline zone persists below the paraphyses, which is interlaced by hyphae connecting the excipulum with the hyphal layer from which the paraphyses are formed. The ascogenous hyphae gather at the base of the paraphyses and form the hypothecium (Fig. 10.18F). At this stage the excipulum has separated into two parts; an inner meristematic dark-stained layer and an outer hyaline layer. In the hyaline layer, the cells have stretched to form a space net of contorted hyphae. This characteristic mode of excipulum differentiation is especially marked in *Omphalodium arizonicum* Tuckerman (Fig. 10.18A).

The hyphae grow inward from the short cells of the meristematic part of the excipulum and connect the excipulum with the hypothecium, thus forming hyphae throughout the hyaline layer. In young apothecia the enlargement formed from the excipulum and the hyaline layer outgrows that of the hymenium and leads to an incurvation of the disc (Fig. 10.18F). The hyaline layer as well as the excipulum are stretched and thinned out as the growth of the hymenium is advanced. Correspondingly, the disc takes on a less concave shape (Figs. 10.17D and

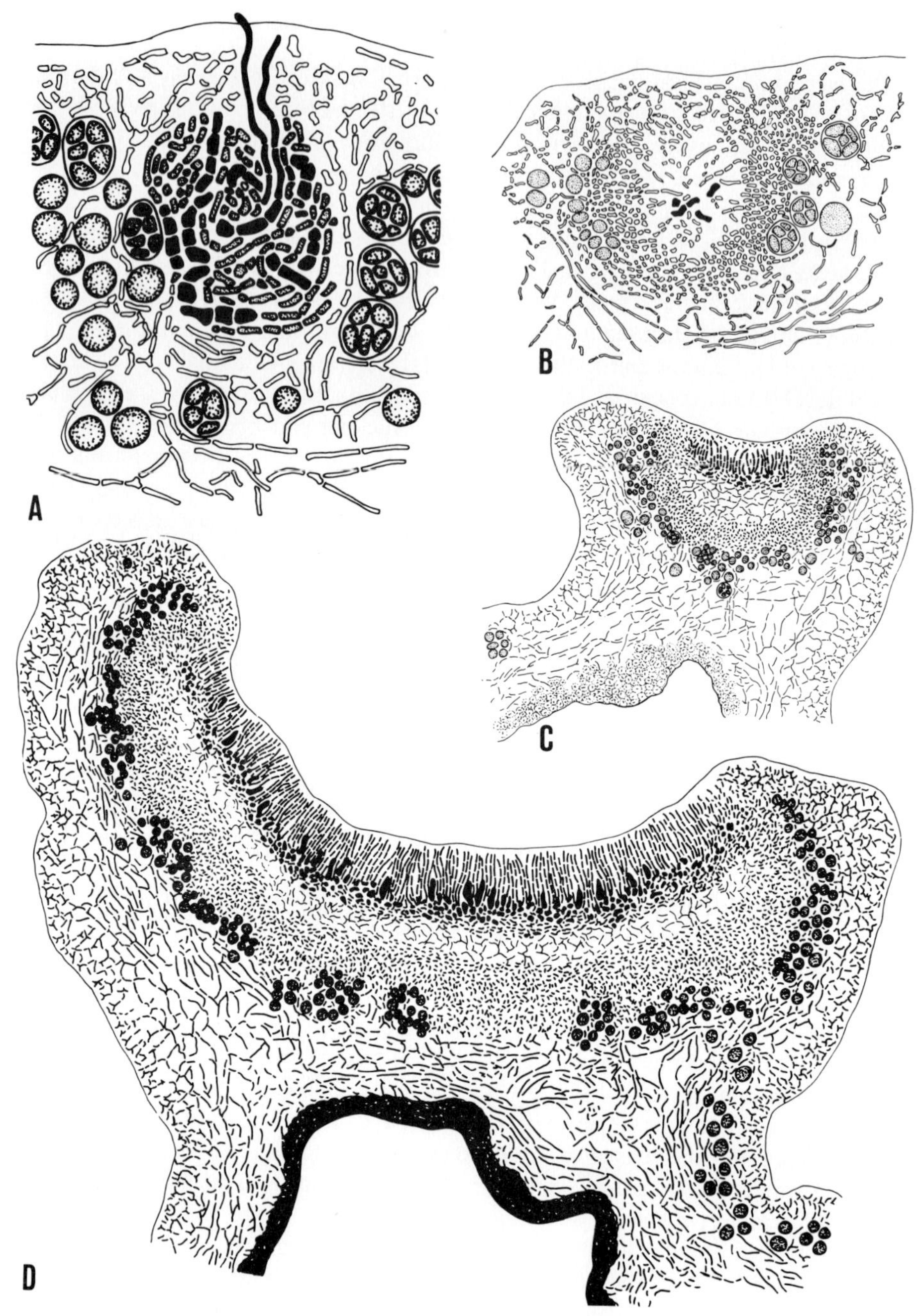

Fig. 10.17 A–D. Development of the apothecium in *Parmelia exasperata*. **A** Generative tissue with ascogonia. **B** Primordium with cupulate excipulum. **C** Young hemiangiocarpous apothecium. The thallus has been grown upwards together with the primordium and forms a margo thallinus. **D** Mature apothecium. (A–D after Henssen and Jahns, 1974.)

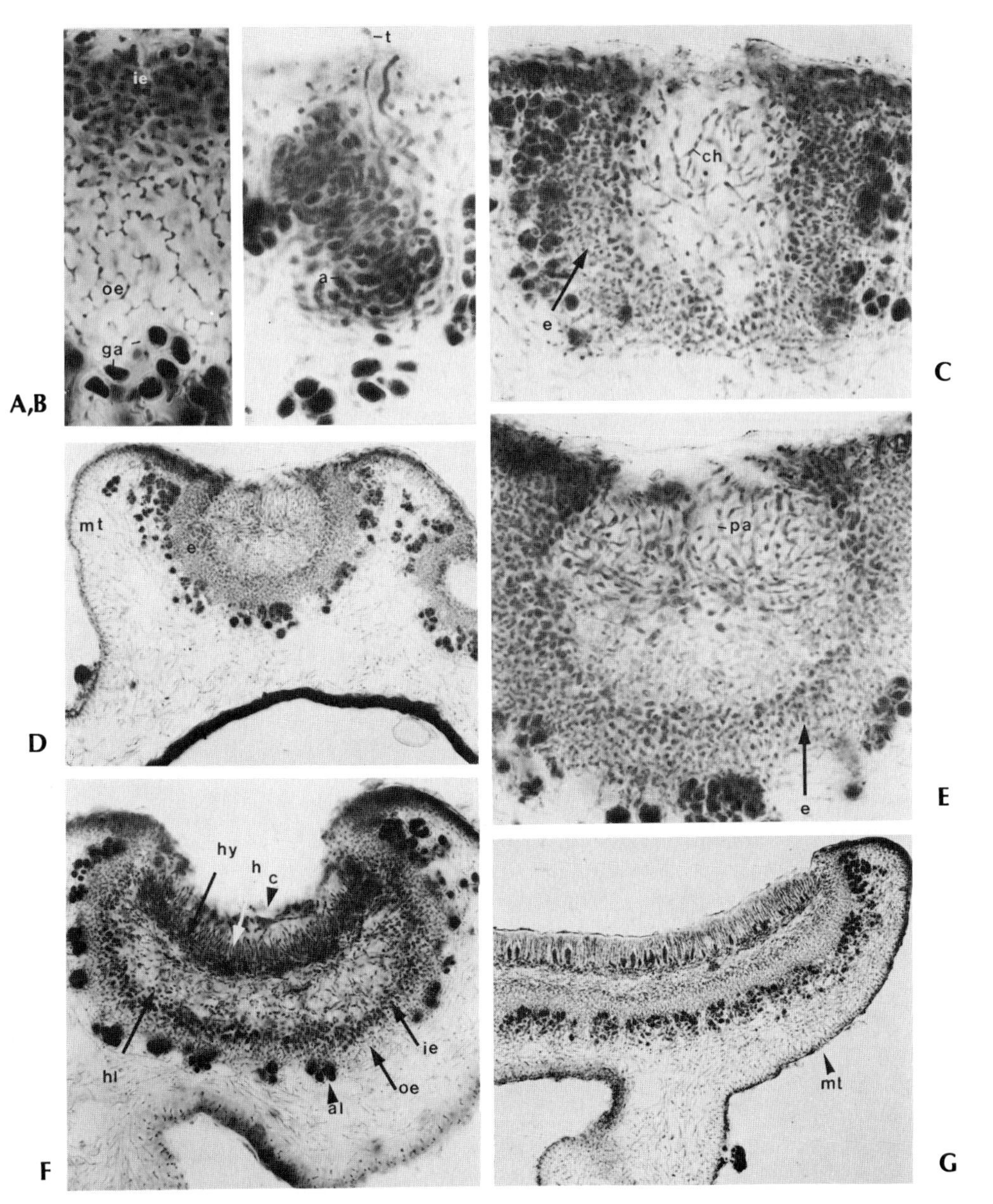

Fig. 10.18 A–G. Development of apothecium in the Parmeliaceae. **A** *Omphalodium arizonicum,* double-structured excipulum. X500. **B–G** *Parmelia exasperata.* **B** Generative tissue with ascogonia. X750. **C** Primordium with meristematic excipulum and connecting hyphae. X380. **D** Young ascocarp surrounded by a thalline margin. X150. **E** Young ascocarp with paraphysoids. X380. **F** Young hemiangiocarpous apothecium with double-structured excipulum. X160. **G** Marginal portion of mature apothecium. X80. *a* ascogonium, *al* algal cells, *c* covering layer, *ch* connecting hyphae, *e* excipulum, *ga* green algae, *h* hymenium, *hl* hyaline layer, *hy* hypothecium, *ie* inner excipulum, *mt* margo thallinus, *oe* outer excipulum, *pa* paraphysoids, *t* trichogyne. (Microtome sections. A after Henssen and Jahns, 1974.)

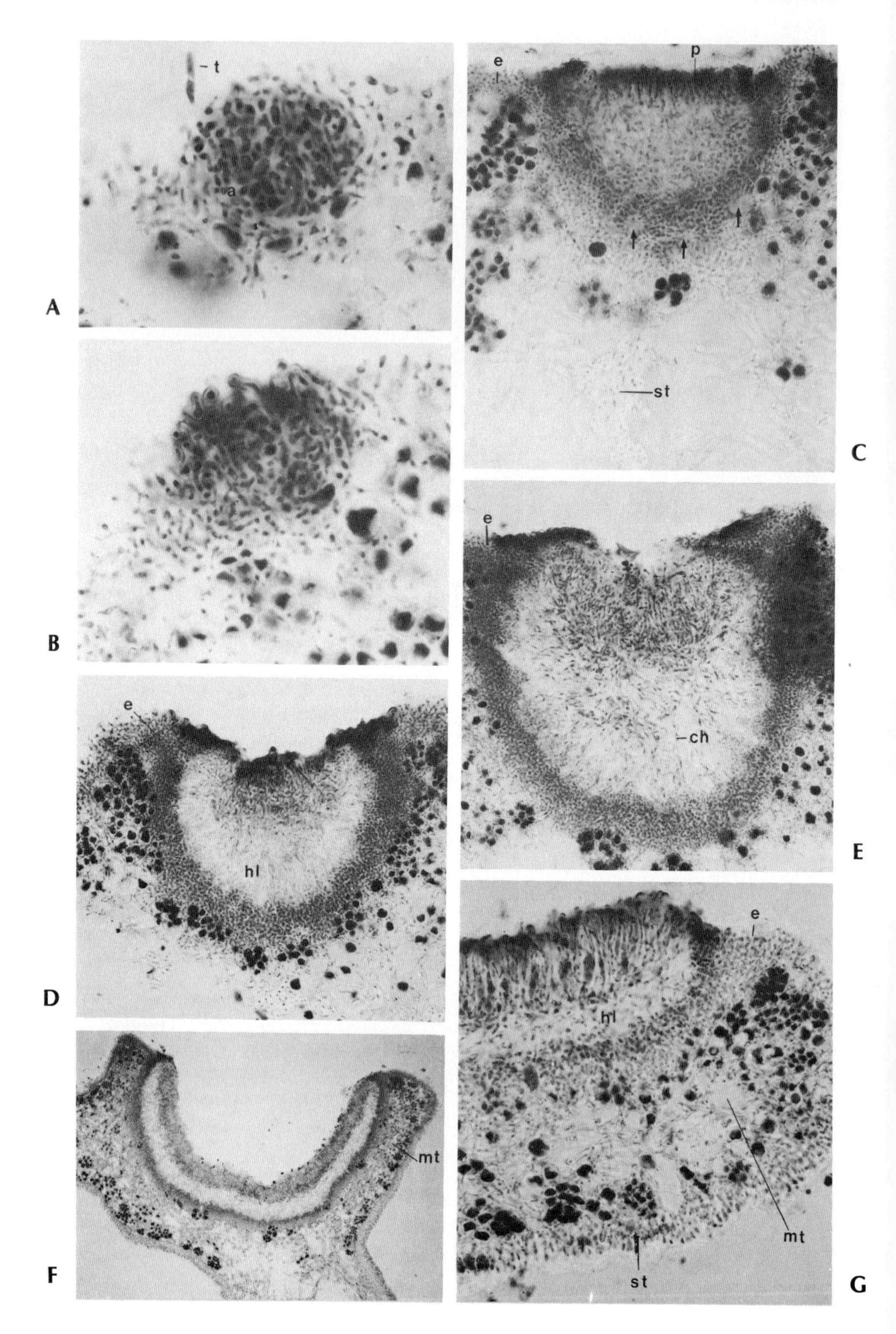

t
a
A
B
e
p
st
C
e
hl
D
e
ch
E
mt
F
e
hl
mt
st
G

10.18G). The ascogenous hyphae extend in a horizontal direction and continue to give rise to asci between the paraphyses. The short-celled paraphyses lie loosely embedded in the hymenial mucilage and are interconnected by anastomoses at their base or higher up. The neighboring parts of the thallus are stimulated to growth by the developing primordium and enlarge to form a thalline margin surrounding the apothecial initial (Figs. 10.17C and 10.18D). On the thalline margin the medullary hyphae form the same type of cortex as is produced on the upper side of the thallus. The algal cells are also stimulated into more active division. Those at the base of the primordium multiply rapidly and proceed to surround the developing fruit body in a saucer-shaped layer. They multiply similarly in the medullary hyphae. In this way, a continuous algal layer is developed extending from the thallus into the margo thallinus and, from there, surrounding the excipulum in a cup-shaped layer. Frequently, however, the continuation of the algal layer in the thallus and margo thallinus can only be seen on one side (Figs. 10.17, C, D and 10.18D). On the other side an interruption occurs when the algal cells are pushed upward by the medullary hyphae in the vigorously vertical growth of the thalline margin.

In *Letharia columbiana* (Nuttall) Thomson, the development of the ascocarp is gymnocarpous. The compact globose primordium lies in the thallus cortex (Fig. 10.19A). The tips of the outer hyphae swell and become dark pigmented (Fig. 10.19B). When the excipulum from the radiating hyphae is formed, the ascogenous hyphae are again difficult to discern; algal cells incorporated in the meristematic zone die away (Fig. 10.20B). Deviating from *Parmelia* Acharius, a layer of true paraphyses with free, dark-pigmented tips is formed first (Figs. 10.19C and 10.20B). The richly branched paraphyses with partly connected tips develop here later on (Figs. 10.19D and 10.20C) and multiply rapidly (Fig. 10.19E). The increased development of the hymenium causes a stretching of the hyaline layer and the excipulum, which gradually become smaller (Figs. 10.19, F, G). The production of the paraphyses and connecting hyphae in a maturing apothecium is seen in detail in Figure 10.20D. Many more paraphyses are formed than connecting hyphae. The paraphyses are short and branched; their tips become free later on.

In *Protousnea magellanica* (Montagne) Krog, the generative tissue is also initiated above the algal layer in the upper part of the thallus as a compact structure enclosing several ascogonia with trichogynes (Fig. 10.21A). The primordium enlarges partly by gelatinization of the hyphal walls and the generative hyphae develop into enlarged and dark-pigmented cells in the upper part and orientate

---

Fig. 10.19 A–G. Development of the apothecium in *Letharia columbiana.* **A** Generative tissue with ascogonium. X570. **B** Somewhat later stage with swelling hyphal tips. X570. **C** Primordium with paraphyses, arrows indicating dead algal cells. X230. **D** Later stage with hyaline layer and connecting hyphae. X230. **E** Young ascocarp with paraphysoids and meristematic excipulum around the hyaline layer. X180. **F** Young apothecium with concave disc. X50. **G** Portion of mature apothecium. X410. *a* ascogonium, *ch* connecting hyphae, *e* excipulum, *hl* hyaline layer, *mt* margo thallinus, *p* paraphyses, *t* trichogyne. (Microtome sections.)

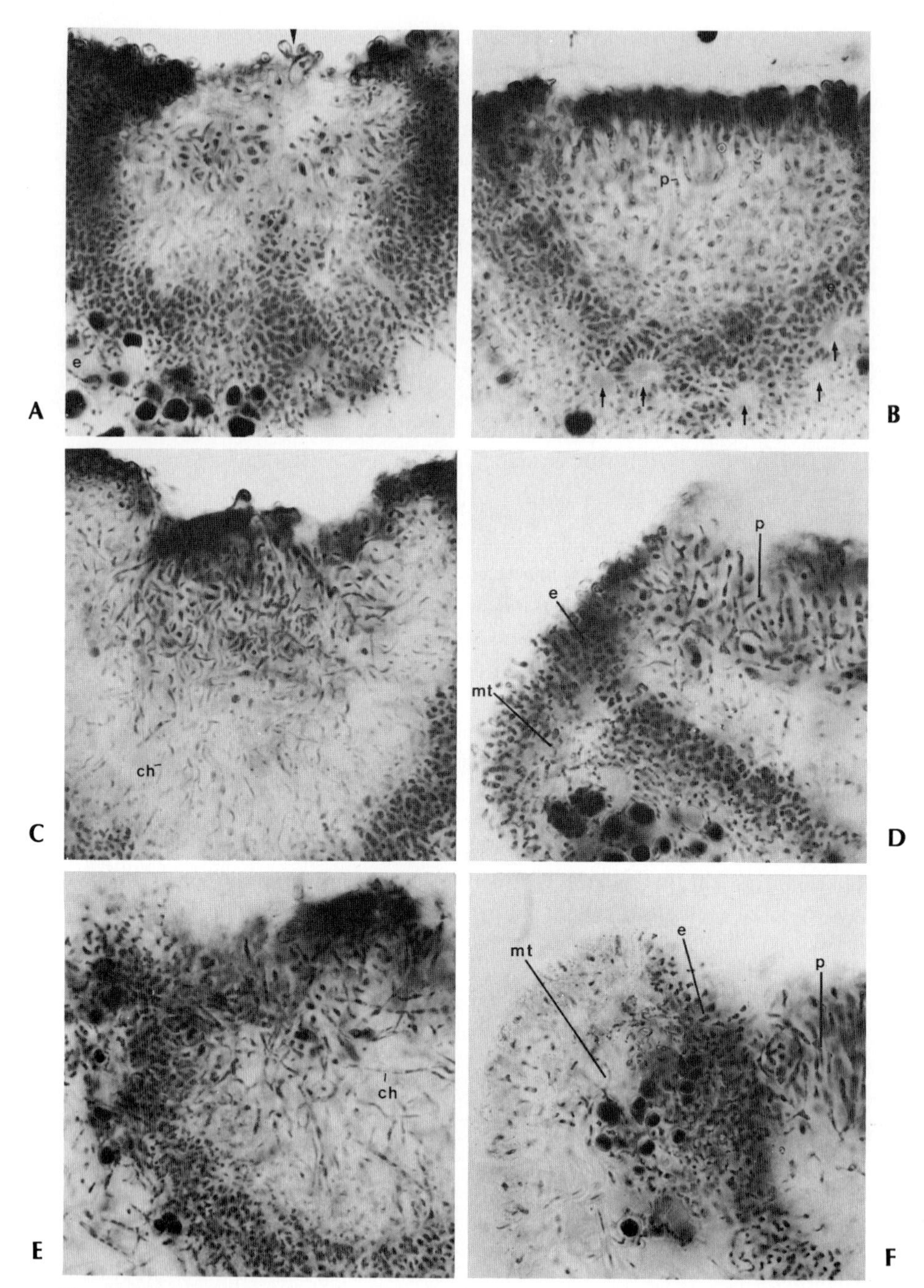

Fig. 10.20 A–F. Development of the apothecium in the Parmeliaceae. **A–D** *Letharia columbiana*. **A** Primordium with meristematic excipulum; arrows indicate swollen and pigmented hyphal tips (inclined sections). **B** Somewhat later stage with true paraphyses;

periclinally around the lower part forming a cup-shaped layer (Fig. 10.21, B, C). Additional ascogonia are produced in these early stages of primordium development. In the following stages the formation of connecting hyphae is especially distinct (Figs. 10.20E and 10.21, D, E). In mature apothecia, the marginal part of the meristematic excipulum is displaced outward, and the paraphyses arise with free tips (Figs. 10.20F and 10.21H). The cortex in the margo thallinus is well developed (Fig. 10.21H). In *Protousnea* and *Letharia,* groups of medullary hyphae with strongly gelatinized walls form supporting structures below the excipulum (Figs. 10.19C and 10.21G).

The complex structure of the apothecium of the Parmeliaceae with a hyaline layer below the hypothecium, interspersed by connecting hyphae and a meristematic excipulum differentiating on two sides, is a unique feature among lichenized ascomycetes. In unlichenized forms, no similar development is known (Letrouit-Galinou and Lallemant, 1970a). The genera of the family studied so far agree fairly well with the main steps of development. The excipulum always consists of anticlinally running and reticulating hyphae with initially relatively large cell lumina. The cells of the excipulum elongate frequently only to a slight degree and form a loose pseudoparenchyma with the lumina more or less closely grouped and embedded in a gelatinous matrix. Connecting hyphae may be more abundant in *Neuropogon* Nees and Flotow and the hemiangiocarpy more pronounced in the species of *Neuropogon* and *Usnea* Browne ex Adams. In general the thalline margin of the apothecium is provided with a cortex consisting of reticulately branched hyphae with angulose cells. More strongly deviating features are found in the structure of the apothecial margin in *Cavernularia lophyrea* (Acharius) Degelius. Here the edge is formed by irregularly interwoven hyphae with thickened pigmented walls. A peculiar type of margin is also found in the genera *Himantormia* Lamb and *Omphalodium* Mayer and Flotow (Lamb, 1964). Here the excipulum is not marginally surrounded by the thalline margin but is sessile and a pedicel is formed by the latter. In the case of *Omphalodium arizonicum* (Tuckerman ex Willey) Tuckerman, the young primordium lies immersed in the thallus, surrounded by a wall produced in the first developmental stage of the thalline margin. In the course of further development, the growth of the thalline margin is retarded. This results in the placement of the excipulum cup on a pedicel. In *Himantormia,* this condition is even more exaggerated. Here the generative tissue is initiated half exogenously—that is to say protruding above the surface of the thallus. As a result, the upward growth of the thalline margin becomes even further delayed.

---

arrows indicate enclosed dead algal cells. **C** Upper portion of primordium with paraphysoids and hyaline layer interspersed with connecting hyphae. **D** Margin of mature apothecium. **E** and **F** *Protousnea magellanica.* **E** Primordium with meristematic excipulum, hyaline layer interspersed by connecting hyphae. **F** Margin of mature apothecium. *ch* connecting hyphae, *e* excipulum, *mt* margo thallinus, *p* paraphyses. (Microtome sections; all photographs X380.)

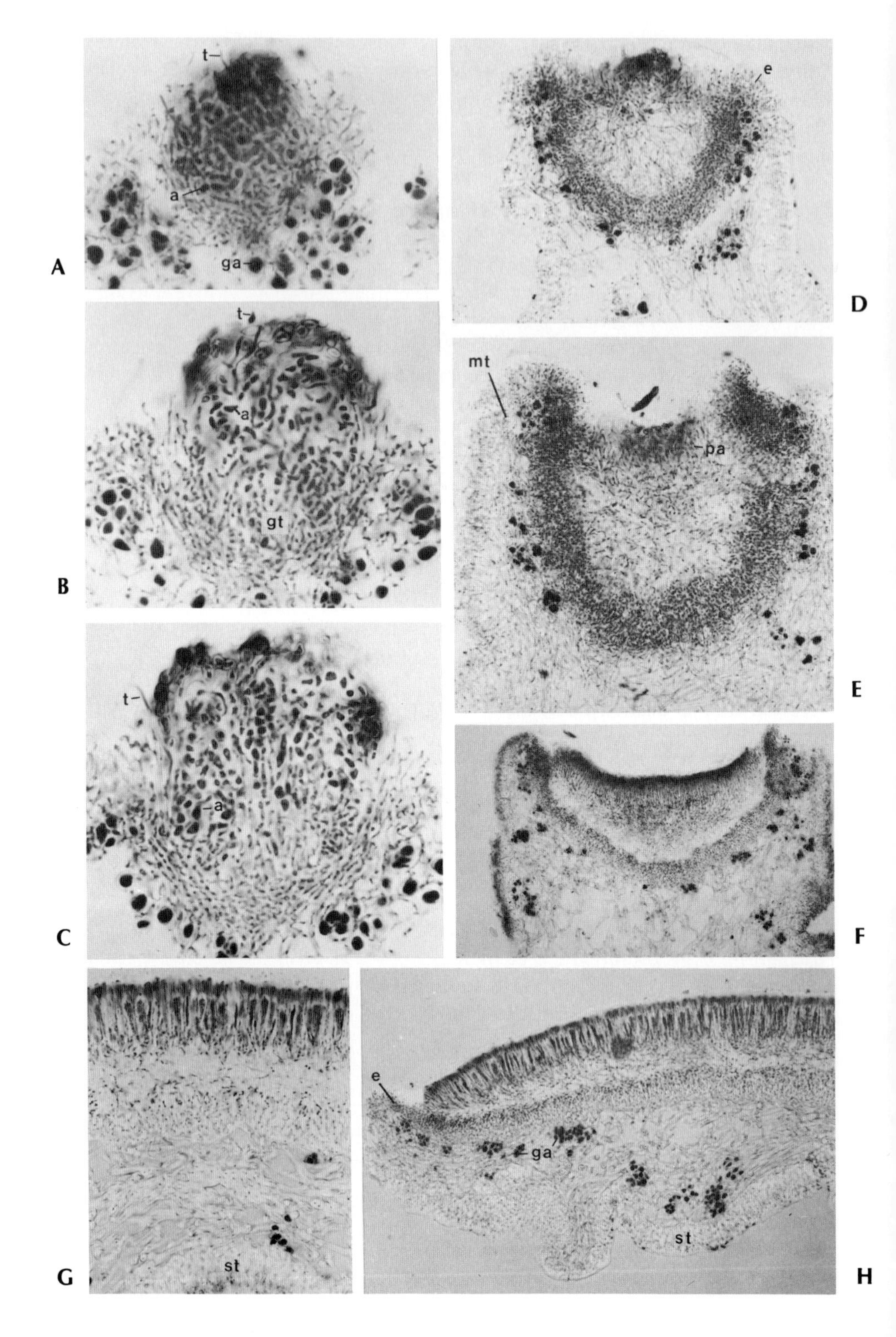
t
a
ga
A
t
a
gt
B
t
a
C
e
D
mt
pa
E
F
G
st
e
ga
st
H

Families with Lecanorine, Lecideine, or Superlecideoid Apothecia

Teloschistaceae

The Teloschistaceae have been placed in the monotypic suborder Teloschistineae on account of a peculiar ascus with a horseshoe-shaped amyloid apical thickening, polaricular spores, and the occurrence of the antraquinone parietin (Henssen and Jahns, 1974). The exceptional position within the Lecanorales has recently been confirmed by electron microscopy studies (Honegger, 1978). The ontogeny of the fruit body corresponds closely to that seen in members of the Lecanoraceae and Lecideaceae. The mature ascocarps are frequently transitional between lecanorine and lecideine apothecium forms. The amount of algal cells included in the thalline margin as well as the production of excipulum hyphae may vary considerably within fruit bodies of one thallus or even in different parts of an apothecium. The varying amount of algal cells in the apothecium of *Caloplaca* Th. Fries species has been noted especially by Poelt and Wunder (1967).

The apothecium is finely lecideine in *Caloplaca ferruginea* (Hudson) Th. Fries and *Caloplaca leucoraea* (Acharius) Branth and remains more or less lecanorine in *Caloplaca cinnamomea* (Th. Fries) Olivier. The first stages of development are the same in all three species. Development begins in the upper part of the algal layer with the formation of generative tissue containing several straight ascogonia, which are composed of relatively few cells and bear long thin-celled trichogynes (Fig. 10.22A). The tips of the trichogynes project above the thallus surface and have been seen covered with conidia (Fig. 10.22, B, C). The hyphae of the generative tissue multiply and form a globose primordium (Figs. 10.22D, 10.24A, and 10.25A). The first true paraphyses grow up as a small intertwined "bush" from the base of the generative tissue, where the ascogenous hyphae congregate and start with the production of asci (Figs. 10.22E, 10.24C, F, and 10.25B). The surrounding parts of the thallus grow upward together with the centrum and develop into a thalline margin (Figs. 10.23, A–C; 10.24B; and 10.25, B, C). In the further course of development, a vertical as well as a horizontal enlargement of the ascocarp takes place. An excipulum becomes differentiated around the hymenium in the usual manner, the formative layer giving rise to paraphyses on its inner side and excipular hyphae on its outer side (Figs. 10.22F and 10.23D). The strongly gelatinized walls of excipular hyphae fuse. The radiating hyphae of the exciple are produced in such abundance in the "lecideine" apothecia that the thalline margin formed earlier is gradually pushed down to the base of the ascocarp (Figs. 10.23, D–G, and 10.25D). The development of excipulum hyphae in

Fig. 10.21 A–H. Development of the apothecium in *Protousnea magellanica*. **A** Generative tissue with ascogonia. X410. **B** and **C** Somewhat later stages with a shell of periclinal hyphae and ascogonia bearing trichogynes. X410. **D** Primordium with excipulum and hyaline layer dispersed by connecting hyphae. X170. **E** Later stage with paraphysoids. X170. **F** Young zeorine apothecium. X90. **G** Median portion of mature apothecium. **H** Marginal part of old apothecium. X180. *a* ascogonium, *e* excipulum, *ga* green algae, *gt* generative tissue, *mt* margo thallinus, *pa* paraphysoids, *st* supporting tissue, *t* trichogyne. (Microtome sections.)

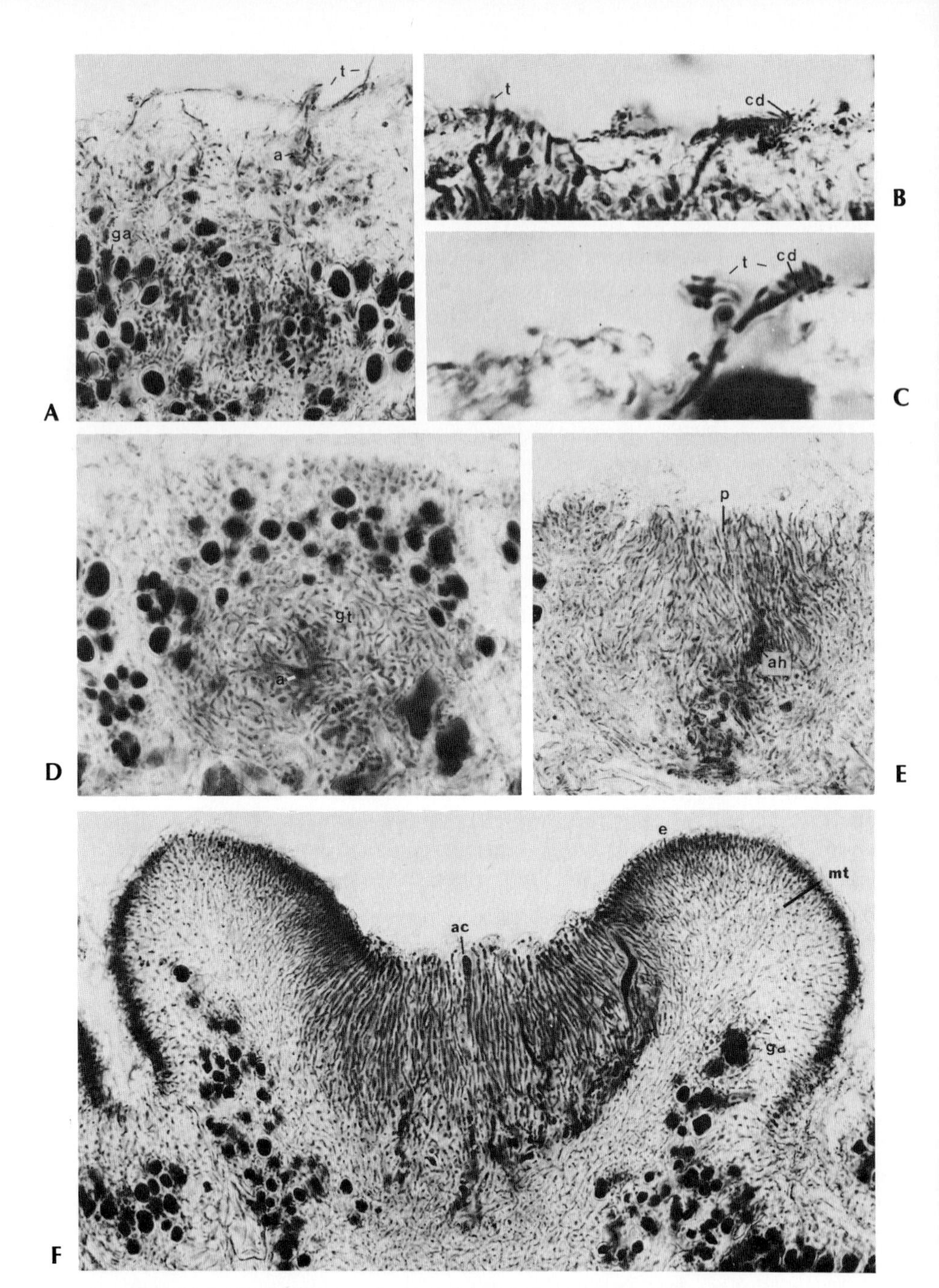

Fig. 10.22 A–F. Development of apothecium in *Caloplaca ferruginea*. **A** Generative tissue enclosing straight ascogonia. X380. **B** and **C** Tips of trichogynes covered by conidia. X620 and X1220. **D** Generative hyphae enclosing ascogenous hyphae and remains of an ascogonium. X400. **E** Primordium with paraphyses and developing ascus. X380. **F** Young apothecium with excipulum and margo thallinus. X290. *a* ascogonium, *ac* ascus, *ah* ascogenous hyphae, *cd* conidia, *e* excipulum, *ga* green algae, *gt* generative tissue, *mt* margo thallinus, *p* paraphyses, *t* trichogyne. (Microtome sections.)

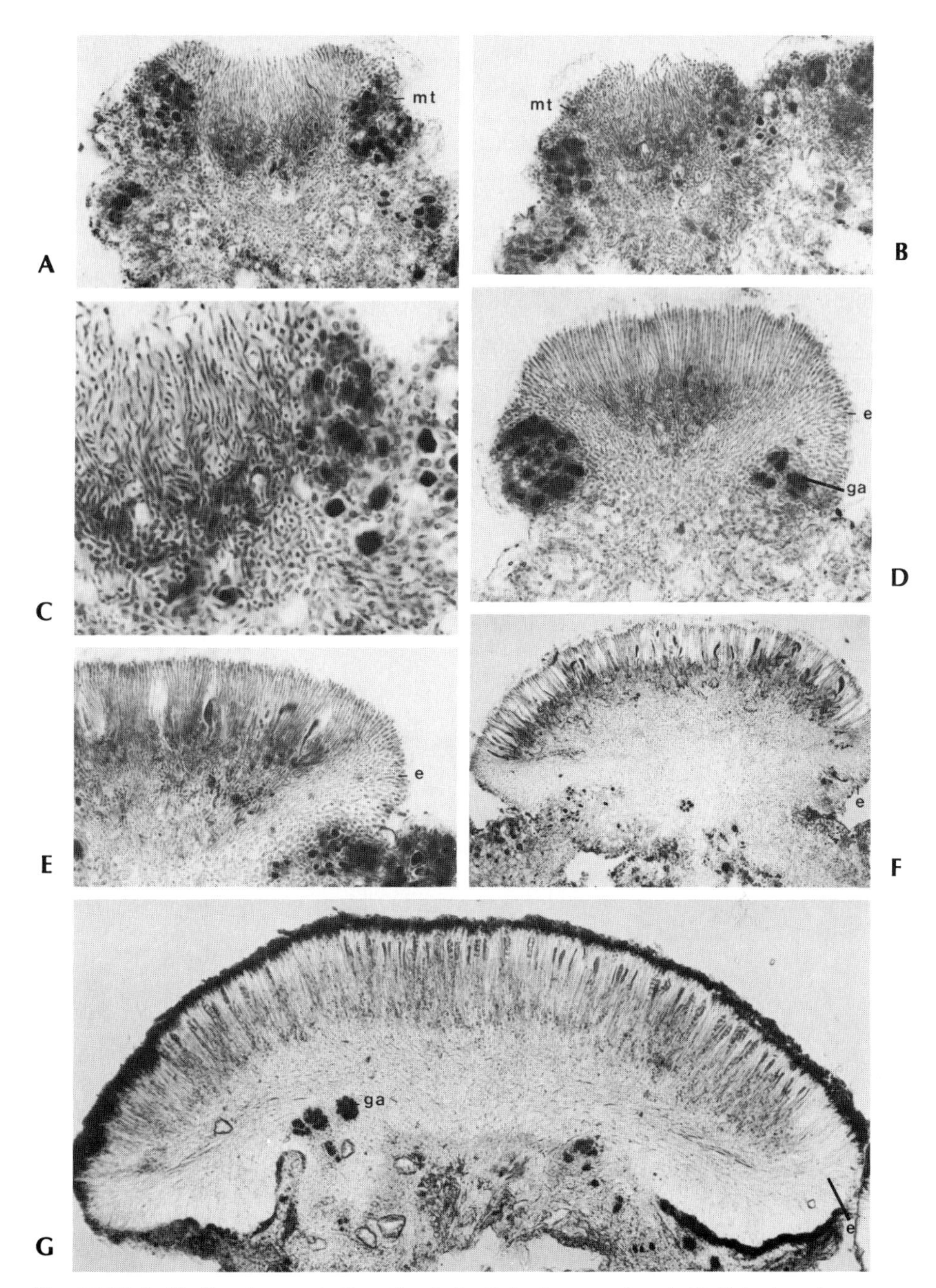

Fig. 10.23 A–G. Development of apothecia in *Caloplaca* species. **A–F** *Caloplaca leucoraea.* **A** and **B** Young ascocarps with thalline margin and paraphyses. X185 and X160. **C** Marginal portion of young ascocarp with paraphyses. X380. **D** Young apothecium with developing excipulum, young asci. X185. **E** Marginal portion of apothecium. X160. **F** Old lecideine apothecium. X80. **G** *Caloplaca ferruginea,* old lecideine apothecium. X90. *e* excipulum, *ga* green algae, *mt* margo thallinus. (Microtome sections.)

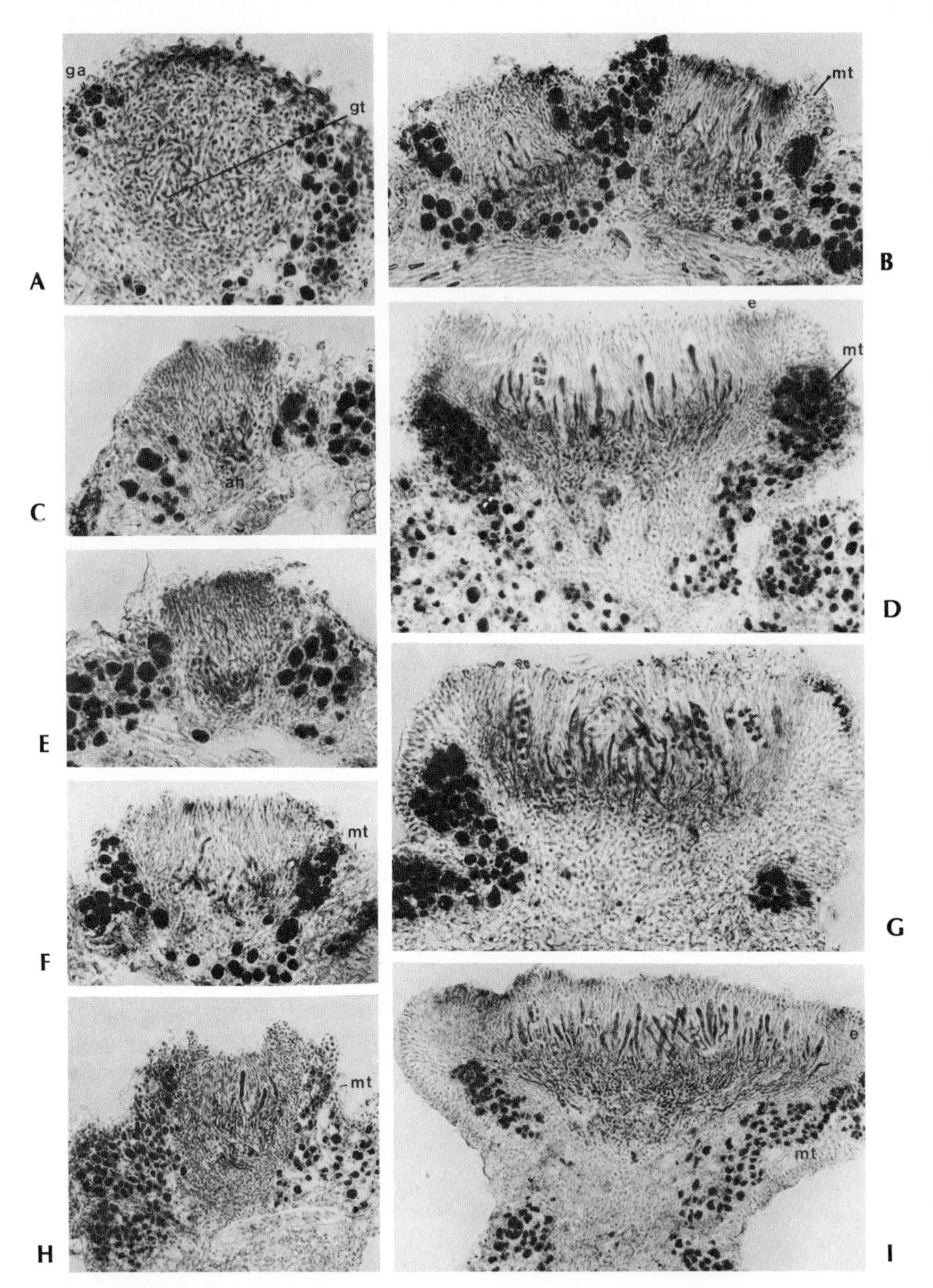

Fig. 10.24 A–I. Development of apothecium in *Caloplaca cinnamomea.* **A** Generative tissue with ascogenous hyphae. X310. **B** Two young apothecia surrounded by a thalline margin. X210. **C** Primordium with paraphysoids. X210. **D** Young lecanorine apothecium. X210. **E** Primordium with paraphysoids. X210. **F** Young stage with paraphyses and thalline margin. X210. **G** Young apothecium with varying amount of algal cells in the thalline margin. X210. **H** Young stage with upgrowing thalline margin. X160. **I** Old apothecium with varying amount of algal cells in the thalline margins. X130. *e* excipulum, *ga* green algae, *gt* generative tissue, *mt* margo thallinus. (Microtome sections.)

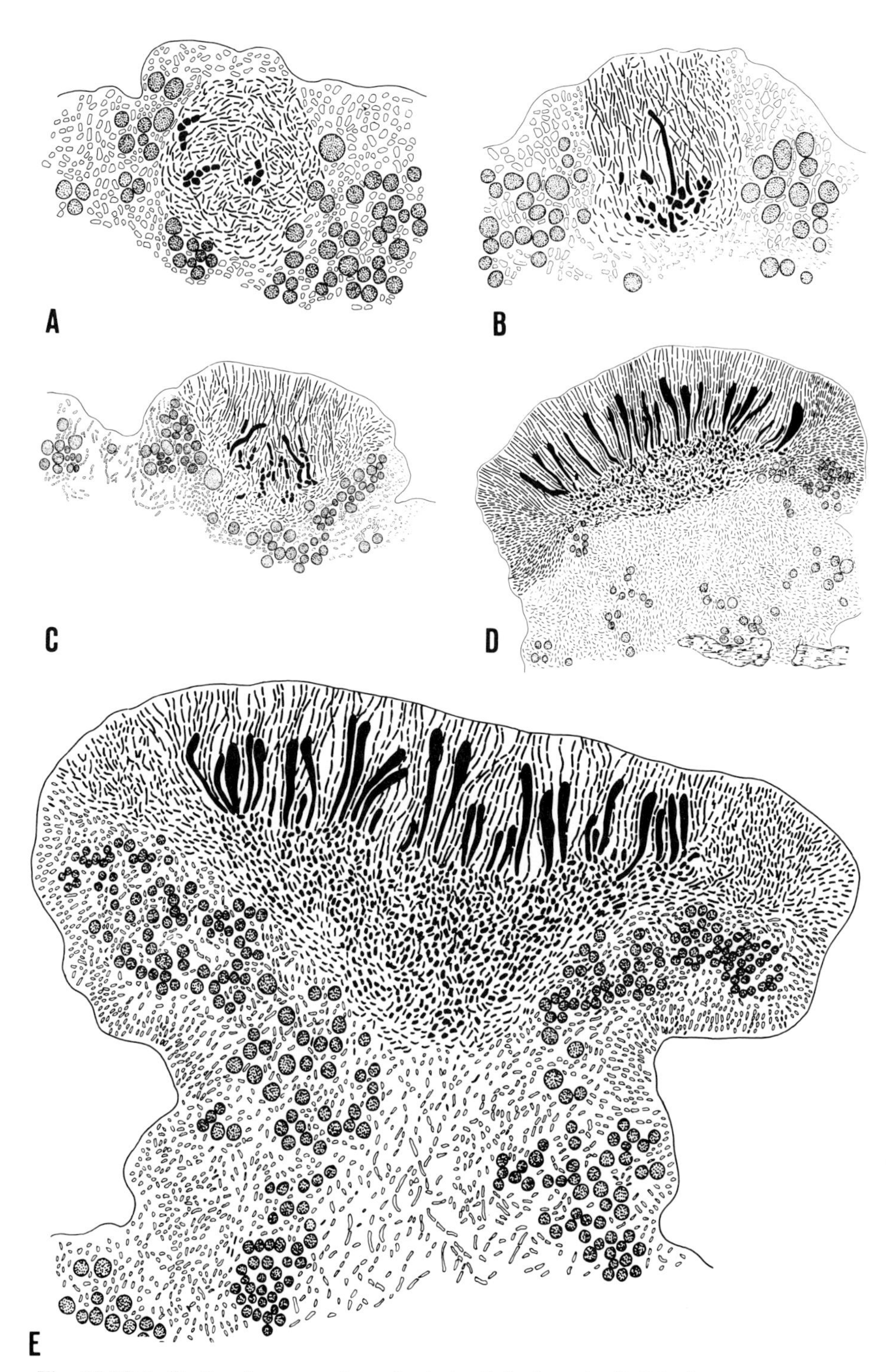

Fig. 10.25 A–E. Development of apothecia in *Caloplaca*. **A–C** *Caloplaca cinnamomea*. **A** Generative tissue with the remains of ascogonia. **B** Beginning of the upward growth of the thallus surrounding the ascocarp. **C** Older stage with asci. **D** *Caloplaca leucoraea*, lecideine apothecium with scattered algal cells in the excipulum. **E** *C. cinnamomea*, mature lecanorine apothecium. (A–E after Henssen and Jahns, 1974.)

the "lecanorine" apothecium is less pronounced and in mature apothecia the thalline margin persists (Figs. 10.24, D, G, I, and 10.25E). The quantity of algal cells present beneath the subhymenium is variable in both types of development. In young apothecia, the algal cells, as in the Parmeliaceae, may form a continuous layer or be lacking in places (Figs. 10.22A and 10.24B, F). The thalline margin develops supportive tissue, which originates from the medullary hyphae and consists of anticlinal, reticulately connected hyphae. This cortical layer is of the same structure as is found in the upper cortex of the thallus of more highly differentiated *Caloplaca* species. Although the number of algal cells in the thalline margin may vary, in old apothecia this margin is always present in *C. cinnamomea,* even when it secondarily appears to be almost destitute of algal cells on one side and consists mainly of the supporting tissue crowned by the excipulum.

The development in the other genera of the family follows a pattern more or less similar to that found in *Caloplaca* Th. Fries. In *Xanthoria* (Fries) Th. Fries and *Teloschistes* Norman, the apothecia are provided with a thalline margin; the fruit bodies of *Xanthopeltis* R. Santesson are immersed. The cortication of the thallus margin always shows the same structure as the neighboring upper thallus cortex. In *Teloschistes,* the excipulum is composed of periclinally interwoven hyphae. In the other genera, it consists of anticlinal, reticulately connected hyphae that may form a loose pseudoparenchyma.

Lecanoraceae

The Lecanoraceae are characterized by having apothecia with a thalline margin. The development of fruit bodies has been studied only in few instances, although this family is one of the largest in the number of species. The most recent investigations are by Letrouit-Galinou (1966), in whose paper the older literature is cited. Baur (1904) stated that the development of the *Lecanora subfusca* (Linnaeus) Acharius is very simple, and the formation of the thalline margin can be interpreted in the same way. The primordium originates below the algal layer and consists of several ascogonia enclosed by some hyphae of the generative tissue (Fig. 10.26A) in *L. subfuscata* H. Magnusson and *Lecanora carpinea* (Linnaeus) Vainio. Baur (1904) observed protruding trichogynes.

The generative tissue hyphae multiply and form the first paraphyses which are paraphysoids in both species. Through enlargement of the primordium, the covering thallus tissue bursts open (Fig. 10.26B). True paraphyses are produced by the formative layer, and are at first intercalary and then predominantly marginal in the part near the formative layer surrounding the hymenium. The thalline margin is formed, as in Collemataceae, Pannariaceae, and Teloschistaceae, by the vertical growth of the thallus surrounding the developing ascocarp (Fig. 10.26C). The enlargement occurs through intercalary growth as well as by the vigorous division of the marginal plasma-rich cells. The excipulum is less distinct than in *C. cinnamomea* and retains the periclinc structure. The exciple may be lacking completely, especially in young apothecia, and the formative layer produces only paraphyses. The ascogenous hyphae keep pace with the vertical growth of the

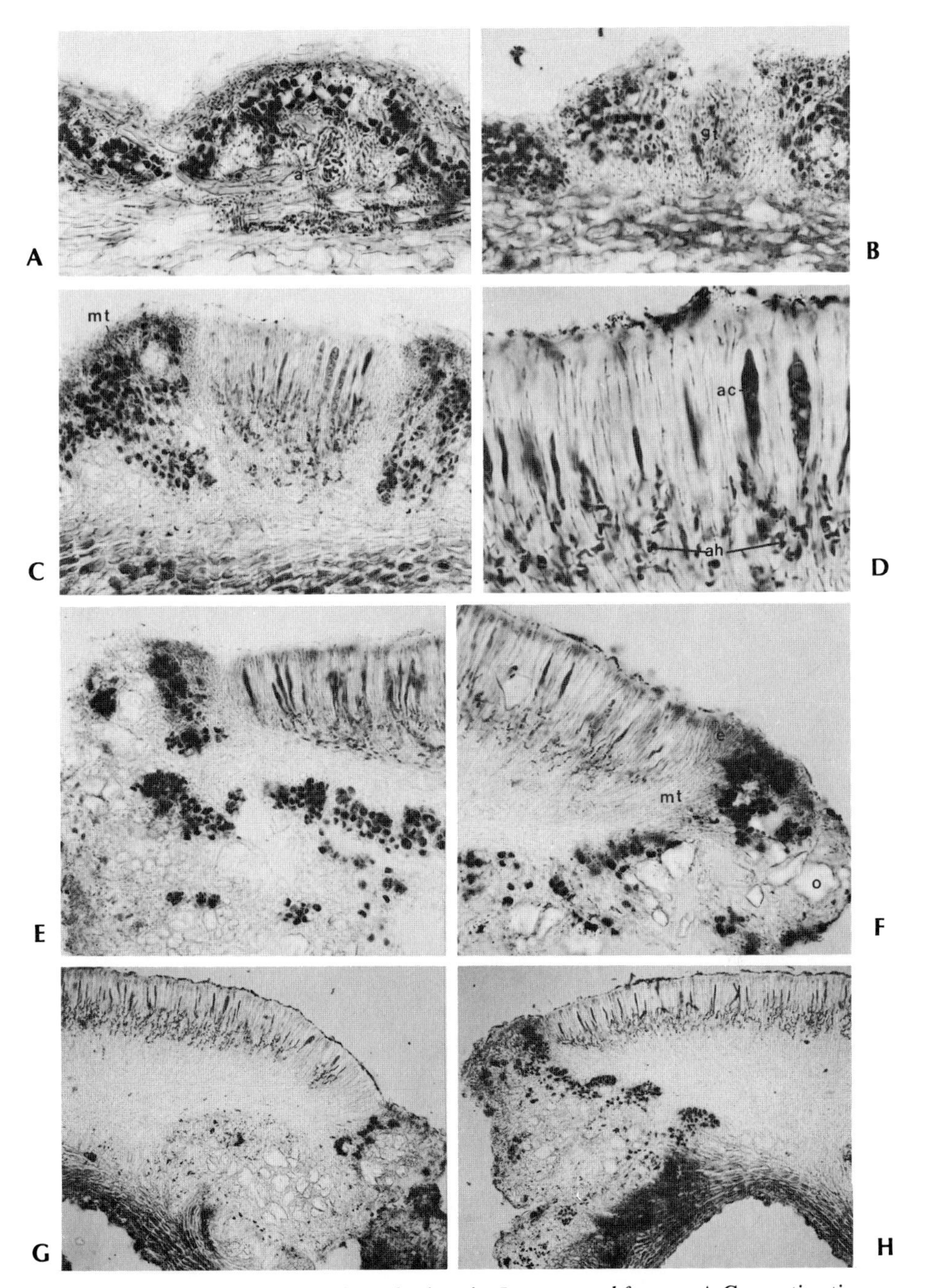

Fig. 10.26 A–E. Development of apothecium in *Lecanora subfuscata*. **A** Generative tissue with ascogonia. X210. **B** Primordium with paraphysoids. X210. **C** Young lecanorine apothecium. X160. **D** Portion of hymenium. X380. **E** and **F** Marginal portions of mature apothecia, the thalline margin interspersed with oxalate crystals. X150. **G** and **H** Marginal parts of very old apothecia. X80. *ac* ascus, *ah* anchoring hyphae, *e* excipulum. *mt* margo thallinus, *o* oxalate crystals. (Microtome sections.)

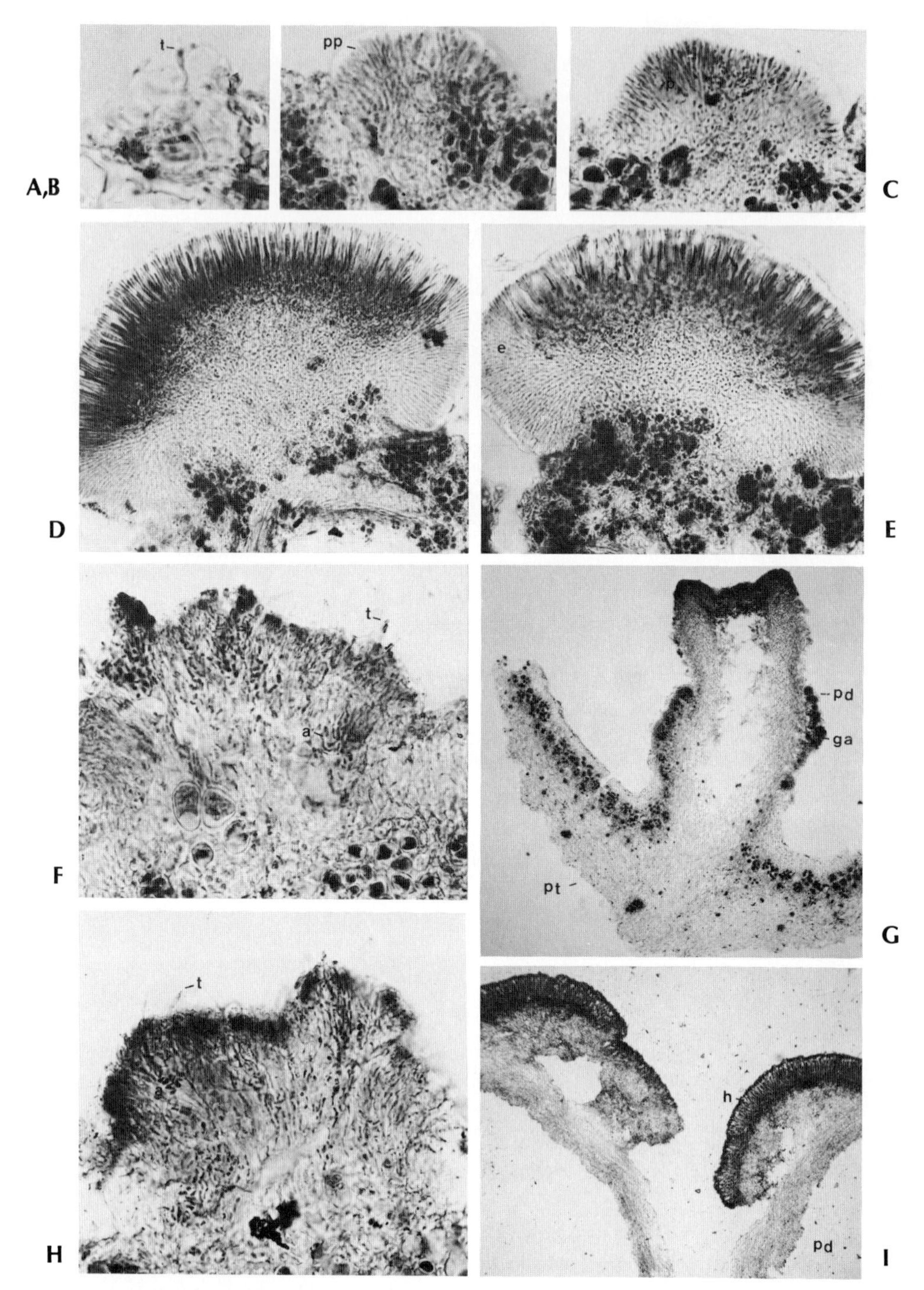

Fig. 10.27 A–I. Development of the fruit bodies in *Lecidea* and *Cladonia*. **A–E** *Lecidea vernalis*. **A** Spirally contorted ascogonium. X1000. **B** Primordium with proparaphyses. X200. **C** Primordium with paraphyses. X410. **D** and **E** Mature apothecia with massively developed excipulum. X210. **F–I** *Cladonia caespiticia*. **F** Primordium with ascogonia and

centrum and surrounding thalline margin (Fig. 10.26, C, D) and grow horizontally in the marginal parts of the maturing fruit bodies (Fig. 10.26E). Part of the hymenium becomes sterile in old degenerating apothecia, and groups of ascogenous hyphae developing asci may lie interspersed within this margin (Fig. 10.26G). Oxalate crystals are deposited in the thalline margin (Fig. 10.26, C, E–H). The amount of algal cells in the margo thallinus varies. In maturing apothecia, supporting tissue is produced at the border of the thalline margin; it is especially well developed in *Lecanora conizaeoides* Nylander ex Crombrie. Strands of hyphae may invade the thalline margin from the formative layer or excipulum (Fig. 10.26, E, F) comparable to the development of anchoring hyphae in the Collemataceae. These strands of hyphae are produced secondarily. In another interpretation (Letrouit-Galinou, 1966), they are regarded as the elements forming the thalline margin or "amphithecium." A comparison of the terminology used here with the French school is given by Keuck (1977).

Lecideaceae

The Lecideaceae differ from the Lecanoraceae in that the apothecia only have a proper margin. Transition forms between *Lecanora* and *Lecidea* Acharius are known that may have an ascocarp development similar to that described for *Caloplaca* species. The family is thought to contain 3000 species and to be approximately three times as large as the Lecanoraceae. The ascocarp ontogeny of some members of the family have been studied by Letrouit-Galinou (1966) and Jahns (1970a). This discussion is primarily based on the work of Jahns. In addition, *Lecidea berengeriana* (Massalongo) Nylander and *Lecidea fuscoatrata* Nylander have been studied as well as some species of *Sporopodium* Montagne—a genus with epithecial algae (Santesson, 1952). The development of epithecial algae in *Sporopodium vainis* or other foliicolous lichens was hitherto unknown, and a study of these is important for a comparison with the development of the hymenial algae in the thallinocarps of the Lichinaceae.

Two lines of development are recognized in the *Lecidea* species so far studied. The development commences in the first type with the formation of a spirally coiled ascogonium as seen in *Lecidea vernalis* (Linnaeus) Acharius which bears a short trichogyne (Fig. 10.27A). The ascogonium gradually becomes surrounded by a hyphal weft of generative tissue. Frequently several ascogonia are enclosed in a cup-shaped structure. The primordium arches up above the surface with further development. The hyphae of the generative tissue are aligned in parallel and protrude from the thallus surface (Fig. 10.27B). These parallel hyphae are termed proparaphyses. The ascogenous hyphae grow in a vertical direction until they reach the upper part of the proparaphyses. More parallel hyphae are produced, and the young ascocarp forms a hemispheric body on the thallus. The hymenium

---

protruding trichogynes. X310. **G** Young fruit body. X80. **H** Generative tissue with ascogonia and ascogenous hyphae. X350. **I** Apothecium that has split into several portions. X50. *a* ascogonium, *e* excipulum, *ga* green algae, *h* hymenium, *pd* podetium, *pp* proparaphyses, *pt* primary thallus, *t* trichogyne. (Microtome sections.)

is developed in the apex of the primordium (Fig. 10.27C) and undergoes horizontal expansion in the usual manner. Paraphyses with gelatinizing walls are formed inward at the marginal growth point, and radiating agglutinating excipular hyphae are formed outward. The growth may continue for a long time so that the fruit body eventually forms a spherical structure on the surface of the thallus. The algal cells, originally surrounding the young primordium (Fig. 10.27B), remain visible around the stipe of the mature ascocarp (Fig. 10.27, D, E).

In *Lecidea granulosa* (Ehrhart) Acharius and *Lecidea berengeriana,* the first stages of development are very similar. A spirally coiled ascogonium is formed and soon surrounded by dark-pigmented hyphae of the generative tissue (Figs. 10.28A and 10.29A). The primordium develops into a globose dark structure with a lighter colored gelatinous center in which ascogenous hyphae are visible (Figs. 10.28B and 10.29B). The primordium then stretches vertically, and paraphysoids arise in the upper part. Through further production of paraphyses and ascogenous hyphae, the centrum enlarges, and the young ascocarp opens gradually (Fig. 10.29C). In *Lecidea berengeriana,* a marginal growth point develops as in *Lecidea vernalis,* and an excipulum of radiating hyphae is correspondingly produced. The hymenium and subhymenium remain, at first, enclosed by a darkly pigmented outer mantle (Fig. 10.28, C, D). In the later course of development, this layer is arched over by the exciple and pressed to the base of the apothecium (Fig. 10.28, E, F). In *Lecidea granulosa,* no radiating excipulum is formed. The hymenium and subhymenium are continuously surrounded by a darkly pigmented layer. The paraphyses are thin, branched, and anastomosing (Fig. 10.29, D, E). A stipe may be developed in both these species.

The development in *Lecidea fuscoatrata* is characterized by a large amount of generative tissue, formed between the thallus squamules, which contains numerous ascogonia (Fig. 10.29, F, G). The formation of the paraphysoids and paraphyses resembles that of *Lecidea granulosa,* aside from the formation of a radiating dark-pigmented excipulum (Fig. 10.29, H, I).

The development of *Sporopodium* is closely connected to the *Lecidea granulosa* type. A darkly pigmented outer layer is formed by the generative tissue surrounding the ascogenous hyphae and young asci (Fig. 10.31, A–C). The upper part of this layer gradually breaks up when the centrum enlarges and a hymenium consisting of paraphysoids and young asci becomes exposed (Figs. 10.30, A–D, and 10.31D). The rest of the tissue covering the paraphysoids gradually dissolves. In the marginal part new paraphysoids arise. The darkly pigmented outer layer develops into an excipular structure of irregularly interwoven hyphae with enlarged cells (Fig. 10.30F). Hairs growing from the outer cells give the excipulum a characteristic appearance (Fig. 10.30F). Epithecial algae occur in the apothecia of *Sporopodium phyllocharis* (Montagne) Massalongo and *Sporopodium xantholeucum* (Müller-Argoviensis) Zahlbruckner (Figs. 10.30F and 10.31, E, F). In the developing primordium algal cells are trapped in the upper part of the generative tissue (Fig. 10.31C) and later on become enclosed in the dark, pigmented mantle layer (Fig. 10.31D). The algal cells are stimulated to increased division and continue to multiply when the ascocarp opens so as to form groups positioned on the hymenial gelatin (Fig. 10.31, E, F). The apothecial cells are

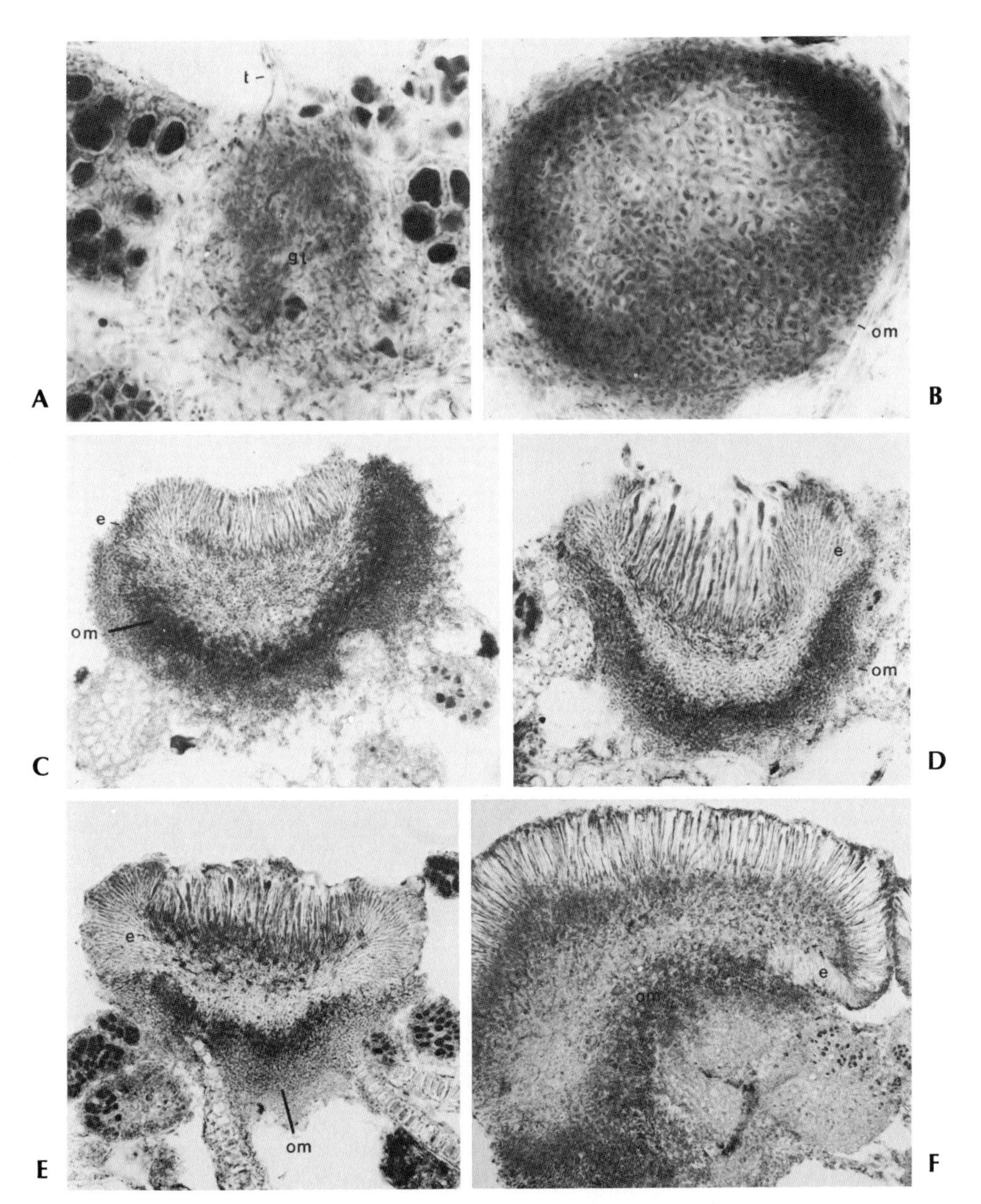

Fig. 10.28 A–F. Development of apothecium in *Lecidea berengeriana* (microtome sections). **A** Primordium composed of generative tissue enclosing remains of ascogonium and trichogyne. X520. **B** Primordium surrounded by dark-pigmented mantle layer. X570. **C** Young apothecium, hymenium, and subhymenium enclosed by the darkly pigmented outer mantle. X160. **D** Later stage demonstrating the development of the excipulum. X210. **E** and **F** Mature apothecia, excipulum well developed and arching over the dark mantle layer. X160 and X130. *e* excipulum, *gl* gelatin mucilage, *om* outer mantle, *t* trichogyne. (Microtome sections.)

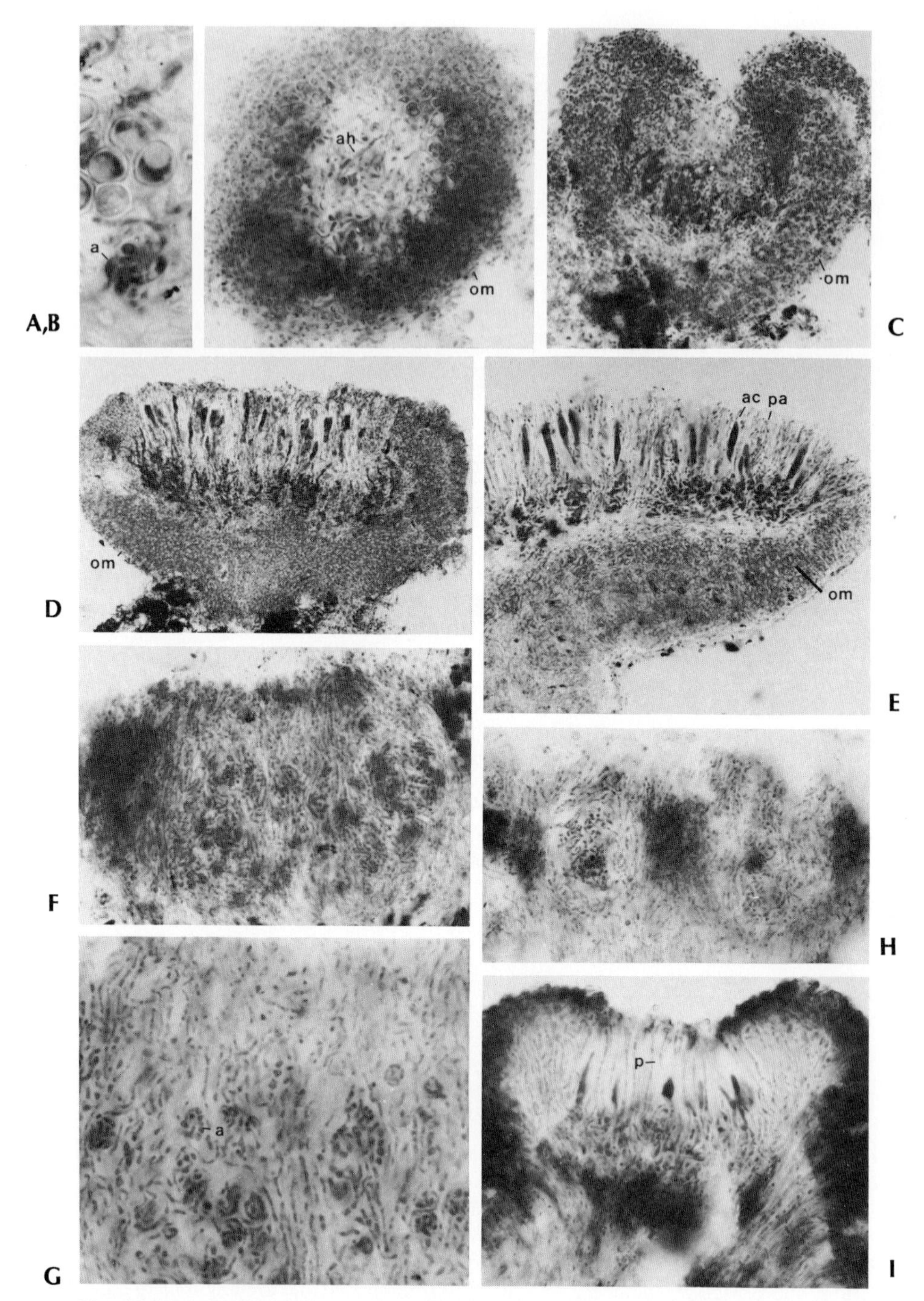

Fig. 10.29 A–I. Development of apothecium in *Lecidea*. **A** *Lecidea granulosa*, spirally contorted ascogonium. X1000. **B** *Lecidea berengeriana*, primordium with ascogenous hyphae and dark pigmented mantle layer. X380. **C–E** *L. granulosa*. **C** Young opening

much smaller than those in the thallus (cf. Fig. 10.31, B, F). This corresponds to the difference in size of the hymenial algae in pyrenocarpous lichens. Whether the apothecial algae in *Sporopodium* are ejaculated together with the muriform ascospores, as are the hymenial algae in *Endocarpon* Hedwig and *Staurothele* Norman, is not known. The main lines of development in *Lecidea* and *Sporopodium* described here are summarized in Figure 10.32.

The first stages of development in *Bacidia* de Notaris in Zahlbruckner (Jahns, 1970a) closely resemble the ascocarp ontogeny of *Lecidea vernalis*. Hyphae develop into a primordium bearing proparaphyses at the apex (Fig. 10.33, A–C). Subsequently, the proparaphyses elongate irregularly to form a large cluster of loose hyphae, which resembles trichogynes projecting above the surface of the thallus (Fig. 10.33D). These hyphae later become closely aggregated again to form a hymenium and a narrow excipulum.

During this development, the underlying vegetative tissue enlarges into a stipe in *Bacidia sabuletorum* Schreber. In *Bacidia sphaeroides* (Dickson) Zahlbruckner, it grows upward and around the fruit body; it forms a pseudoexcipulum that does not contain algal cells (Fig. 10.33, E–G). Such a development type is strongly reminiscent of that seen in the family Stereocaulaceae. It has to be stressed, however, that the border between the excipulum and the pseudoexcipulum is difficult to distinguish in such small stipitate apothecia.

In the genus *Gomphillus* Nylander, the small fruit bodies appear to consist of a black apothecium borne on a stalk, and for this reason the genus was formerly included in the Cladoniaceae. In reality, however, the fruit bodies are not truly stipitate; instead they consist entirely of a high, narrow apothecium in which the height of the hymenium exceeds its diameter (Jahns, 1970a). It is the laterally developed pale excipulum that gives the spurious appearance of a fruit body stalk. The course of development is schematically illustrated in Figure 10.34. The spores are transversely multiseptate and equal the entire fruit body in length.

The developmental type with the production of proparaphyses is found not only in species of *Lecidea* and *Bacidia* but also in the Cladoniaceae and the Stereocaulaceae. The second developmental type characterized by a primordium with a darkly pigmented outer mantle has been observed so far only in species of *Lecidea* and in *Sporopodium*.

Cladoniaceae

The genera of the Cladoniaceae sensu stricto (Jahns, 1970a) are characterized by a duplex type of thallus, consisting of a thallus horizontalis and a thallus verticalis; it is the latter that bears the lecideine, globose apothecia. According to recent

---

ascocarp. X200. **D** Young apothecium. X200. **E** portion of old stipitate apothecium. X210. **F–I** *Lecidea fuscoatrata*. **F** and **G** Generative tissue with numerous spirally contorted ascogonia. X150 and X310. **H** Primordium with paraphysoids. X160. **I** Young apothecium. X380. *a* ascogonium, *ac* ascus, *ah* anchoring hyphae, *om* outer mantle, *p* paraphyses, *pa* paraphysoids. (Microtome sections.)

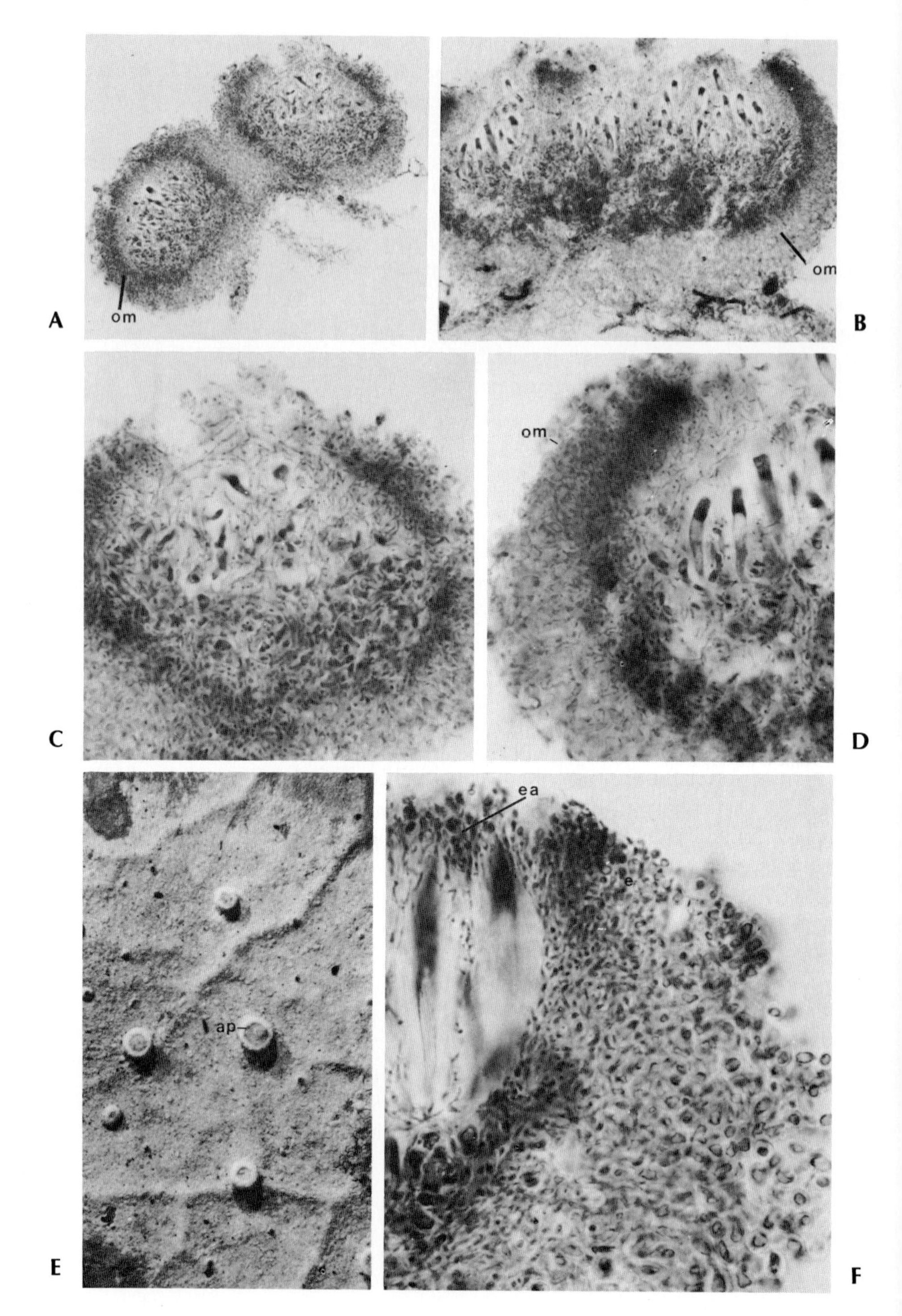
om
A
om
B
C
om
D
ap
E
ea
e
F

electron microscopical studies of *Cladonia macilenta* Hoffmann, the amyloid ascus is a variant of the *Lecanora* type and possibly provides a link to the *Peltigera* type (Honegger, 1978). The Cladoniaceae is considered to be derived from the Lecideaceae (Jahns, 1970a). In regard to the development of the ascocarp, they have in common the occurrence of proparaphyses. The formation of stipitate apothecia in *Bacidia* de Notaris is considered as a precursor of the podetium fruit body stalk found in the Cladoniaceae.

In most of the genera, the thallus verticalis is part of the fruit body, originating from the generative tissue, and can therefore be regarded as a podetium, although *Cladia* Nylander and *Thysanothecium* Berkeley and Montagne have pseudopodetia (Jahns, 1970a). The thallus horizontalis is often not persistent, and the podetia take over the function of the primary thallus. They always contain an algal layer or at least groups of algal cells and may or may not be corticated. The assimilative surface is frequently increased by the production of phyllocladia.

The Cladoniaceae can be derived from the Lecideaceae. Two homogeneous groupings, the Baeomycetaceae and the Stereocaulaceae, have been separated from the Cladoniaceae sensu lato, but the remaining genera show no great measure of uniformity among themselves and probably do not represent a single natural group (Jahns, 1970a; Henssen and Jahns, 1974). For example, it is the union of genera with podetia and those with pseudopodetia in the same family that is questionable.

The development of the podetia and fruit bodies in the species of *Cladonia* Wiggers has been repeatedly studied, leading to contradictory results (Jahns, 1970a; Henssen and Jahns, 1974).

In the genus *Cladonia,* two different types of fruit body development can be distinguished (Jahns, 1970a). In the first type the ascogonia are produced immediately by the generative tissue, which then extends at a second stage to form the podetium. In the second type a podetium is formed first, and then the ascogonia arise later on in its apex. The first type may be exemplified by *Cladonia caespiticia* (Persoon) Flörke. In this species, generative tissue originates as a cluster of strongly stainable parallel hyphae which breaks through the cortex and forms a small protuberance of the thallus horizontalis. The ascogonia with trichogynes are formed inside this primordium (Figs. 10.27, F, H, and 10.35, A–D). A conidium fused with the tip of a trichogyne has been seen (Fig. 10.37B). The generative tissue develops into an elongated podetium, inside which the ascogenous hyphae

Fig. 10.30 A–F. Development of apothecia in *Sporopodium.* **A–D** *Sporopodium leprieurii.* **A** Young ascocarps with dark-pigmented mantle layer. X170. **B** Young apothecium filled with paraphysoids and asci. X170. **C** Young ascocarp with bursting mantle layer. X400. **D** Marginal part of young apothecium in excipulum formation. X400. **E** *Sporopodium phyllocharis* var. *flavescens,* habit photograph of thallus and apothecia. X8. **F** *Sporopodium xantholeucum,* marginal portion of apothecium with massively developed excipulum and some cells of epithecial algae. X220. *ap* apothecium, *e* excipulum, *ea* epithecial algae, *om* outer mantle, *pa* paraphysoids. (All except E are microtome sections.)

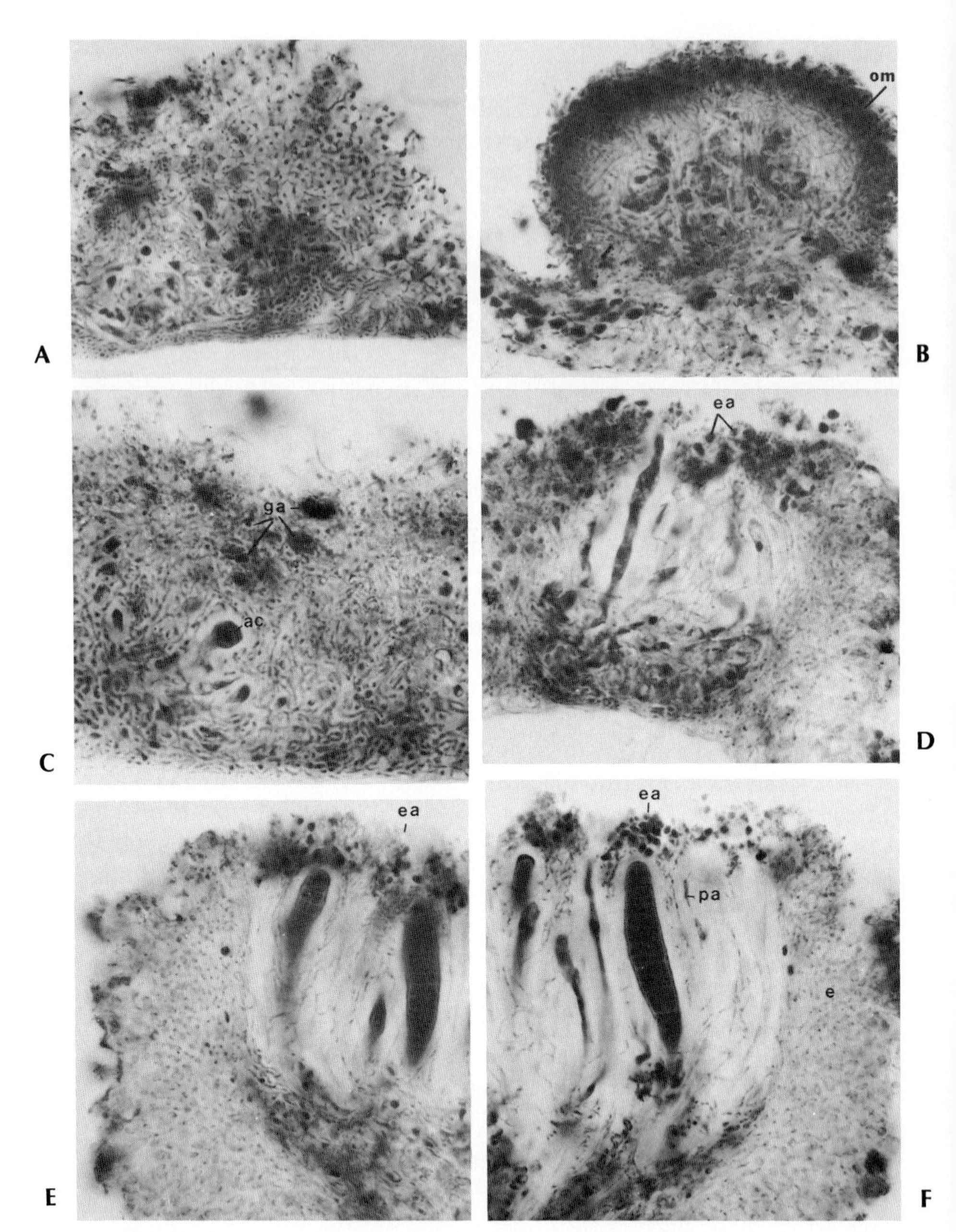

Fig. 10.31 A–F. Development of epithecial algae in *Sporopodium phyllocharis*. **A** Generative tissue with ascogenous hyphae. X400. **B** Primordium with dark-pigmented outer mantle layer and a paraphysoid centrum. X400. **C** Generative tissue with enclosed algal cells. X550. **D** Young ascocarp. X400. **E** and **F** Marginal portions of mature apothecia, the epithecium with scattered cells of epithecial algae. X400. *ac* ascogonium, *e* excipulum, *ea* epithecial algae, *ga* green algae, *om* outer mantle, *pa* paraphysoids. (Microtome sections.)

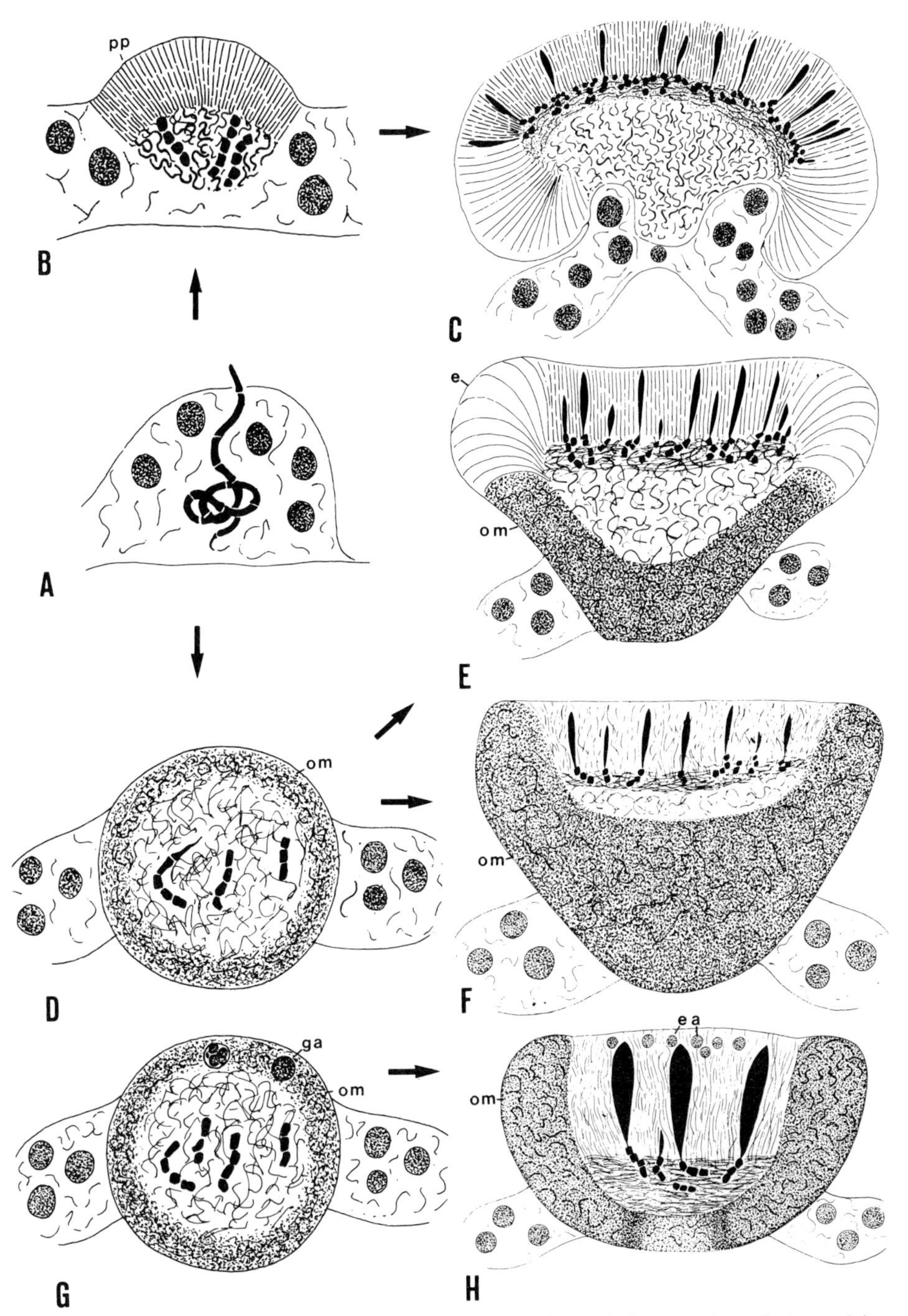

Fig. 10.32 A–H. Development of apothecia in *Lecidea* and *Sporopodium* (schematic). **A** Spirally contorted ascogonium. **B** Primordium with proparaphyses, **C** Mature apothecium of *Lecidea vernalis.* **D** Primordium with outer mantle layer. **E** Mature apothecium of *Lecidea berengeriana.* **F** Mature apothecium of *Lecidea granulosa.* **G** Primordium of *Sporopodium* with outer mantle layer enclosing algal cells. **H** Mature apothecium of *Sporopodium* with algal cells in epithecium. *e* excipulum, *ea* epithecial algae, *ga* green algae, *om* outer mantle, *pp* proparaphyses.

proliferate upward simultaneously. At the apex of the podetium, when it is fully grown, they arrange themselves in a layer and, at the same time, the paraphyses of the hymenium are formed above them (Figs. 10.27G and 10.37, A, C). They are distinctly septate when mature, and the asci grow up between them (Figs. 10.35, F, G, and 10.37, D, E). Occasionally, the podetia are branched and an apothecium is produced in each of the tips (Fig. 10.35E). In the normal development, the hymenium also becomes secondarily divided. The inner tissue of the podetium ruptures by the elongation process, and further enlargement causes the split of the fertile parts into different hymenia (Fig. 10.27, G, I). The podetium is corticated at the base (Fig. 10.37F). Algal cells may be pushed up by the vertical growth of the generative tissue and enclosed by these hyphae, but parts of

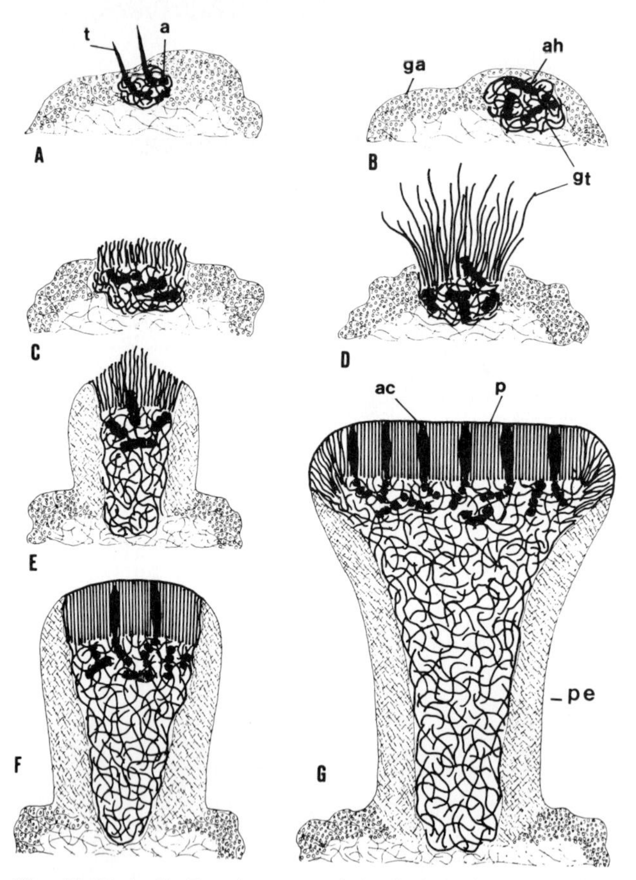

Fig. 10.33 A–G. Development of the fruit body in *Bacidia sphaeroides* (schematic). **A** Generative tissue with ascogonia. **B** Generative tissue with ascogenous hyphae. **C** Stage with proparaphyses. **D** Elongation of proparaphyses. **E** Stipitate fruit body with hymenium. **F** and **G** Young and mature stipitate apothecium. *a* ascogonium, *ac* ascus, *ah* ascogenous hyphae, *ga* green algae, *gt* generative tissue, *p* paraphyses, *pe* pseudoexcipulum, *t* trichogyne. (A–G after Jahns, 1970a.)

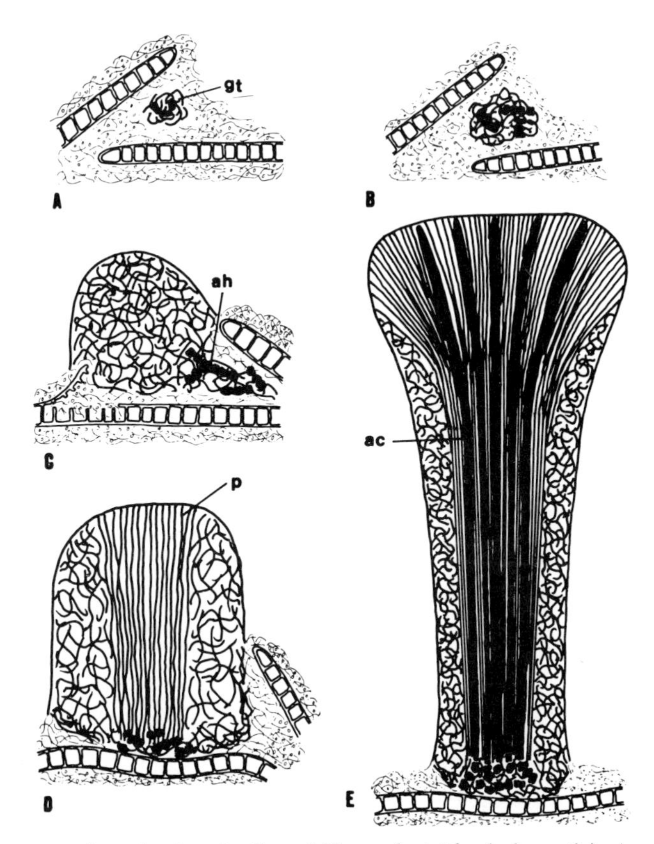

Fig. 10.34 A–E. Development of apothecium in *Gomphillus calycioides* (schematic). **A** Generative tissue with remains of ascogonium. **B** Primordium with ascogenous hyphae. **C** Upward growing primordium. **D** Young ascocarp. **E** Mature ascocarp. *ac* ascus, *ah* ascogenous hyphae, *gt* generative tissue, *p* paraphyses. (A–E after Jahns, 1970a.)

the podetium may be also entirely free of algal cells (Figs. 10.37, F, G, and 10.39A).

Examples of the second type of development are *Cladonia floerkeana* (Fries) Sommerfeldt (Fig. 10.36) and *Cladonia hypoxantha* Tuckerman (Fig. 10.38). In these species, the generative tissue is formed in the same manner as in *C. caespiticia*. The difference lies in the fact that no ascogonia are produced in the young primordium, which initially undergoes elongation to form a fully developed podetium. Ascogonia are first produced at the tip of the podetium (Figs. 10.36D and 10.38B). A limited vertical growth of the podetium tip and the simultaneous proliferation of the ascogenous hyphae follow the formation of ascogonia in *C. hypoxantha* (Fig. 10.38D). The ascogenous hyphae then grow horizontally, and the terminal hyphae develop into paraphyses (Fig. 10.38F). The final stages of development are the equivalent of the first type.

The podetium in *C. hypoxantha* has the same structure as in *C. caespiticia*. A cortical structure is only formed at the podetium base (Fig. 10.38, A–E). The pushing up of algal cells by the vertically growing hyphae of the generative tissue is illustrated in Figure 10.38A. The algal cells are stimulated to multiply and then become surrounded by hyphal branches. The hyphae may develop into phyllocladia having the same dorsiventral structure with a well-developed upper cortex as the squamules of the primary thallus (Figs. 10.38E and 10.39, C, F, G). In other species, such as *Cladonia crispata* (Acharius) Flotow, the podetium may have a well-developed outer cortex consisting of reticulately connected hyphae as well as an inner layer of deviating structure (Fig. 10.39B). In *Cladonia digitata* (Linnaeus) Hoffmann, the podetium not only takes over the assimilative function of the primary thallus but also that of the vegetative propagation by the production of soredia (Fig. 10.39, D, E). In most cases, the algal layer of the podetia probably develops from algal cells of the thallus horizontalis, as described above. Scattered algal cells are carried upward together with the young generative tissue. In a few cases, however, it appears that free-living algae from the environment may also be entrapped and incorporated (Jahns and Beltman, 1969).

The genus *Cladia* was long regarded as a subgenus of *Cladonia*. However, in contrast to the latter, it has pseudopodetia, which develop from a granulose or squamulose thallus horizontalis. One of its noteworthy characters is the produc-

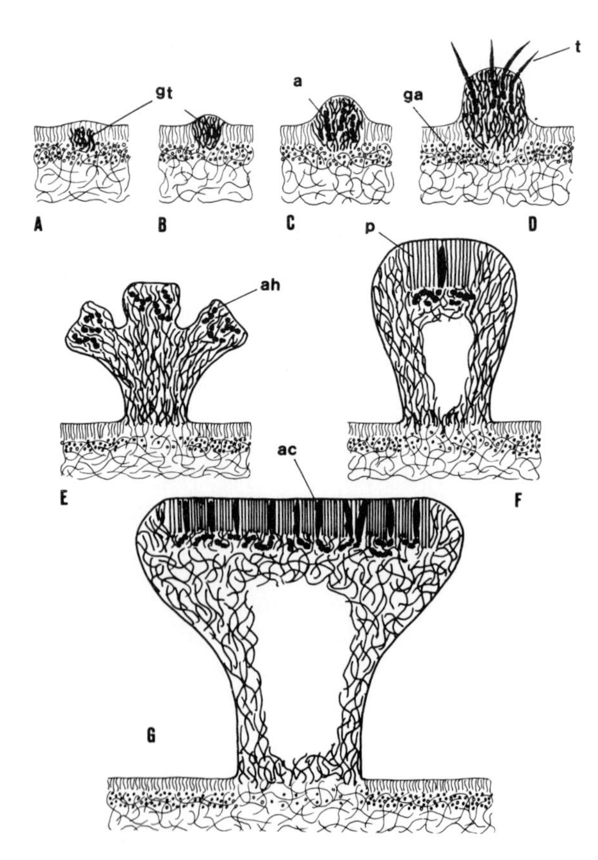

Fig. 10.35 A–G. Development of fruit body in *Cladonia caespiticia* (schematic). **A** Generative tissue formed above the algal layer. **B** Somewhat later stage. **C** Primordium with ascogonia in differentiation. **D** Primordium containing ascogonia with trichogynes. **E** Divided podetium. **F** Young fruit body. **G** Mature apothecium upon hollow podetium. *a* ascogonium, *ac* ascus, *ah* ascogenous hyphae, *ga* green algae, *gt* generative tissue, *p* paraphyses, *t* trichogyne. (A–G after Jahns, 1970a.)

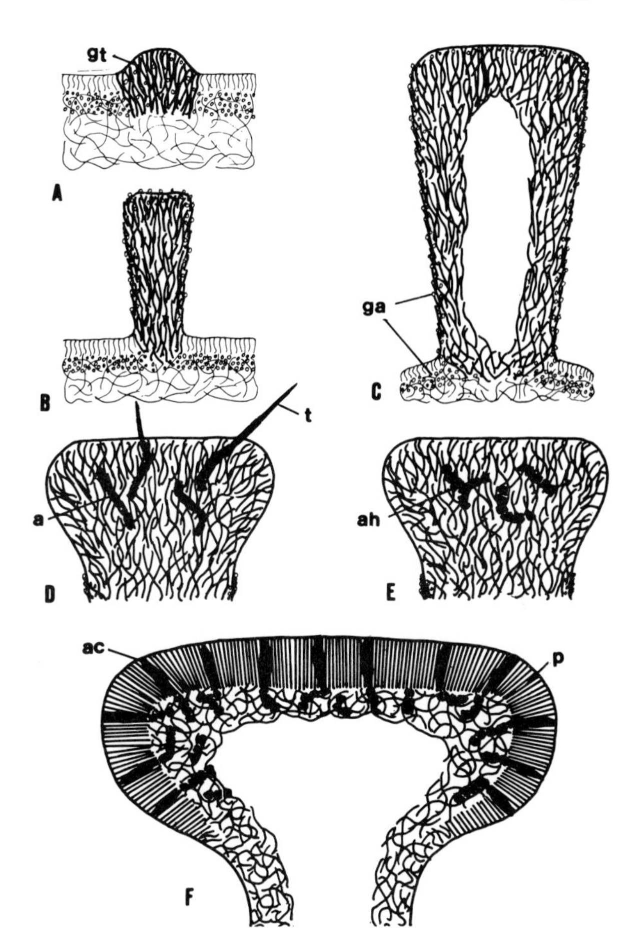

Fig. 10.36 A–F. Development of apothecia in *Cladonia floerkeana* (schematic). **A** Generative tissue. **B** Development of podetium. **C** Hollow podetium. **D** Ascogonia with trichogynes formed in the podetium tip. **E** Tip of podetium with ascogenous hyphae. **F** Mature apothecium. *a* ascogonium, *ac* ascus, *ah* ascogenous hyphae, *ga* green algae, *gt* generative tissue, *p* paraphyses, *t* trichogyne. (A–F after Jahns, 1970a.)

tion of proliferating apothecia (Fig. 10.40). They are found on the apices of the pseudopodetium branches. In *Cladia aggregata* (Swartz) Acharius, the development begins with a weft of generative tissue that protrudes through the thallus cortex (Jahns, 1970b). The hyphae orientate vertically to form an apothecium-like structure comprised of palisade hyphae. Ascogonia are differentiated first at this stage; their trichogynes project above the level of the palisade layer. The ascogenous hyphae developing from the ascogonial cells lie among the palisade hyphae. A paraphysate hymenium arises above the ascogenous hyphae, and the asci grow up between the paraphyses. The differentiation of the final hymenium is restricted to these areas only if parts of the palisade layer contain ascogenous hyphae, and then small apothecia are produced upon the primarily formed disc (Fig. 10.40B). However, if no ascogonia are formed, a second fruit body of palisade hyphae emerges from the first. This process may be repeated until the sexual system appears. The ascocarp ontogeny of *C. aggregata* is of special interest because the apothecia are formed by the generative tissue without any stimulation by the sexual system.

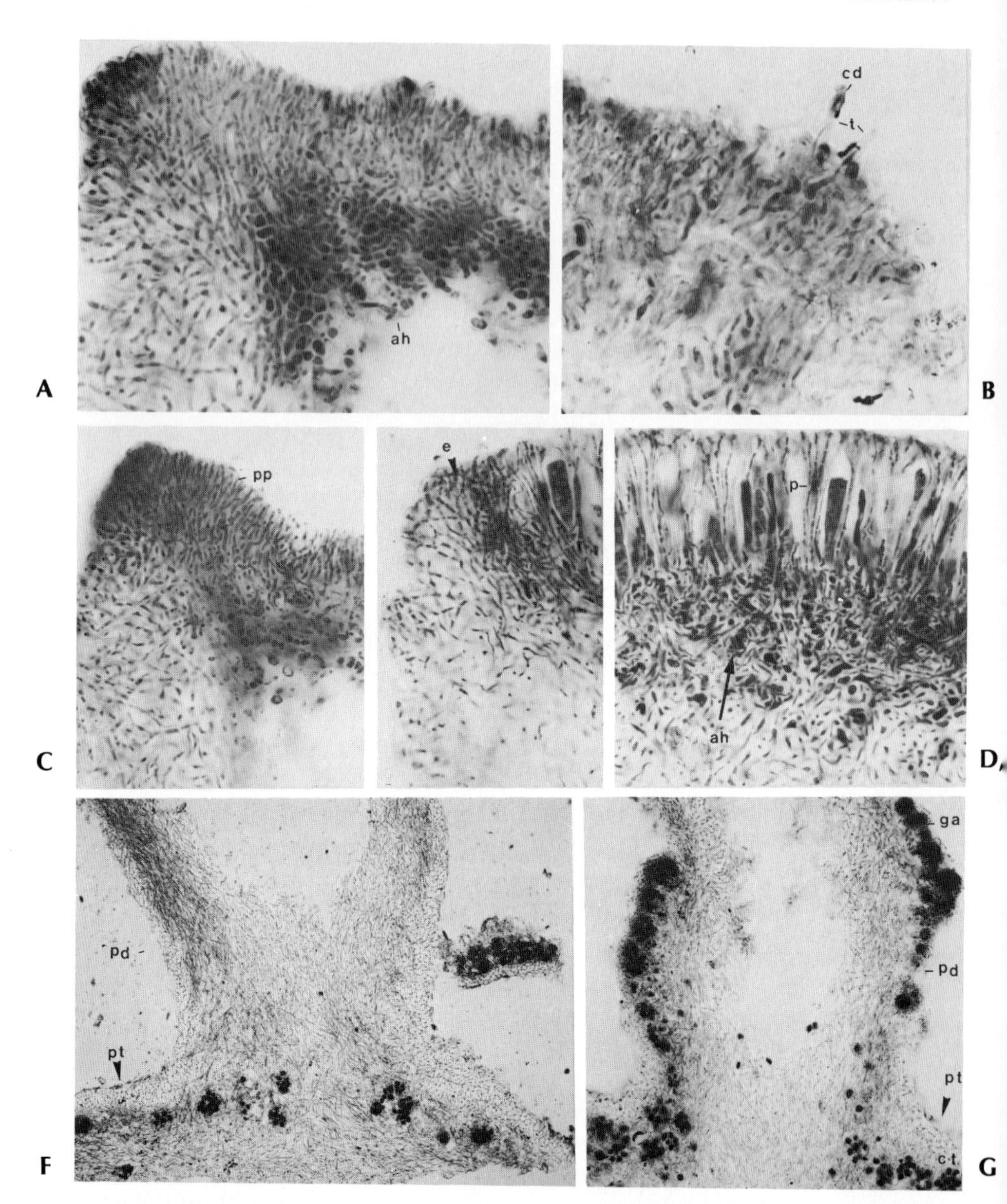

Fig. 10.37 A–G. Development of apothecium in *Cladonia caespiticia*. **A** Tip of podetium with differentiating paraphyses and ascogenous hyphae. X270. **B** Primordium with ascogonia; a conidium fused with the tip of a trichogyne. X220. **C** Tip of podetium with proparaphyses and ascogenous hyphae. X270. **D** Margin of mature apothecium. X270. **E** Portion of apothecium with hymenium and hypothecium. X270. **F** Primary thallus and lower part of hollow podetium free of algal cells. X75. **G** Podetium with algal layer. X110. *ah* anchoring hyphae, *cd* conidia, *ct* cortex, *e* excipulum, *ga* green algae, *p* paraphyses, *pd* podetium, *pp* proparaphyses, *pt* primary thallus. (Microtome sections.)

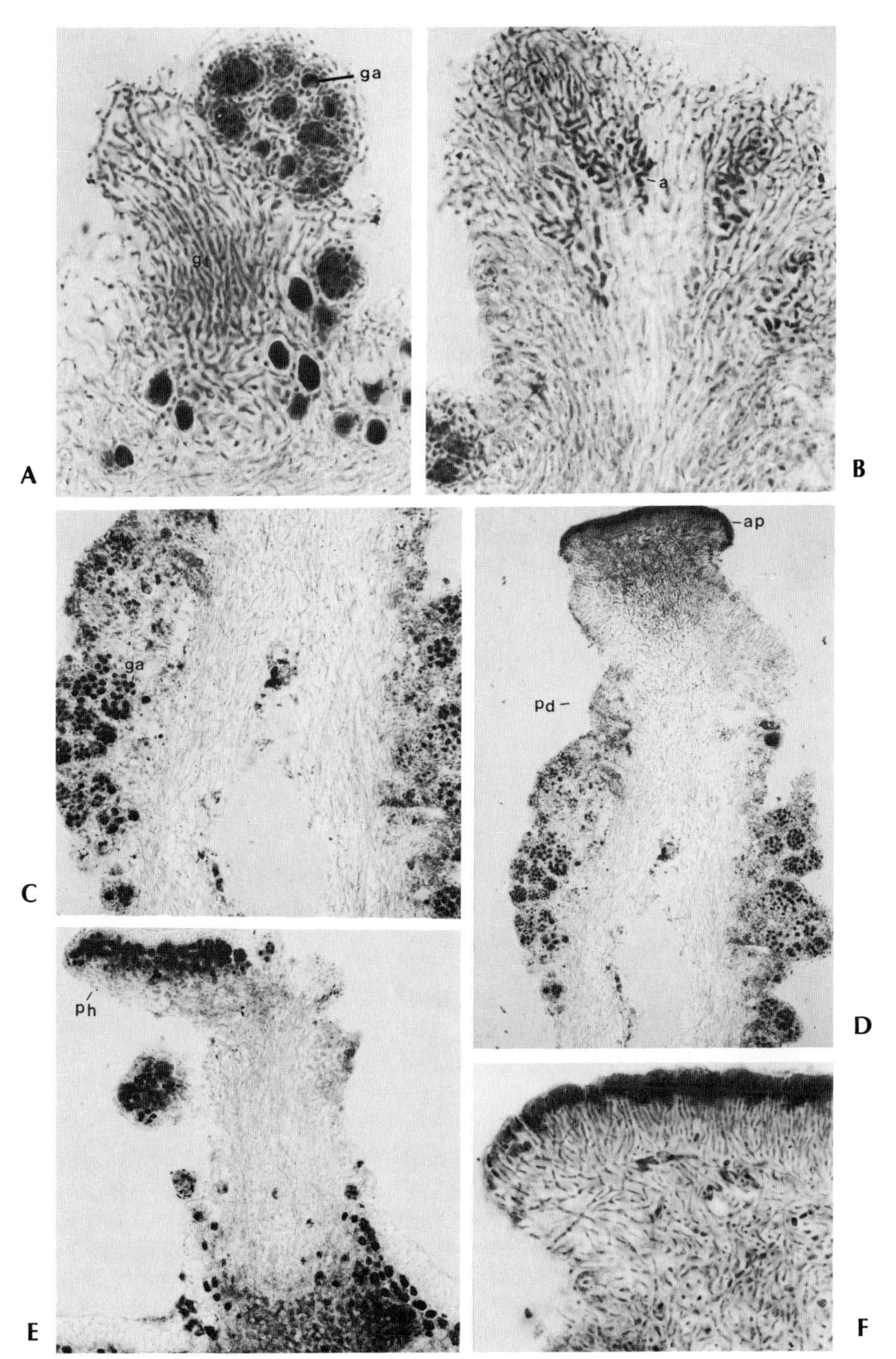

Fig. 10.38 A–F. Development of fruit body in *Cladonia hypoxantha*. **A** Generative tissue pushing up algal cells. X340. **B** Tip of podetium with ascogonia. X340. **C** Central part of hollow podetium. X115. **D** Upper part of podetium with young ascocarp. X55. **E** Fragment of podetium bearing phyllocladium. X115. **F** Young ascocarp with differentiating paraphyses and horizontally extending ascus. X300. *ap* apothecium, *ga* green algae, *pd* podetium, *ph* phyllocladium. (Microtome sections.)

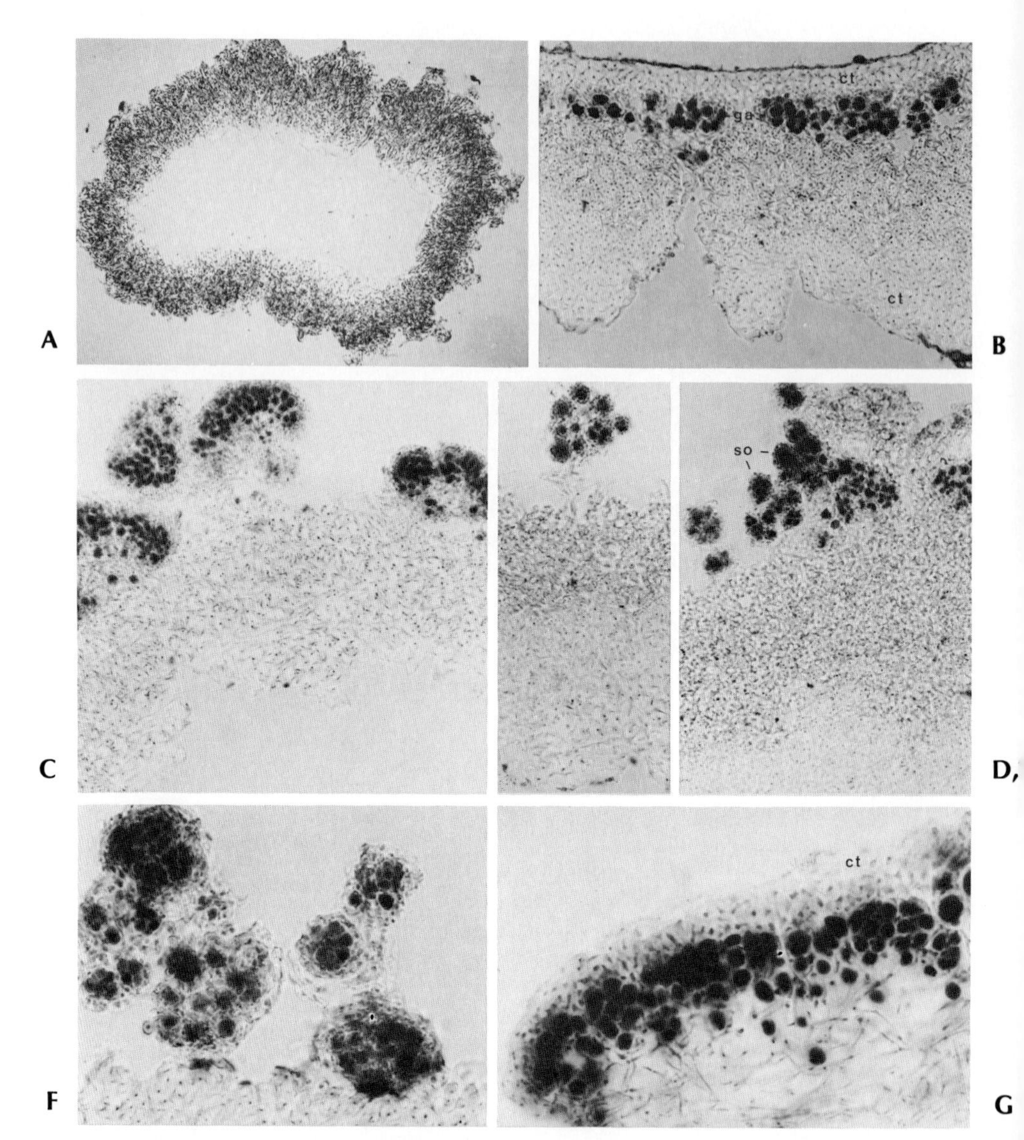

Fig. 10.39 A–G. Structure of podetia in *Cladonia* species. **A** Tangential section of *Cladonia caespiticia,* no cortex and no algal layer developed. X60. **B** Tangential section of *Cladonia crispata,* podetium with outer and inner cortical structure and algal layer. X120. **C** Tangential section of *Cladonia hypoxantha,* early stages of the development of phyllocladia, no cortical structure present. X130. **D** and **E** Tangential section of *Cladonia digitata,* production of soredia. X130 and X120. **F** and **G** Tangential section of *C. hypoxantha.* **F** Soredia. X260. **G** Phyllocladium. X300. *ct* cortex, *ga* green algae, *so* soredium.

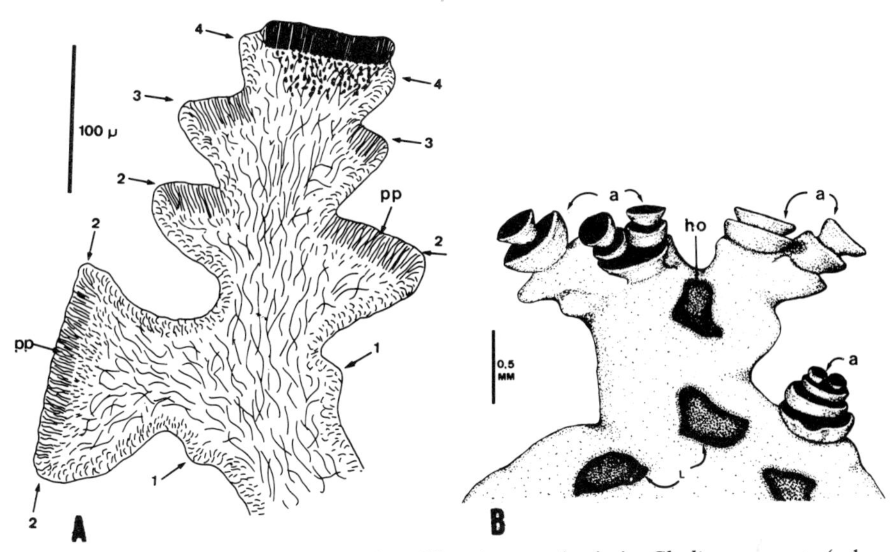

Fig. 10.40 A and B. Development of proliferating apothecia in *Cladia aggregata* (schematic). **A** Longitudinal section of thallus bearing proliferating apothecia. **B** General view of proliferating podetia. *a* ascogonium, *ap* apothecium, *ho* hollow, *pp* proparaphyses, *1–4* levels of proliferation. (A–B after Jahns, 1970b.)

Gymnoderma Nylander has small podetia (Jahns, 1970a). Generative tissue develops in the lower part of thallus near the tip of a lobe (Fig. 10.41). Ascogonia with trichogynes are formed. The generative tissue extends to a podetium, which in *Gymnoderma lineare* (Evans) Yoshimura and Sharp is finally surrounded by a cortical structure.

Glossodium Nylander and *Thysanothecium* are very similar in habitus but differ in their ontogeny. *Glossodium* (Fig. 10.42) has true podetia, whereas *Thysanothecium* develops pseudopodetia (Fig. 10.43). In both genera, the apothecia are perpendicularly formed at the apex of the thallus verticalis and become displaced to the side of the stalk as growth proceeds so that the thallus verticalis assumes a spatulate appearance.

Stereocaulaceae

The Stereocaulaceae are characterized by a duplex type of thallus consisting of a primary thallus horizontalis from which a thallus verticalis bearing the apothecia develops secondarily (Jahns, 1970a; Henssen and Jahns, 1974). The apothecia are surrounded by a pseudoexcipulum (superlecideoid apothecia in the sense of Frey, 1936) or contain no excipular structure at all. The thallus verticalis is always a solid pseudopodetium. All genera of this family have a simultaneous symbiotic relationship with green and blue-green algae. In most of the genera, the latter are enclosed in external cephalodia situated on the thallus horizontalis or on pseudopodetia. Only in *Compsocladium* M. Lamb do the filamentous blue-green algae lie within the pseudopodetia among the medullary hyphae. The Ster-

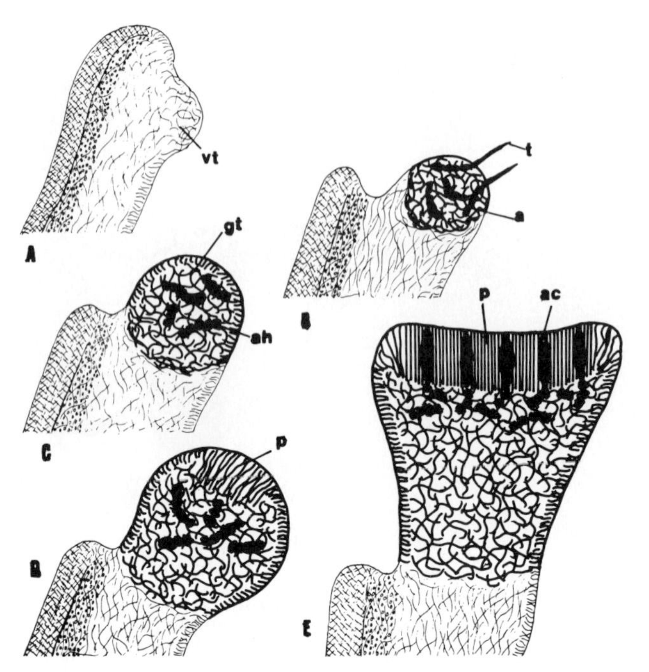

Fig. 10.41 A–E. Development of stipitate fruit body in *Gymnoderma linearis* (schematic). **A** Swelling of vegetative tissue of the lower side of the thallus. **B** Generative tissue with ascogonia. **C** Primordium with ascogenous hyphae. **D** Primordium with proparaphyses. **E** Mature fruit body on short podetium. *ac* ascus, *ah* ascogenous hyphae, *gt* generative tissue, *p* paraphyses, *t* trichogyne, *vt* somatic tissue. (A–E after Jahns, 1970a.)

eocaulaceae shows a phylogenetic connection with the Lecideaceae with the presence of proparaphyses, as does the Cladoniaceae.

The pseudopodetia of the Stereocaulaceae may be formed according to two different modes of development, which are not particularly correlated with distinct genera (Lamb, 1951; Jahns, 1970a). The first mode, called the enteropodial type, is found in the species of *Stereocaulon* Hoffmann that occur in the northern hemisphere and in the genus *Pilophorus* Th. Fries. Most of the *Stereocaulon* species of the southern hemisphere and the genera *Argopsis* Th. Fries and *Compsocladium* have the second development type, which is termed holostelidial.

In lichens with the enteropodial type (Fig. 10.44, F–H), the pseudopodetium derives from the squamuliform portion of the thallus horizontalis. Part of the medullary tissue beneath the algal layer (see Fig. 10.47A) begins to expand and grow upward. This medullary tissue is ordinary thalline rather than generative tissue in contrast to the podetium formation. The pseudopodetial initial thus formed breaks through the thalline squamule, divides it, and raises the fragmented portions of it up to its summit and sides. The algal layer and the squamulose phyllocladia of the pseudopodetia are derived from these remains of the thallus horizontalis.

A granular portion of the thallus horizontalis grows upward uniformly and in its entirety in lichens with the holostelidial type of pseudopodetial development

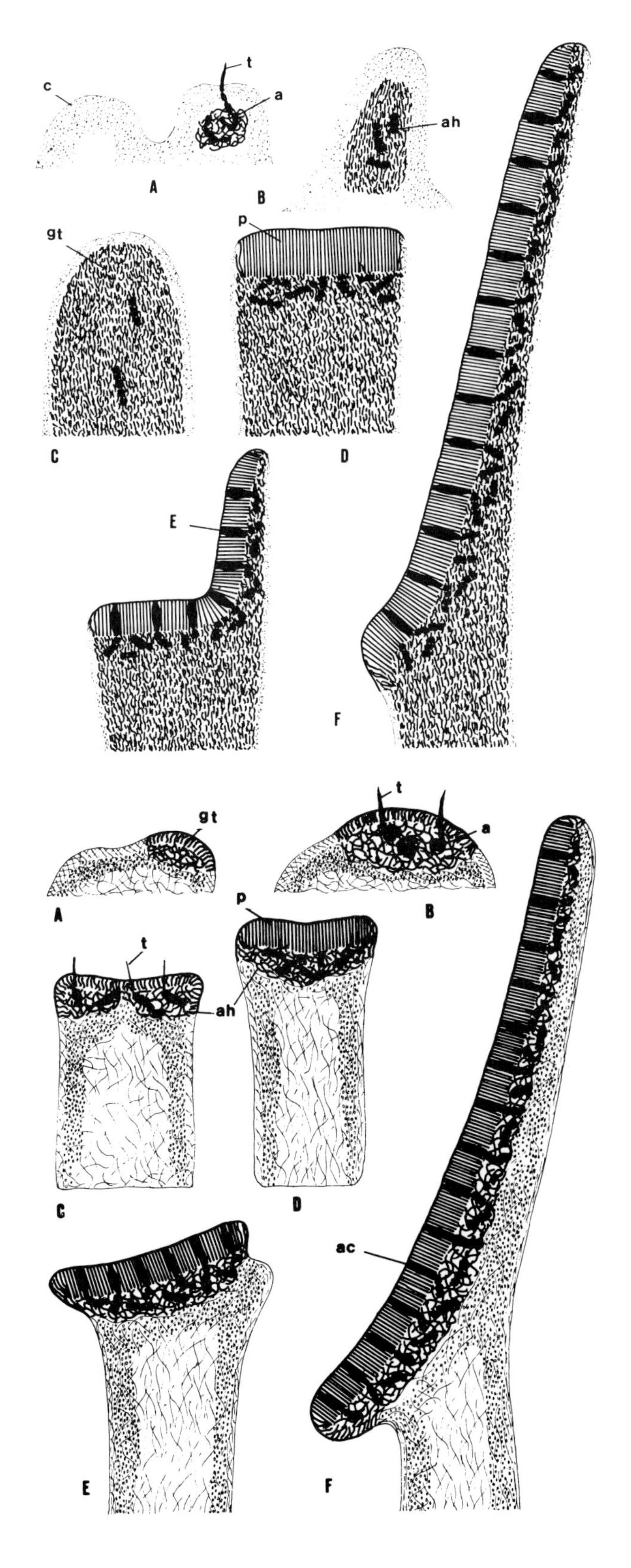

Fig. 10.42 A–F. Development of tongue-shaped apothecium in *Glossodium aversum* (schematic). **A** Generative tissue with ascogonia. **B** Primordium with ascogenous hyphae. **C** Young podetium with ascogenous hyphae growing up. **D** Tip of podetium with layer of paraphyses. **E** Young, and **F** old apothecium on tip of podetium. *a* ascogonium, *ah* ascogenous hyphae, *c* covering layer, *gt* generative tissue, *p* paraphyses, *t* trichogyne. (A–F after Jahns, 1970a.)

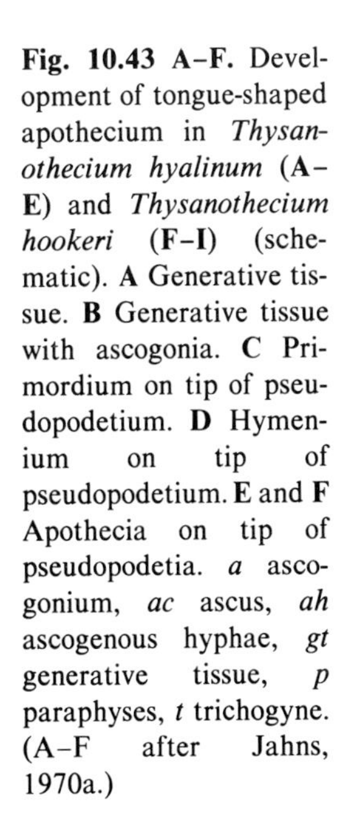

Fig. 10.43 A–F. Development of tongue-shaped apothecium in *Thysanothecium hyalinum* (**A–E**) and *Thysanothecium hookeri* (**F–I**) (schematic). **A** Generative tissue. **B** Generative tissue with ascogonia. **C** Primordium on tip of pseudopodetium. **D** Hymenium on tip of pseudopodetium. **E** and **F** Apothecia on tip of pseudopodetia. *a* ascogonium, *ac* ascus, *ah* ascogenous hyphae, *gt* generative tissue, *p* paraphyses, *t* trichogyne. (A–F after Jahns, 1970a.)

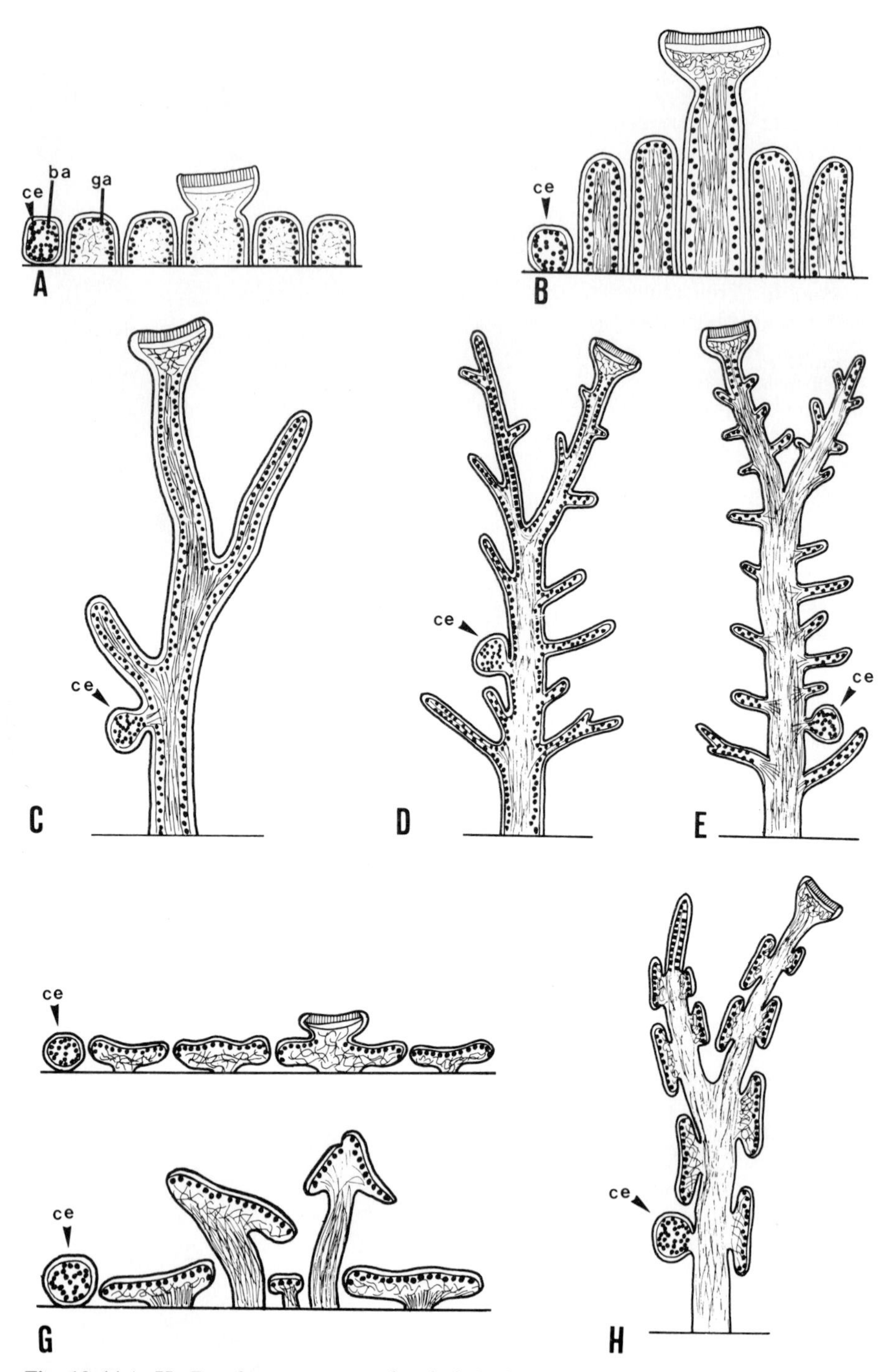

Fig. 10.44 A–H. Development types of podetia in *Stereocaulon* (schematic). **A–E** Holostelidial type of development. **F–H** Enteropodial type of development. For details see text. *ba* blue-green algae, *ce* cephalodium, *ga* green algae. (A–H after Lamb, 1951.)

(Fig. 10.44, A–D). A cylindrical pseudopodetium is produced that is completely clothed from the beginning with an algal layer. The phyllocladia are not squamulose in form in this type of development but terete and cylindrical, being short lateral branchlets of the pseudopodetium.

The development of the Stereocaulaceae has been studied by Jahns (1970a). The first stages of apothecial development are found in fully grown pseudopodetia. They occur in the apices of branches or in the phyllocladia. The subgenera *Stereocaulon* and *Holostelidium* Lamb of the genus *Stereocaulon* show the same development pattern of the ascocarp in spite of a basic difference in the formation of the pseudopodetia (Figs. 10.45 and 10.46). The ascogonia are formed first and only subsequently become enveloped by the generative tissue. The ascogonia are spirally coiled and the trichogynes relatively thin. In *Stereocaulon dactylophyllum* Flörke, the ascogonia may be deeply immersed in the thallus and the ascocarp then is hemiangiocarpous. Proparaphyses are formed preceeding the differentiation of the paraphyses. The cortex of the thallus begins to grow up around the sides of the primordium initial, as the hymenium is being formed, and in this way, the young fruit body is surrounded by a pseudoexcipulum that superficially resembles a proper margin. In *Stereocaulon condensatum* Hoffmann, for example, the vertical growth of cortical tissue is pronounced, and a pedicel is formed that elevates the apothecium for some distance above the surface of the thallus (Fig. 10.45I). In old apothecia, the hymenium may be so strongly developed that it covers the pseudoexcipulum; in such fruit bodies the proper margin is no longer visible. In *Stereocaulon implexum* Th. Fries, the apothecium is surrounded by a massively developed plicated pseudoexcipulum (Fig. 10.46H). The two species of the section *Redingeria* Lamb differ by first producing the generative tissue and then the ascogonia. In the genus *Argopsis* the ascocarp develops as in *Stereocaulon,* and the mature flat apothecium is surrounded by a thick pseudoexcipulum. *Compsocladium* differs in having globose fruit bodies in which the pseudoexcipulum is only poorly developed.

In the genus *Pilophorus,* neither excipulum nor pseudoexcipulum is formed. The generative tissue develops before the ascogonia. The latter consist of straight rows of cells in contrast to the spirally twisted ascogonia found in the other genera. Up to a hundred or more ascogonia bearing trichogynes may be produced in one primordium (Jahns, 1973).

In *Pilophorus robustus* Th. Fries, the fusion of conidia with swollen tips of trichogynes has been observed (Jahns, 1970a). The course of development in *P. robustus* is illustrated in Figure 10.47. The generative tissue becomes darkly pigmented as the ascogenous hyphae develop and as paraphyses differentiate around the hemispherical primordium. A distinction between vegetative and generative tissue can be made by the pigmentation (Fig. 10.47F). The hymenium enlarges considerably and projects on all sides beyond the pseudopodetium. The outermost rim is composed of sterile paraphyses. An excipulum is not developed. The vegetative tissue enlarges simultaneously to form a columella. The delimitation between the subhymenial layers and the vegetative tissue of the columella always remains distinct. In other species, a pigmentation or an additional particular texture defines a distinct edge (Jahns, 1970c). In these instances, a columella may or may not be present (Fig. 10.48).

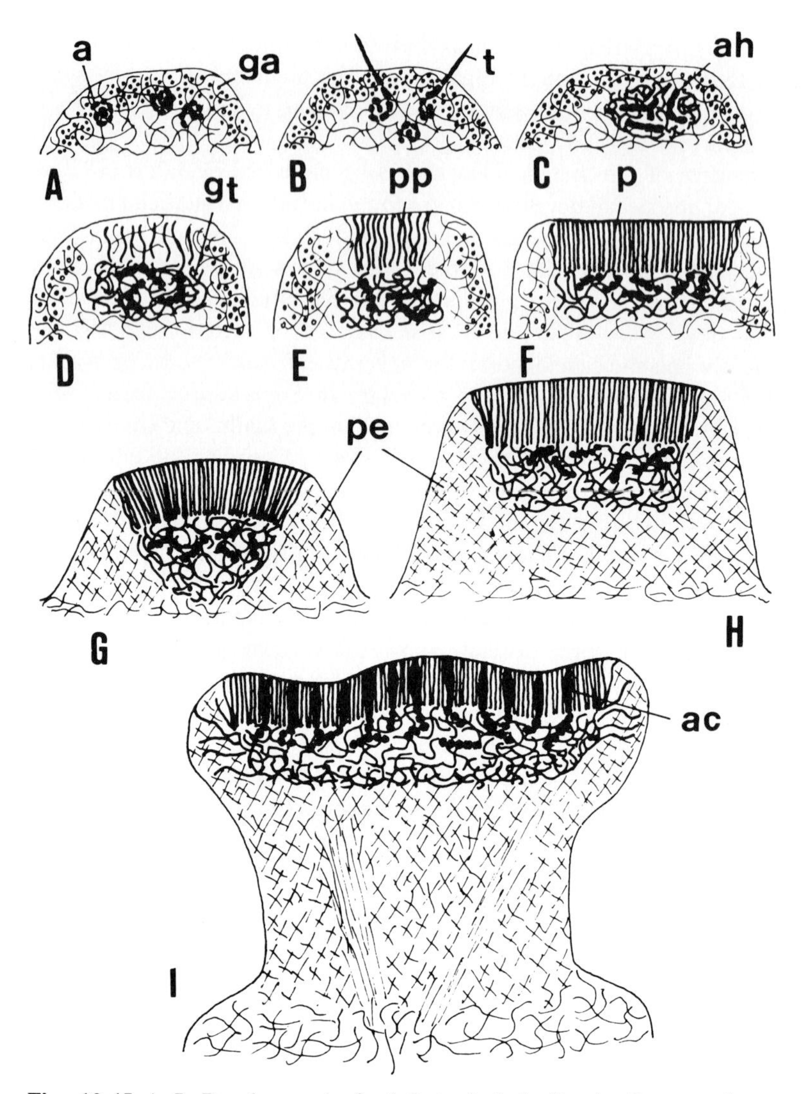

Fig. 10.45 A–I. Development of stipitate fruit bodies in *Stereocaulon condensatum* (schematic). **A** and **B** Spirally contorted ascogonia between thallus hyphae. **C** Primordium with generative tissue enclosing ascogenous hyphae. **D** and **E** Formation of proparaphyses. **F** Young hymenium with paraphyses. **G** and **H** Young ascocarp in tip of pseudopodetium. **I** Mature stipitate apothecium. *a* ascogonium, *ac* ascus, *ah* ascogenous hyphae, *ga* green algae, *gt* generative tissue, *p* paraphyses, *pe* pseudoexcipulum, *pp* proparaphyses, *t* trichogyne. (A–I after Jahns, 1970a.)

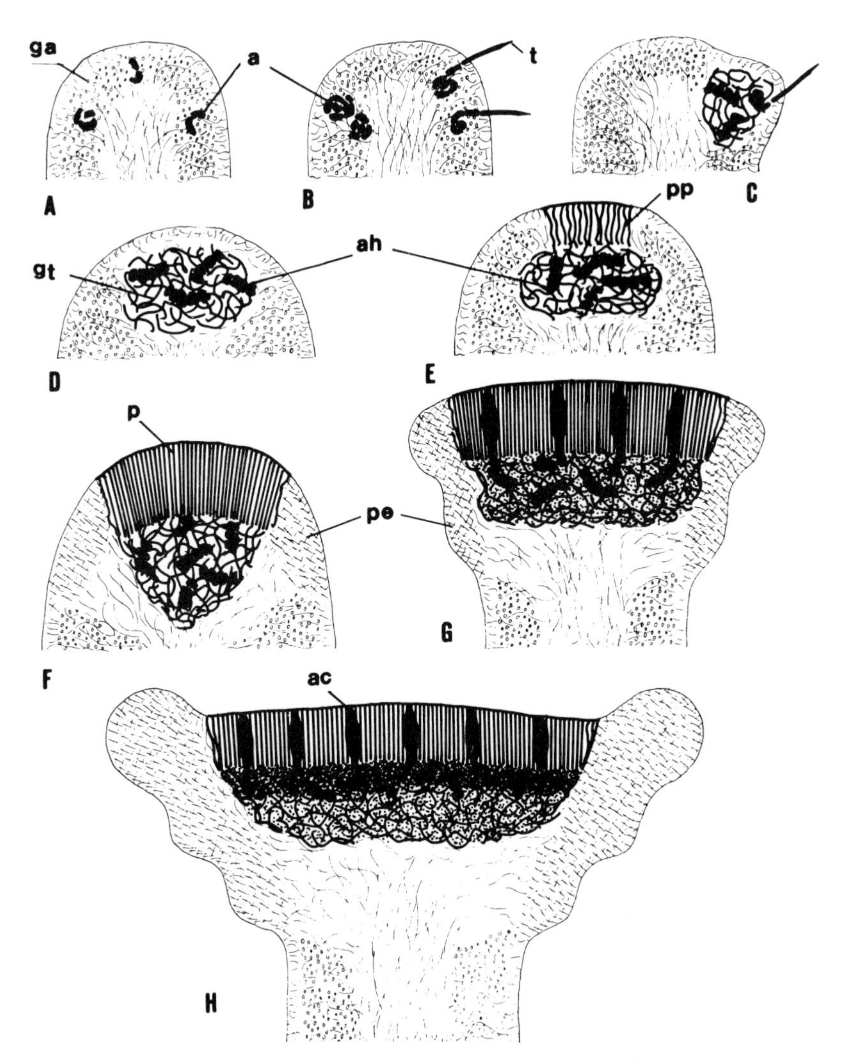

Fig. 10.46 A–H. Development of apothecium in *Stereocaulon implexum* (schematic). **A–C** Formation of ascogonia. **D** Generative tissue enclosing ascogenous hyphae. **E** Primordium with proparaphyses. **F** and **G** Young apothecia with pseudoexcipulum. **H** Mature fruit body with pseudoexcipulum. *a* ascogonium, *ac* ascus, *ah* ascogenous hyphae, *ga* green algae, *gt* generative tissue, *p* paraphyses, *pe* pseudoexcipulum, *pp* proparaphyses. (A–H after Jahns, 1970a.)

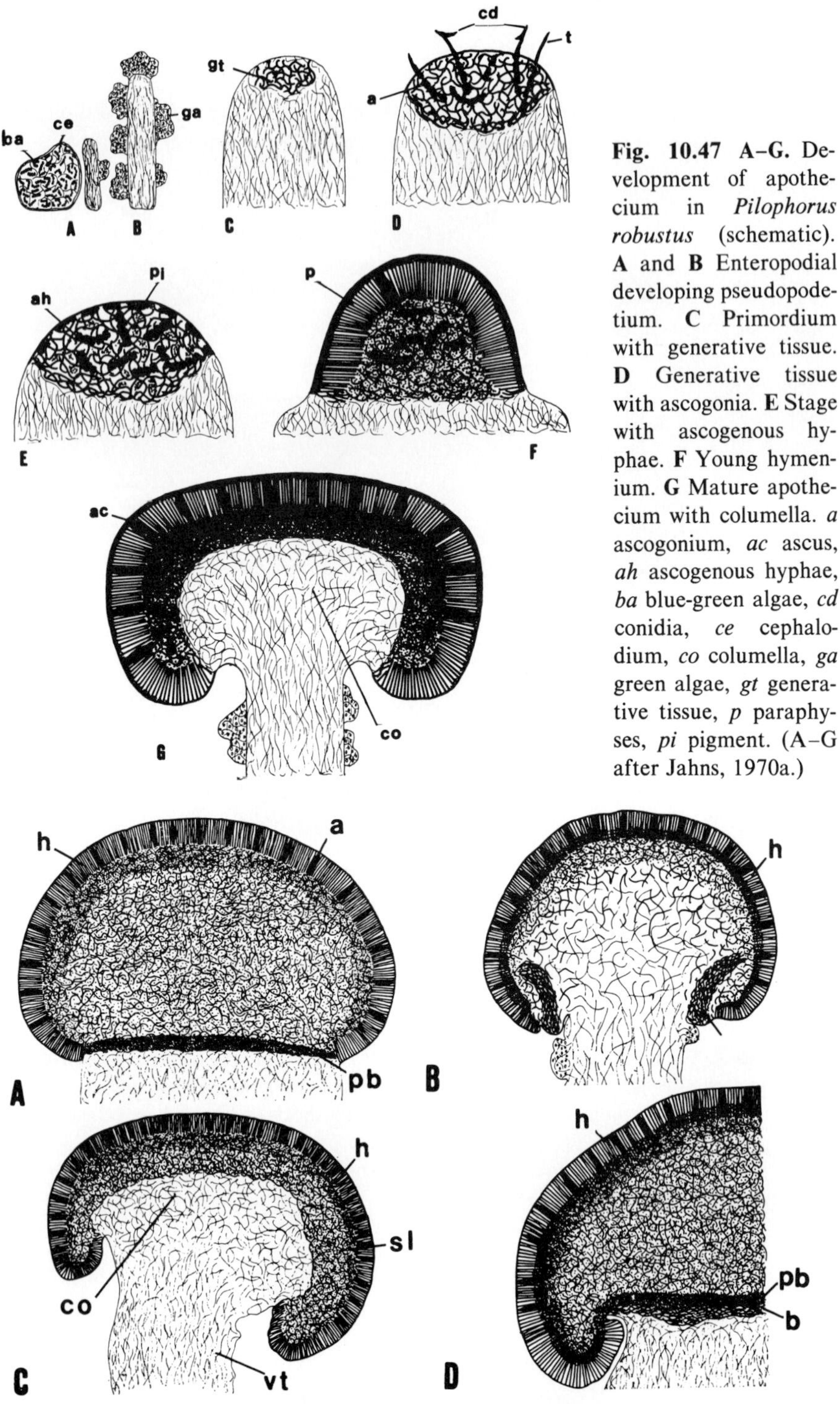

Fig. 10.47 A–G. Development of apothecium in *Pilophorus robustus* (schematic). **A** and **B** Enteropodial developing pseudopodetium. **C** Primordium with generative tissue. **D** Generative tissue with ascogonia. **E** Stage with ascogenous hyphae. **F** Young hymenium. **G** Mature apothecium with columella. *a* ascogonium, *ac* ascus, *ah* ascogenous hyphae, *ba* blue-green algae, *cd* conidia, *ce* cephalodium, *co* columella, *ga* green algae, *gt* generative tissue, *p* paraphyses, *pi* pigment. (A–G after Jahns, 1970a.)

Fig. 10.48 A–D. Apothecium structure in *Pilophorus* species. (schematic). **A** *Pilophorus cereolus.* **B** *Pilophorus acicularis.* **C** *Pilophorus robustus.* **D** *Pilophorus fibula. a* ascogonium, *b* boundary tissue, *co* columella, *h* hymenium, *pb* pigment boundary, *sl* subhymenial layer, *vt* somatic tissue. (A–D after Jahns, 1970c.)

Umbilicariaceae

The Umbilicariaceae remain isolated by their particular development of the fruit body. In the majority of the species, the discs of the lecideine or superlecideoid apothecia are provided with a central umbo or with ridges arranged either concentrically or in a star-shaped formation (Figs. 10.50, E–G). The family has been included in the suborder Lecanorineae (Henssen and Jahns, 1974) or has been placed in a separate suborder, the Umbilicariineae (Poelt, 1974). Based on the apothecial morphology, several genera have been distinguished (Scholander, 1934; Llano, 1950). This separation, however, is not justified in terms of the ascocarp ontogeny (Henssen, 1970), and the taxonomy of Frey (1936) is followed, which unites all species within the one genus, *Umbilicaria* Hoffmann.

The development of the pseudoexcipulum is best seen in *Umbilicaria rigida* (Du Rietz) Frey, a species in which the apothecial disc is undivided. Here, the development commences somewhat below the level of the algal layer with the production of generative tissue with several ascogonia (Fig. 10.49A). The uppermost hyphae of the generative tissue extend to form paraphyses (Figs. 10.49C and 10.50A). The neighboring hyphae of the thallus cortex begin to grow in a vertical direction (Figs. 10.49, B, D, E, and 10.51B). The development of the pseudoexcipulum may be asymmetrical at one side of the ascocarp, while at the other side sterile hyphae are growing upward from the generative tissue (Fig. 10.50B). The pseudoexcipulum has the same structure as the thalline cortex, consisting of reticulately branched hyphae with strongly gelatinizing walls. The ascogenous hyphae grow upward simultaneously with the elongating generative tissue (Fig. 10.50B). In the second phase of development, a growth zone arises as usual around the hymenium in which the formative layer produces paraphyses inward and a limited amount of radiating excipulary hyphae outward (Figs. 10.49F and 10.51C). The excipulum hyphae integrate with those of the pseudoexcipulum (Fig. 10.50, C, D). The hymenium may already be composed of fertile and sterile parts (Fig. 10.49, B, D) in *U. rigida* which has a flat disc. This developmental aspect is found combined in other species, with cessation of growth in some parts that leads to umbonate or gyrose apothecia. The main steps of such a developmental pattern in the apothecia of *Umbilicaria muehlenbergii* (Acharius) Tuckerman are illustrated in Fig. 10.51, D–F. In early stages, when the first paraphyses have been formed, the central part of these hyphae ceases to grow actively and becomes surmounted by the neighboring portions (Fig. 10.51D). This process is repeated several times and is accompanied by a considerable increase in height, as can be seen by comparing young and old apothecia (Fig. 10.51, D–F). Ascogenous hyphae and asci are restricted to the growing portions of the hymenium (Fig. 10.51E). Finally, the mature apothecia are deeply divided by grooves which separate the different gyri from each other (Fig. 10.51F). The hymenium is surrounded by a black excipulum in *U. muehlenbergii*. Apothecia of other species are provided with a well-developed annular excipulum consisting of radiating hyphae. The annular excipulum is later on connected by a texture, developed mainly by the formative layer in the lower part of the subhymenium in *Umbilicaria decussata* (Villars) Frey and *Umbilicaria cylindrica* (Linnaeus) Delise. The texture

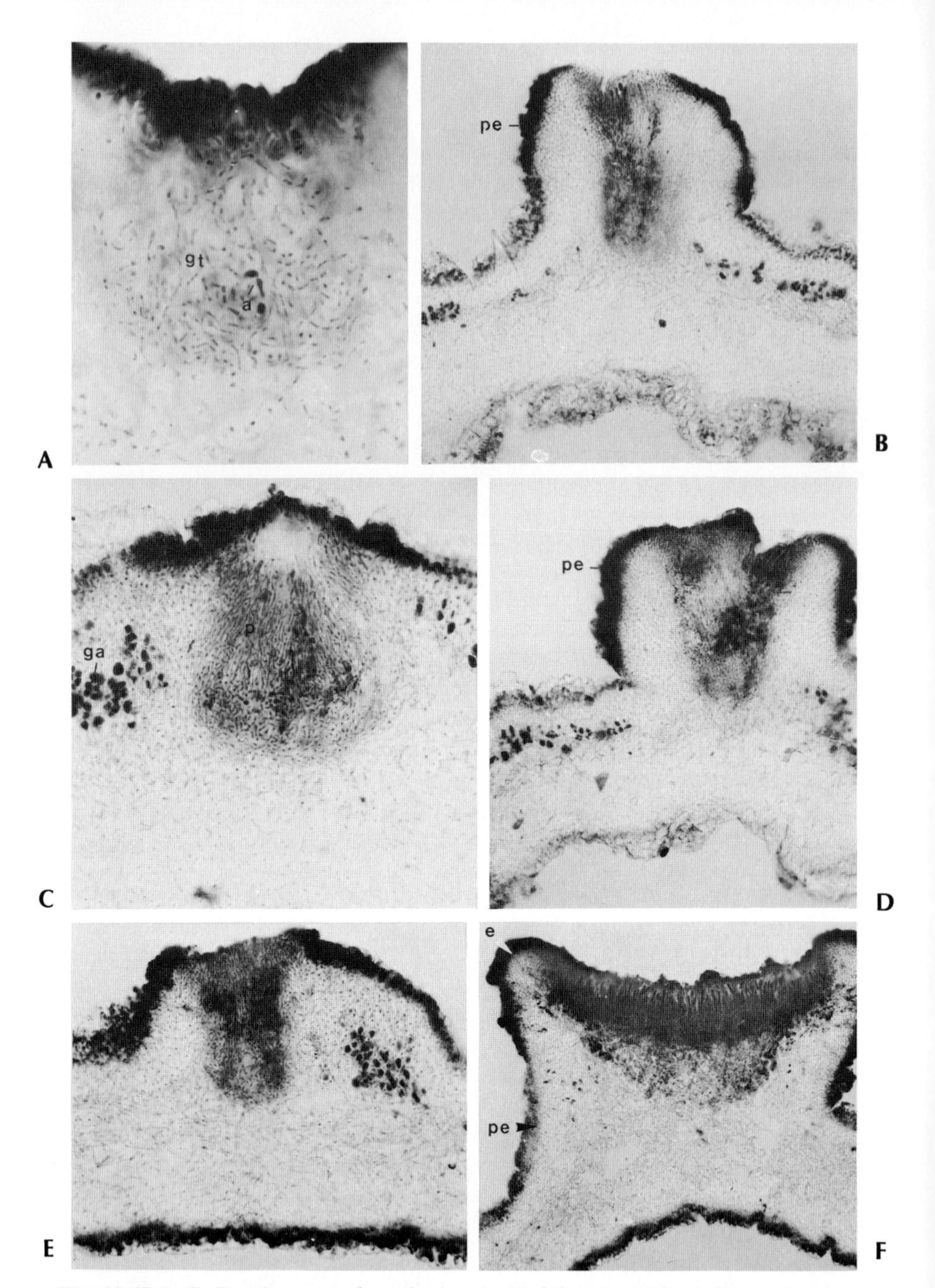

Fig. 10.49 A–F. Development of apothecium in *Umbilicaria rigida*. **A** Generative tissue with remains of ascogonium. X410. **B** Young asoccarp with pseudoexcipulum. X110. **C** Primordium with paraphyses. X230. **D** Young ascocarp with segregated parts of ascogenous hyphae. X110. **E** Primordium enclosed by vertical growing thallus cortex. X180. **F** Young apothecium. X90. *a* asocgonium, *ga* green algae, *gt* generative tissue, *p* paraphyses, *pe* pseudoexcipulum. (Microtome sections.)

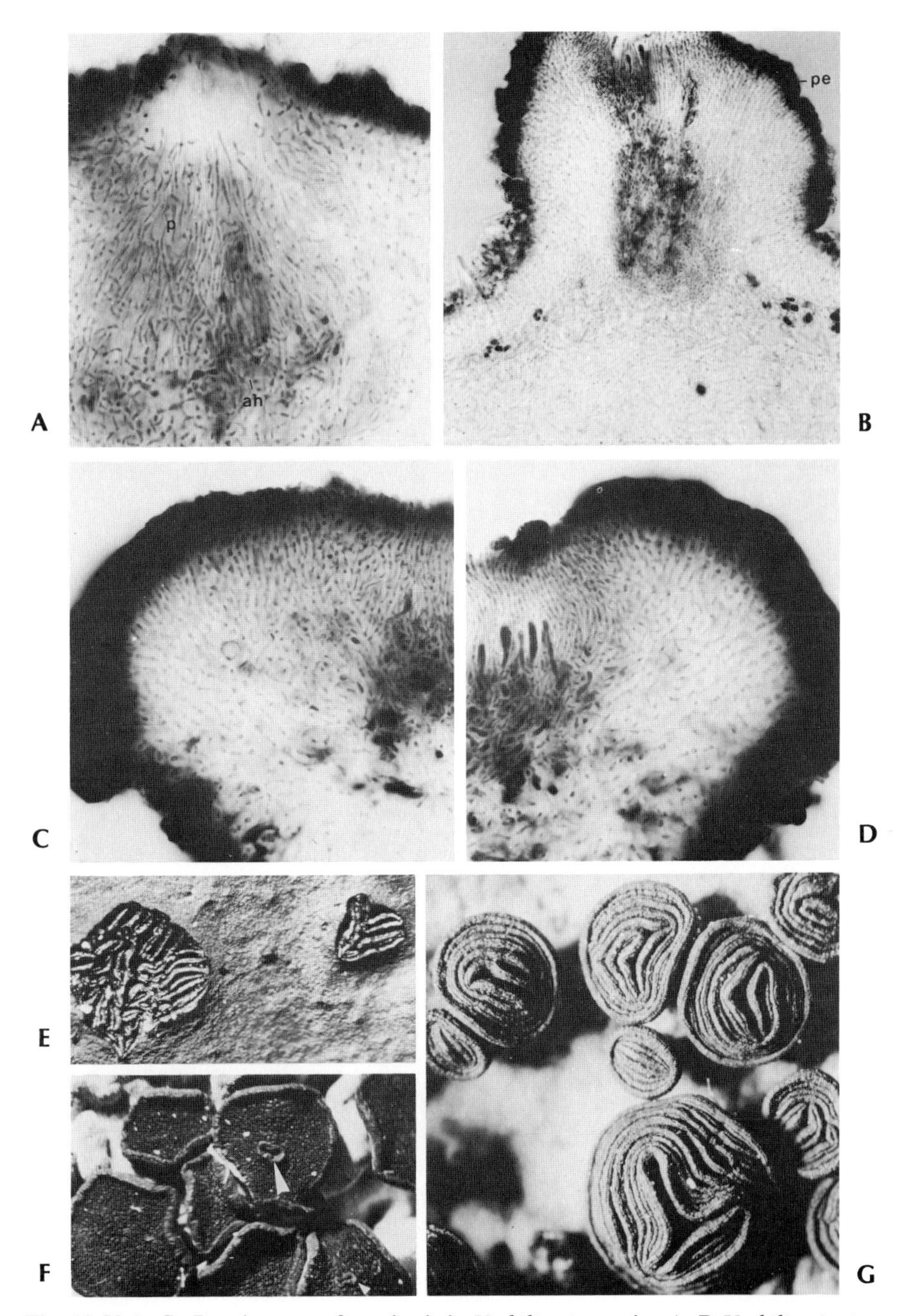

Fig. 10.50 A–G. Development of apothecia in *Umbilicaria* species. **A–D** *Umbilicaria rigida*. **A** Primordium with paraphyses and ascogenous hyphae. X380. **B** Young ascocarp with one-sided pseudoexcipulum. X260. **C** and **D** Marginal part of apothecium. X380. **E** *Umbilicaria muehlenbergii*, disc with actinogyri. X14. **F** *Umbilicaria virginis*, apothecia partly with umbonate disc (arrowheads indicate sterile columella). X10. **G** *Umbilicaria cylindrica*, gyrophorous apothecium. X16. *ah* anchoring hyphae, *p* paraphyses, *pe* pseudoexcipulum. (A–D microtome sections.)

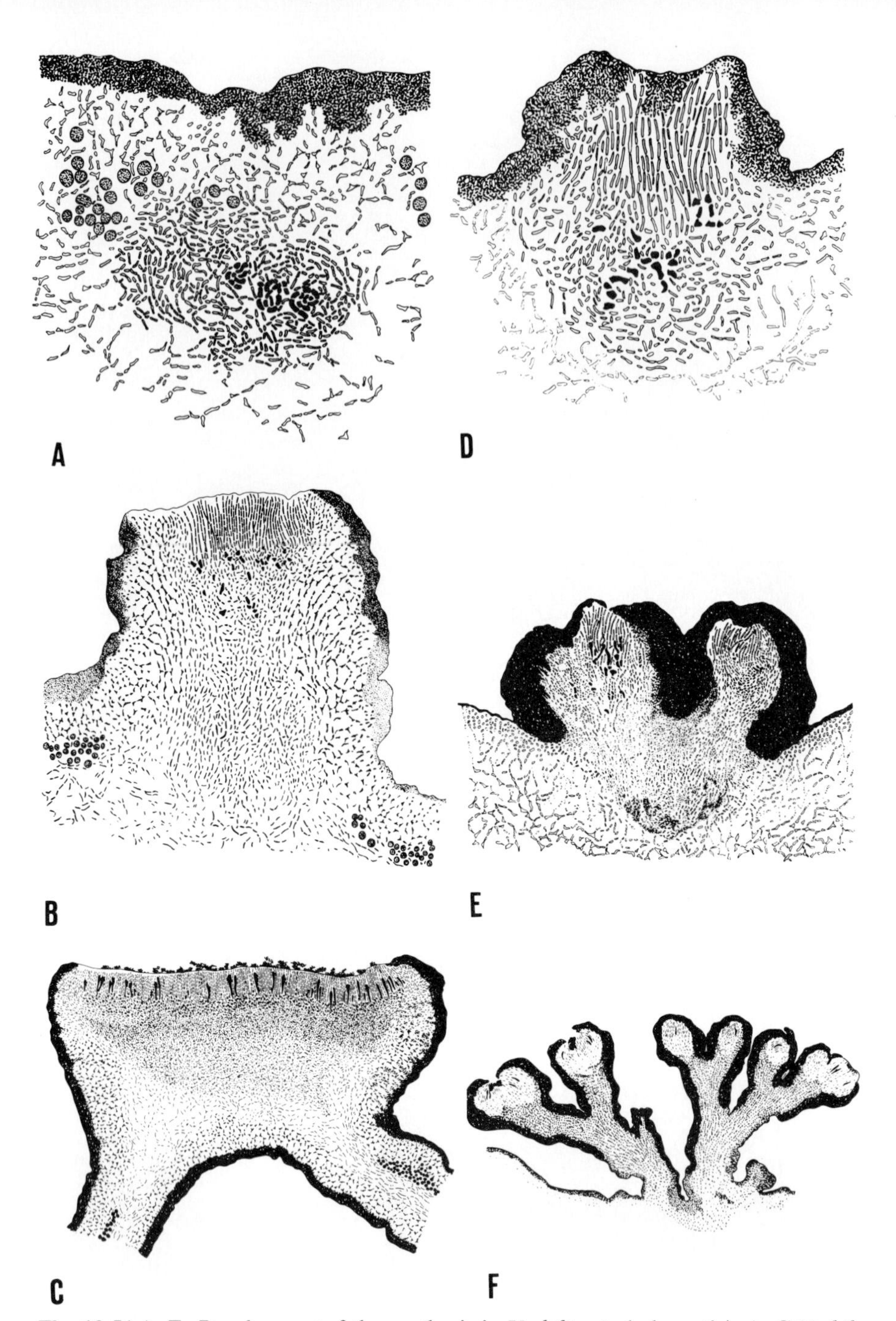

Fig. 10.51 A–F. Development of the apothecia in *Umbilicaria* (schematic). **A–C** *Umbilicaria rigida*. **A** Generative tissue with ascogonia. **B** Young apothecium with pseudoexcipulum. **C** Mature apothecium with asci. **D–F** *Umbilicaria muehlenbergii*. **D** Primordium with first stage in the division of the hymenium. **E** Young apothecium with two gyri. **F** Old, deeply gyrose apothecium. See text for further explanations. (A–F after Henssen and Jahns, 1974.)

gradually develops from the margin to the center and finally gives rise to the formation of a cupular excipulum. This development is comparable to the secondarily developed cupular excipulum in *Ramalodium;* the base part of the excipulum in *Umbilicaria,* however, does not have the same structure as the marginal part; instead, it has an anatomy comparable to the cortex of the thallus.

In general, the spirally coiled ascogonia lie in homogeneous generative tissue; more rarely the primordium consists of a sterile central column surrounded by a ring of generative hyphae enclosing the ascogonia (Fig. 10.52 B). The different lines of development are illustrated in Figure 10.52. The formation of pseudopodetia in some species is worth noting (Fig. 10.52I).

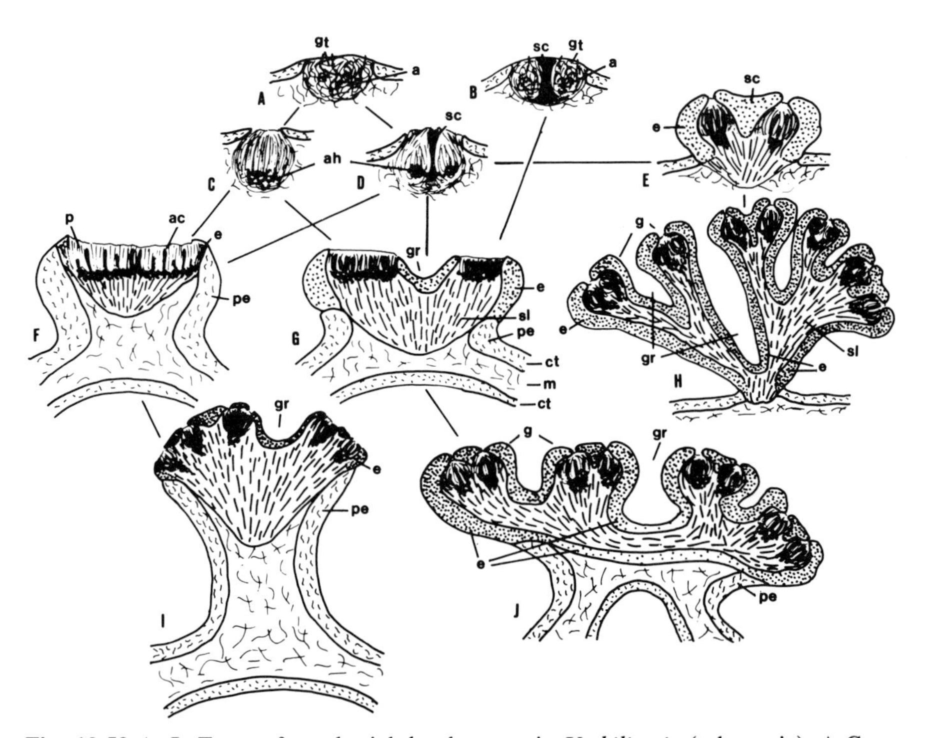

Fig. 10.52 A–J. Types of apothecial development in *Umbilicaria* (schematic). **A** Generative tissue with ascogones. **B** Generative tissue with sterile columella. **C** Primordium with paraphyses and ascogenous hyphae. **D** The same stage around a sterile central part. **E** Young ascocarp with two gyri and excipulum. **F** Apothecium with undivided disc (leiodisc) and pseudoexcipulum. **G** Omphalodisc apothecium with two gyri. **H** Deeply gyrose apothecium with excipulum. **I** Stipitate apothecium (on pseudopodetium). **J** Gyrose apothecium with cupulate excipulum. *a* ascogonium, *ac* ascus, *ah* ascogenous hyphae, *ct* cortex, *e* excipulum, *g* gyri, *gr* groove, *gt* generative tissue, *m* medulla, *p* paraphyses, *pe* pseudoexcipulum, *se* sterile column, *sl* subhymenial layer. (A after Henssen, 1970.)

Coccocarpiaceae

The family Coccocarpiaceae is polymorphic and includes genera with minutely filamentous fruticose, small placoid foliose, or umbilicate thalli. The apothecia develop from an aggregation of cells being continuously replenished by the adjacent hyphae of the vegetative thallus. The morphology of development in *Coccocarpia* Persoon, *Spilonema* Bornet, and *Steinera* Zahlbruckner has been studied by Henssen (1963a, 1975) and Keuck (1977). The genus *Peltularia* R. Santesson is a new genus of the family. True paraphyses are the only form of interascal filaments; they are often branched and have acuminate or enlarged end cells.

The development of the fruit bodies is well illustrated by *Spilonema paradoxum* Bornet. The apothecia, almost spherical in the mature condition, are laterally sessile on the filamentous branches. The phycobiont is a *Stigonema* species. The formation of the apothecia commences with the spread of generative tissue on the outside of a filament. This tissue consists of an aggregation of short cells forming a loose pseudoparenchyma in which numerous ascogonia consisting of straight cell rows and provided with trichogynes are formed (Fig. 10.53, A, B). The marginal cells of the primordium extend to form paraphyses. The ascogenous hyphae lie in groups at the base of the layer of paraphyses (Fig. 10.53C). The surrounding thallus hyphae continuously add cells to the formation layer as long as the fruit body is growing. As a result of the continued division of the cells of the generative tissue, the ascocarp initial protrudes outward in a hemispherical form (Fig. 10.53D). The asci grow in among paraphyses that are continuously replenished from the cells of the subhymenium. The outermost cell rows remain short celled and no structurally distinct excipulum is developed (Fig. 10.53D).

In *Coccocarpia parmelioides* (Hooker) Trevisan, generative tissue is formed in the algal layer and always includes algal cells (Figs. 10.54A and 10.55A). The tissue originates from medullary hyphae that become richly branched and abstrict short cells. The ascogonia are straight or slightly curved; the tips of the trichogynes project only insignificantly above the thallus surface. The primordium enlarges by division of its cells; continuous addition of new cells is by abstriction from neighboring medullary hyphae (Fig. 10.54, B, C). A dark pigment is deposited in the hymenial gelatin, which obscures the formation of the ascogenous hyphae in the primordium center and the differentiation of the paraphyses in the marginal part (Figs. 10.54D and 10.55B). The formative layer in the developing apothecium has a hemispherical or conical shape; the participation of the medullary hyphae in building up this layer is distinctly seen, especially in this genus of the family (Figs. 10.54, F, G, and 10.55, C, D). The margin of the mature apothecium lies closely appressed to the thallus. The excipulum consists of radiating hyphae and is poorly developed in *C. parmeliodes*. In *Coccocarpia epiphylla* (Fée) Müller-Argoviensis, the excipular hyphae are not closely adnate. The hyphal cells are enlarged and produce outgrowing hairs (Fig. 10.54H).

The monotypic *Peltularia* has an umbilicate thallus. Ascogonia have not been seen in the type specimen studied. The primordium consists of densely aggregated cells, including ascogenous hyphae (Fig. 10.56B). Development precisely follows the pattern of the Coccocarpiaceae. The formative layer has a conical shape and encloses algal filaments as in *Coccocarpia;* adjacent hyphae continuously add isodiametric cells to the developing fruit body (Fig. 10.56, C–G). The apothecium remains immersed for a long time within the thallus. The branched paraphyses are relatively long and often have acuminated or enlarged end cells (Fig. 10.56, F, G). The medullary hyphae may also grow upward around the apothecium, carrying groups of algal cells with them and forming a small thalline margin, which is in a later stage depressed by the overgrowing marginal part of the hymenium (Fig. 10.56, D, F). A similar formation of a thalline margin is seen in species of the placoid genus *Steinera* (Henssen, 1975; Henssen and James, 1980). Such secondary development of a margo thallinus is a characteristic feature of the Pannariaceae.

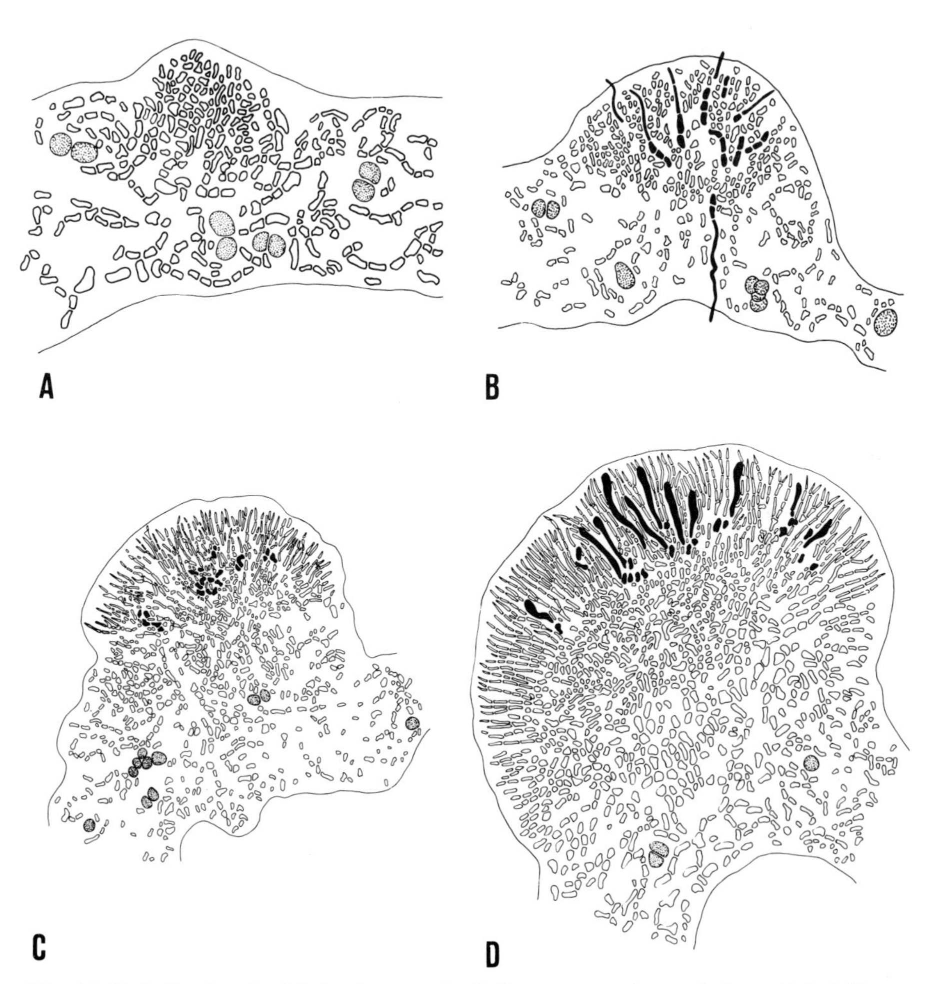

Fig. 10.53 A–D. Apothecial development in *Spilonema paradoxum* (schematic). **A** Generative tissue. **B** Primordium with ascogonia. **C** Young ascocarp with ascogenous hyphae and paraphyses. **D** Mature apothecium. (A–D after Henssen and Jahns, 1974.)

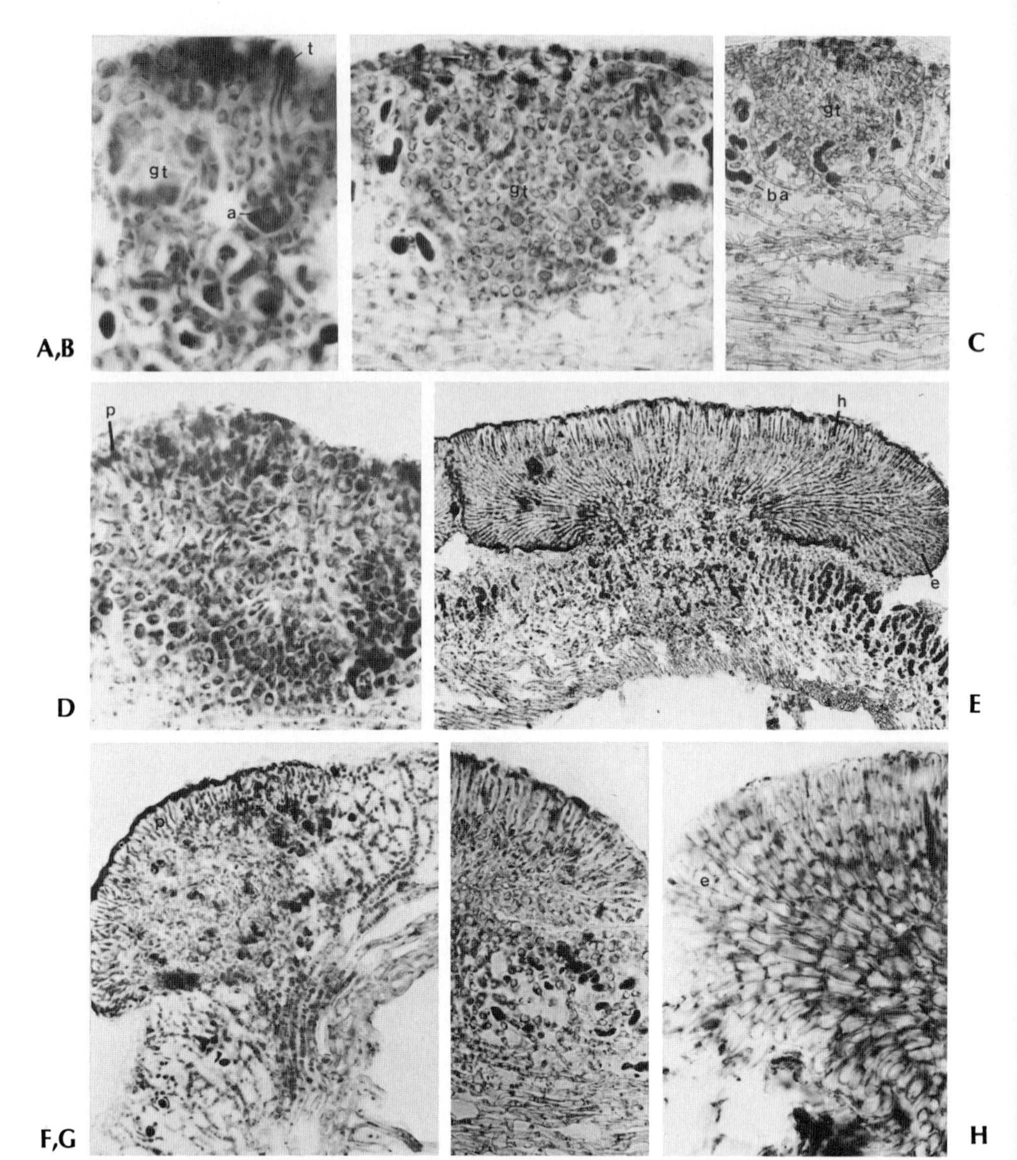

Fig. 10.54 A–H. Development of apothecia in *Coccocarpia*. **A–G** *Coccocarpia parmeliodes*. **A** Ascogonia in generative tissue. X380. **B** Compact generative tissue. X380. **C** Generative tissue replenished from medullary hyphae. X240. **D** Primordium with first paraphyses. X380. **E** Old apothecium adnate to the thallus. X100. **F** Young apothecium, the surrounding medullary hyphae dividing into small cells. X180. **G** Marginal portion of old apothecium, short cells formed by medullary and cortex hyphae. X240. **H** *Coccocarpia epiphylla*, excipulum. X380. *a* ascogonium, *ba* blue-green algae, *e* excipulum, *gt* generative tissue, *h* hymenium, *p* paraphyses, *t* trichogyne. (Microtome sections.)

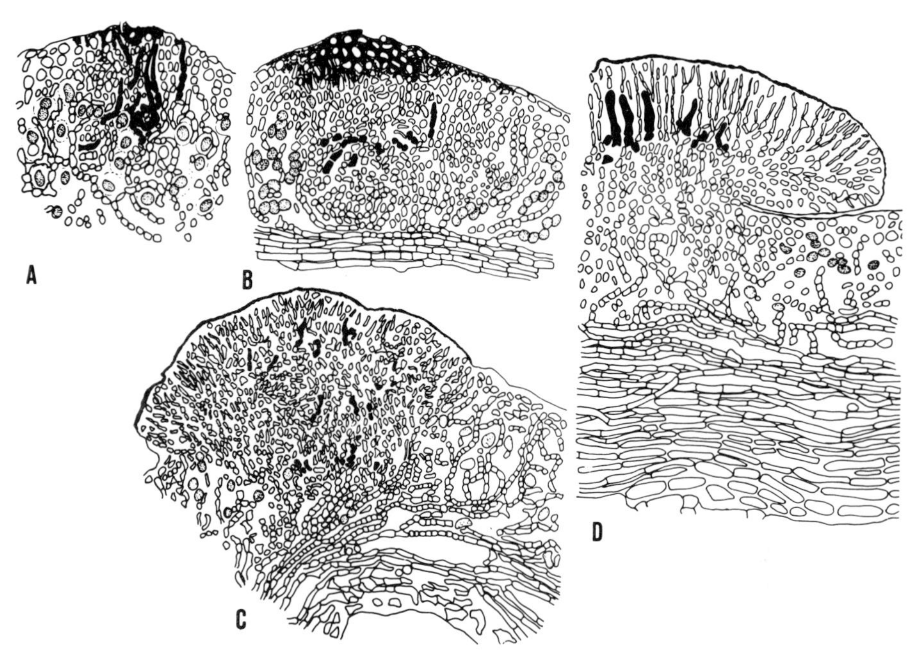

Fig. 10.55 A–D. Apothecial development in *Coccocarpia* (schematic). **A** Primordium with ascogonia. **B** Primordium with ascogenous hyphae, upper part dark pigmented. **C** Young apothecium with paraphyses, the surrounding vegetative hyphae producing small cells. **D** Marginal portion of mature apothecium.

Pannariaceae

The apothecia of the Pannariaceae are provided either with a massively developed annular excipulum, a corticated thalline margin, or a double margin consisting of an excipulum and a margo thallinus that is partly confluent with it (Fig. 10.57). A peculiar secondary development of the margo thallinus in the vicinity of the developing apothecium is a characteristic feature of the Pannariaceae (Henssen 1969; Henssen and Jahns, 1974). The structure of the ascus is not uniform throughout the genera. In some species, the ascus is provided with an amyloid cap, in others, with a distinct amyloid ring, and in some *Parmeliella* Müller-Argoviensis species, a limited expansion of the inner ascus layer has even been observed (Keuck, 1977).

The ascocarp ontogeny has been studied by Henssen (1963a, 1969) and Keuck (1977). The ascogonia are developed from medullary hyphae at the division between the algal layer and the thallus cortex. They occur in groups and may have partly branched trichogynes (Figs. 10.58A and 10.60, A, B). Either the ascogonia are formed first and secondarily surrounded by generative tissue or they both form together at the same time (Fig. 10.60, A, B). The hyphae of the generative tissue develop into paraphysoids (Figs. 10.60C and 10.61B). The primordium enlarges primarily in a vertical direction and breaks through the thallus

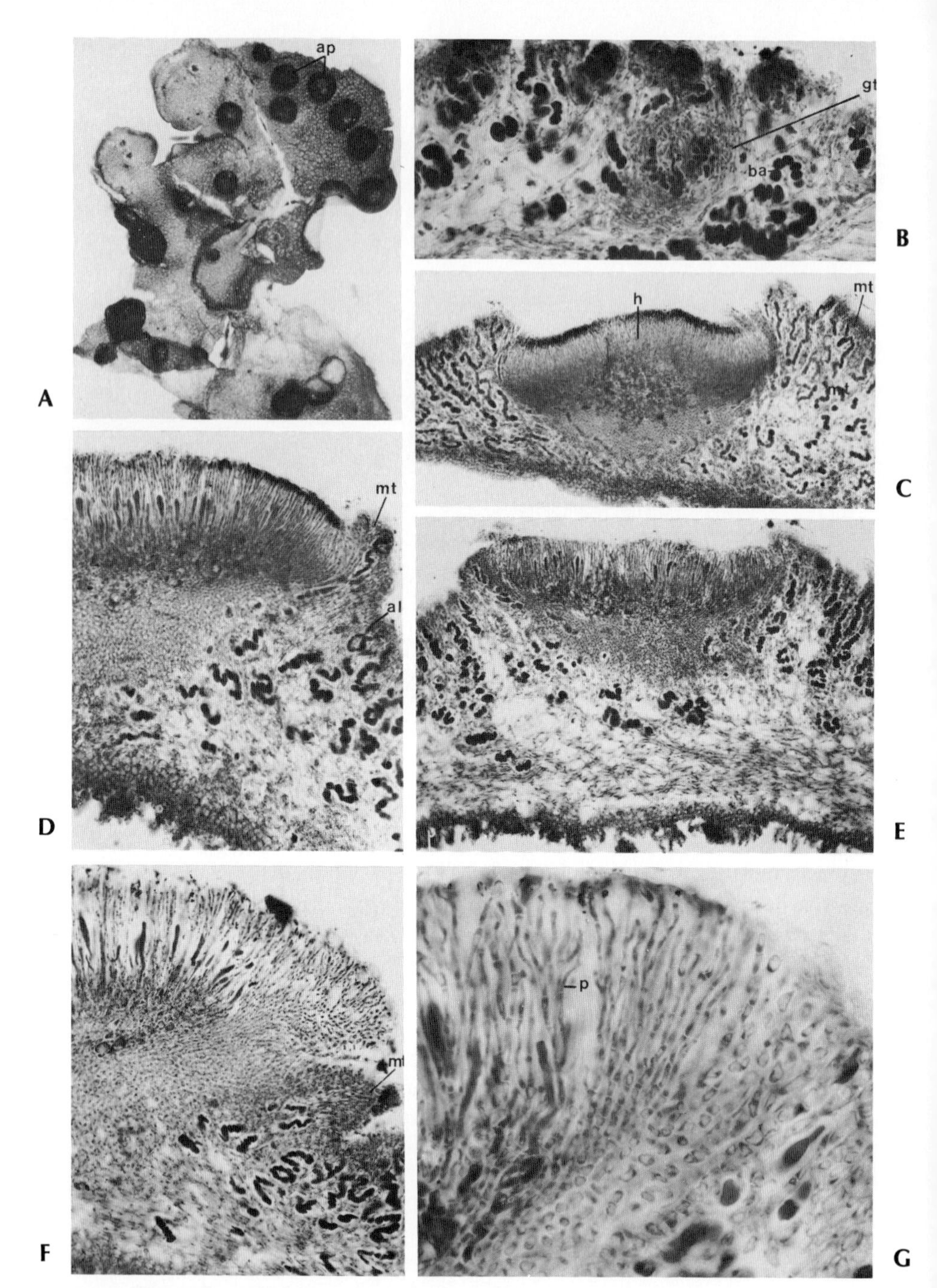

Fig. 10.56 A–G. *Peltularia gyrophoroides.* **A** Habit photograph of the holotype. X50. **B** Generative tissue with ascogenous hyphae. X210. **C** Formation of thalline margin. X100. **D** Young apothecium with thalline margin and small-celled subhymenium. X210. **E** Immersed young apothecium. X130. **F** Margin of old apothecium, the margo thallinus pressed down by the overlying hymenium. X240. **G** Development of paraphyses. X210. *al* algal cells, *ap* apothecium, *ba* blue-green algae, *gt* generative tissue, *h* hymenium, *mt* margo thallinus, *p* paraphyses. (B–G microtome sections.)

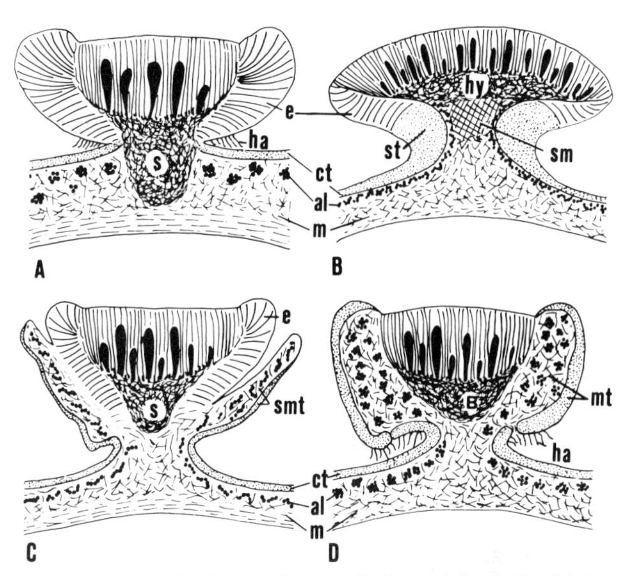

Fig. 10.57 A–D. Types of apothecia in the Pannariaceae (schematic). **A** Lecideine apothecium with stipe. **B** Apothecium with proper margin composed of excipulum and pseudoexcipulum, supporting tissue below hymenium formed by medullary hyphae. **C** Apothecium with double margin. **D** Lecanorine apothecium, the margin corticated by supporting tissue. *al* algal cells, *ct* cortex, *e* excipulum, *ha* hairs, *hy* hypothecium, *m* medulla, *mt* margo thallinus, *s* stipe, *sm* supporting tissue medulla, *smt* secondary margo thallinus, *st* supporting tissue. (A–D after Keuck, 1977.)

cortex (Fig. 10.58B). The vertical growth continues to a greater or lesser extent and the mature apothecia lie flattened on the surface of the thallus (Fig. 10.58, C, D) or are distinctly pedicellate (Fig. 10.59B). The first stages of development are very similar in the different genera of the Pannariaceae; the further course of development, however, differs considerably, and four developmental types may be distinguished (Fig. 10.57).

1. In *Parmeliella plumbea* (Lightfoot) Müller-Argoviensis (Fig. 10.58), a massively developed proper margin is formed by an annular excipulum which is typically constructed of radiating hyphae. The base of the excipulum margin ultimately produces hairs. Below the apothecium, the algae are nearly absent because of the voluminose stipe, which may reach the horizontally orientated lower medullary hyphae.
2. The apothecia in species of *Pannaria* Delise and *Psoroma* (Acharius) Michaux bear a corticated margo thallinus. This margin is derived secondarily from strands of medullary hyphae that grow up between the hymenium and the cortex of the thallus carrying groups of algal cells with them (Fig. 10.59C). The thalline margin, formed in this way by intrusion of medullary hyphae and algal cells, remains always somewhat in advance of the hymenium in its vertical growth, expanding horizontally at the same time on the outer side. A supporting tissue with a structure similar to the thallus cortex is differentiated

from the medullary hyphae of the margo thallinus; the only difference is that the cells of the newly formed cortex are slightly smaller and the marginal cells produce hairs at the base. The boundary between the old and the new cortical layers can be distinguished by a difference in cell size (Fig. 10.59D). The individual portions of the apothecial thalline margin formed in this way vary in length and width. As seen from above, the thallus margin presents a crenulated appearance, which is a characteristic feature of the lecanorine apothecia in the Pannariaceae. In some species the crenulate margin is developed all around; in other species it is lacking in places and is replaced by an excipulum of periclinal hyphae. The degree of development of the thalline margin may also be subject to variation within the same species or even on the same thallus.

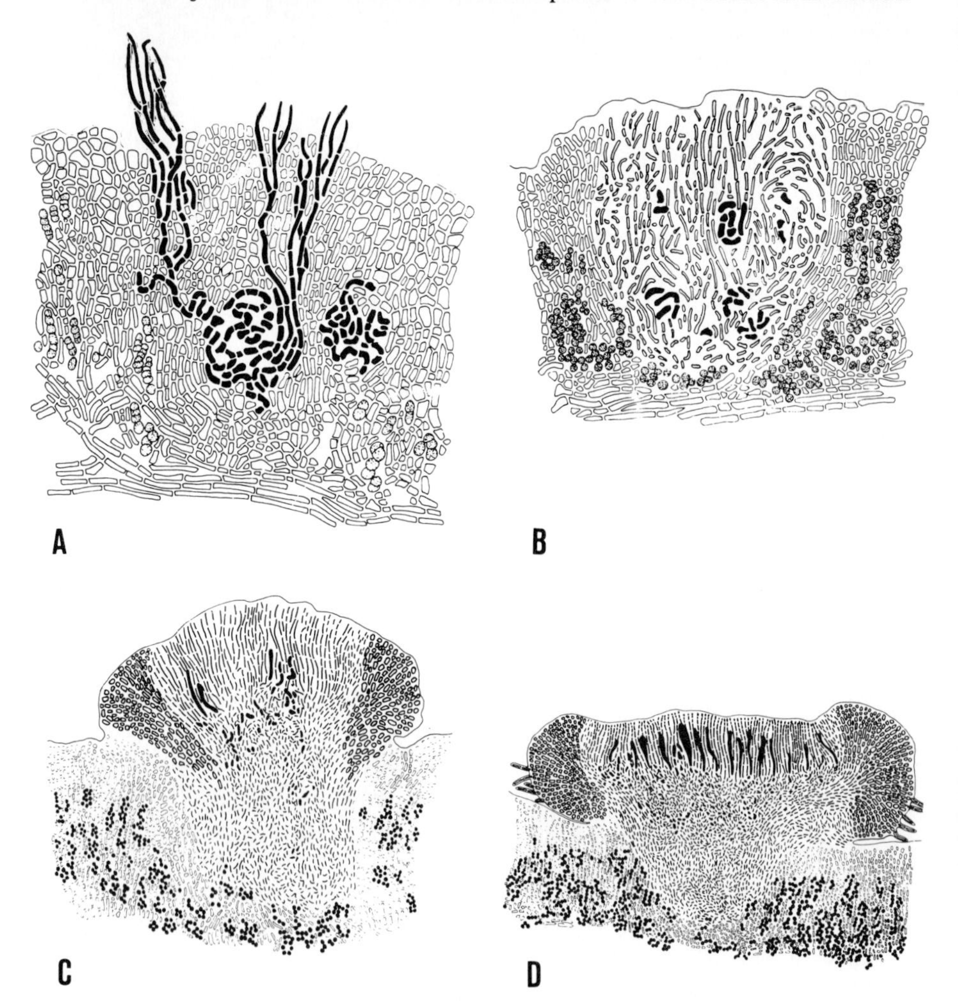

Fig. 10.58 A–D. Development of apothecium in *Parmeliella plumbea* (schematic). **A** Ascogonia with branched trichogynes. **B** Primordium with paraphysoids. **C** Young apothecium. **D** Mature apothecium, hairs developing from the excipulum. (A–D after Henssen and Jahns, 1974.)

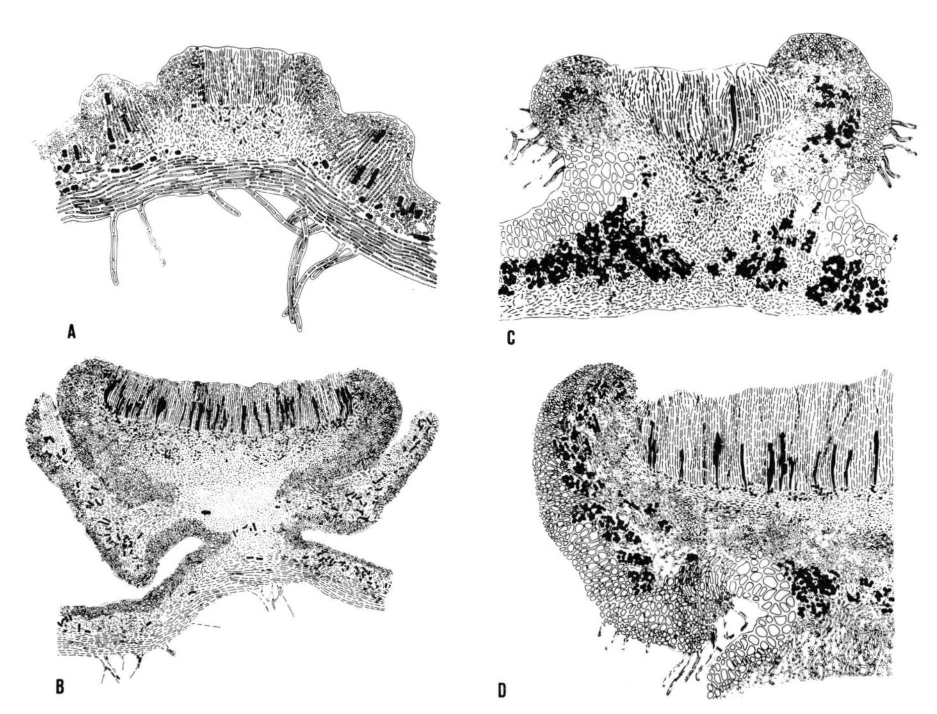

Fig. 10.59 A–D. Types of apothecia in the Pannariaceae (schematic). **A** and **B** *Parmeliella duplomarginata.* **A** Development of the thalline margin. **B** Mature apothecium with double margin. **C** and **D** *Pannaria rubiginosa.* **C** Secondarily formed thalline margin of a young apothecium. **D** Old apothecium with fully differentiated thalline margin. The cortex of the thallus and the later developed cortex of the margo thallinus have cells of different size. (A–D after Henssen and Jahns, 1974.)

In *Psoroma* species, the thalline margin may continue to develop until it overgrows areas of the disc in part. Such apothecia bear patches of thallus tissue on their discs. More rarely, patches of sterile thallus tissue develop upon the apothecial disc by the proliferation of a central sterile columella, including algal cells (James and Henssen, 1975). This process is analogous to that described for *Umbilicaria.*

3. In *Parmeliella duplomarginata,* P. James and Henssen (Henssen and James, 1980) the apothecium is surrounded by a double margin. The first stage of development is the formation of an excipulum of anticlinal hyphae around the young apothecium corresponding to the development in *Parmeliella plumbea.* Simultaneously, medullary hyphae grow a short distance upward to surround the excipulum, thereby forming a thalline margin (Fig. 10.59A). Both the apothecium and the thalline margin grow in a vertical direction and in later stages may become confluent in places in the lower part (Fig. 10.59B). Seen from above, the crevice between the thalline margin and the proper is distinctly visible. In mature apothecia, a supporting tissue is formed by medullary hyphae corresponding to the structure in type four (Henssen and James, 1980).

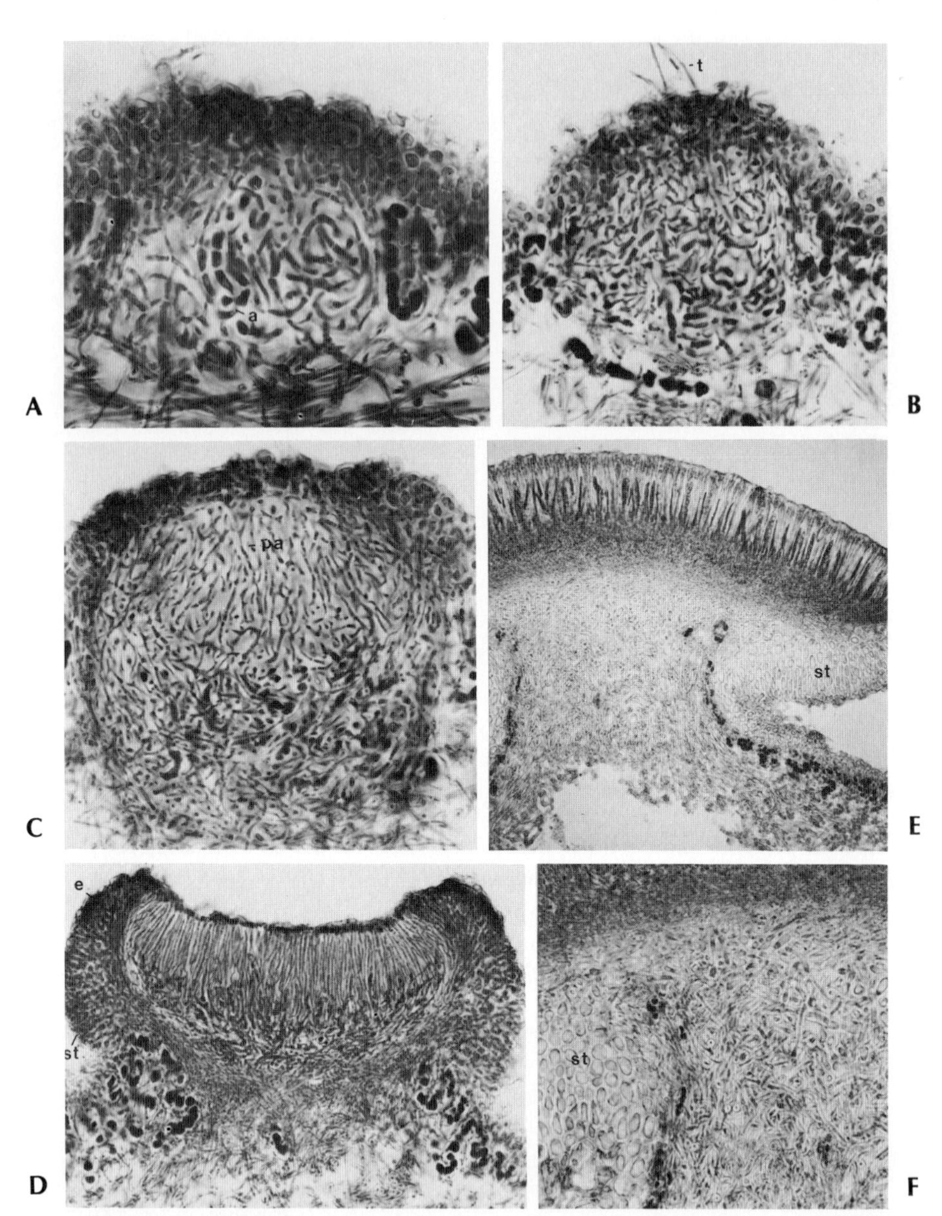

Fig. 10.60 A–F. Development of apothecium in *Parmeliella pycnophora*. **A** and **B** Generative tissue with ascogonia. X520 and X380. **C** Primordium with paraphysoids. X380. **D** Young apothecium with composed proper margin. X210. **E** Older apothecium with supporting tissue in the lower part of the apothecial margin and below the subhymenial layers. X80. **F** Portions of marginal and basal supporting tissue. X230. *a* ascogonium, *e* excipulum, *pa* paraphysoids, *st* supporting tissue, *t* trichogyne. (Microtome sections. A–D after Keuck, 1977.)

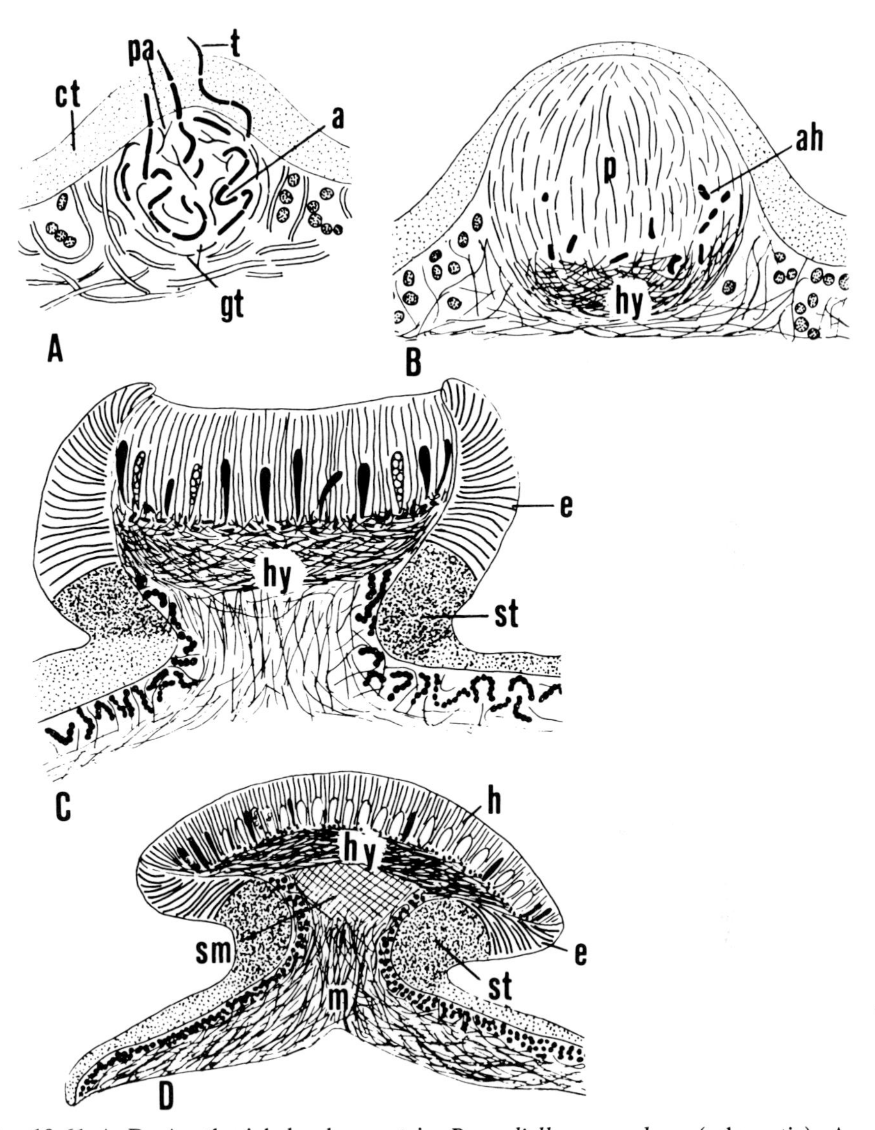

Fig. 10.61 A–D. Apothecial development in *Parmeliella pycnophora* (schematic). **A** Generative tissue with ascogonia. **B** Primordium with paraphysoids. **C** Young apothecium, the margin consisting of excipulum and supporting tissue. **D** Later stage with supporting tissue formed below the hypothecium. *a* ascogonium, *ah* ascogenous hyphae, *ct* cortex, *e* excipulum, *gt* generative tissue, *h* hymenium, *hy* hypothecium, *m* medulla, *p* paraphyses, *pa* paraphysoids, *sm* supporting tissue medulla, *st* supporting tissue, *t* trichogyne. (A–D after Keuck, 1977.)

4. The fourth type of development (Keuck, 1977), occurring in the *Parmeliella pycnophora* (Nylander) R. Santesson group and in the species of *Erioderma* Fée, is characterized by an apothecial margin composed of an upper part formed by the excipulum and of a lower part corresponding to the thallus margin (Figs. 10.60D and 10.61C). The upper part consists of radiating hyphae with small cells; the lower part is thicker and originates from medullary hyphae secondarily producing a cortex. A second peculiarity of this type is the supporting tissue formed by the medullary hyphae below the subhymenium. This texture consists of interwoven hyphae with gradually thickened walls (Figs. 10.60, E, F and 10.61D). No conspicuous stipe is developed in these lichens. Secondarily produced structures below the subhymenium occur also in *Ramalodium, Pilophorus,* and *Umbilicaria,* but in these genera the textures have derived, at least predominantly, from the formative layer (generative tissue) and are not a part of the vegetative lichen thallus.

Families with Hemiangiocarpous Apothecia

Peltigerineae

The suborder Peltigerineae is characterized by a hemiangiocarpous development of the fruit body. The upper part of the paraphysoid tissue derived from the generative hyphae ruptures and a cavity filled with mucilage arises between the covering layer and the layer of paraphyses. The thickness of the covering layer and the size of the cavity may vary and, correspondingly, the hemiangiocarpy is more or less marked in different genera or species. Some subtypes may be distinguished according to the details of developmental morphology. The lichens included in this suborder differ greatly in habit and thallus anatomy. Foliose lichens with a highly differentiated thallus belong to this group in addition to small species with a simple structure comparable to that of the Pannariaceae. The asci also vary greatly in structure. An amyloid apical ring or amyloid caps may be developed, and in some genera a limited expansion of the endoascus has been observed (Keuck, 1977). The bitunicate structure of the *Peltigera* ascus has been demonstrated in electron microscope studies by Honegger (1978). The amyloid ring distinctly seen in this genus becomes everted during the expansion of the inner ascus layer. Three or four families have been distinguished in the suborder depending on the characteristics used as delimiting features (Henssen and Jahns, 1974; Poelt, 1974; Keuck, 1977). The developmental morphology in a number of genera is at present too incompletely known to allow a satisfactory delimitation of the different varying family groups.

Placynthiaceae

A number of hemiangiocarpous genera with small thalli have been suggested for inclusion in this family (Henssen, 1963b, 1969; Henssen and Jahns, 1974). A simple thallus structure does not always correlate with the developmental subtypes. In ascocarp ontogeny, the genus *Massalongia* Koerber, for instance, mostly resembles members of the Peltigeraceae in the strict sense (Henssen, 1963b), and *Vestergrenopsis* Gyelnik mostly resembles the Stictaceae (Kueck, 1977).

In the genus *Placynthium* S. Gray, the hemiangiocarpy is more or less pronounced in the different species (Henssen and Keuck, unpublished results). The ascogonia are large-celled and slightly bent. In *Placynthium flabellosum* (Tuckermann) Zahlbruckner, they lie free within the thallus above the algal layer, being formed by ordinary vegetative hyphae (Fig. 10.62A). Generative tissue is especially developed above the ascogenous hyphae as a paraphysoid structure that ruptures in the uppermost part (Fig. 10.62, C, D). The covering layer and cavity are inconspicuous in this species and the hemiangiocarpy is hardly seen. The thallus enclosing the primordium first keeps pace with the enlargement but eventually bursts above the covering layer. The young ascocarp is surrounded by a thalline margin, which in the later course of development is pressed downward by the massively developing excipulum. The excipulum grows in the marginal growing zone around the hymenium in the usual way, being formed by hyphae that produce outwardly radiating excipulum hyphae and inwardly directed paraphyses (Fig. 10.62, E, F). Pseudoparenchyma is formed in the basal part of older apothecia (Fig. 10.62B).

The young ascocarp of *Placynthium* seems to resemble a corresponding stage in the Pannariaceae. An important difference lies in the formation of the margo thallinus. Whereas in the Pannariaceae the apothecial thalline margin is formed as a new structure by the vertical growth of medullary hypha, in the Placynthiaceae it originates by enlargement from the part of the thallus surrounding the apothecial primordium. The margo thallinus is replaced in most of the *Placynthium* species by a well-developed proper margin. An exception is *Placynthium stenophylla* (Tuckermann) Fink in which the excipulum remains rudimentary, and the mature apothecium is correspondingly lecanorine (Henssen, 1963c).

Peltigeraceae

The Peltigeraceae sensu stricto include *Peltigera* Willdenow, *Hydrothyria* Russell and *Solorina* Acharius—genera with a distinct amyloid ring in the ascus apex. The fissitunicate structure of the *Peltigera* ascus has been recently demonstrated by Honegger (1978). The development of the hemiangiocarpous apothecia has been repeatedly studied in *Peltigera* and *Solorina* (literature summarized in Letrouit-Galinou and Lallement, 1971; Keuck, 1977).

The development pattern prevailing in this family is best recognized in the genus *Hydrothyria.* The foliose thallus has a single-layered cortex on the upper and lower side. The apothecia are initiated marginally or subterminally and, as in the case of *Peltigera,* become displaced later on to the upper side of the thallus. In the first stage of development (Fig. 10.63A), a complex of straight or slightly bent ascogonia is situated between a paraphysoid tissue of generative tissue, and a small layer of medullary hyphae enclosing algal cells lies underneath. The lower cortex has started to form a supporting tissue. One gets the impression that true paraphyses with free tips have grown in between the stretched hyphae of the paraphysoid texture (Fig. 10.63B). In the further course of development, the true

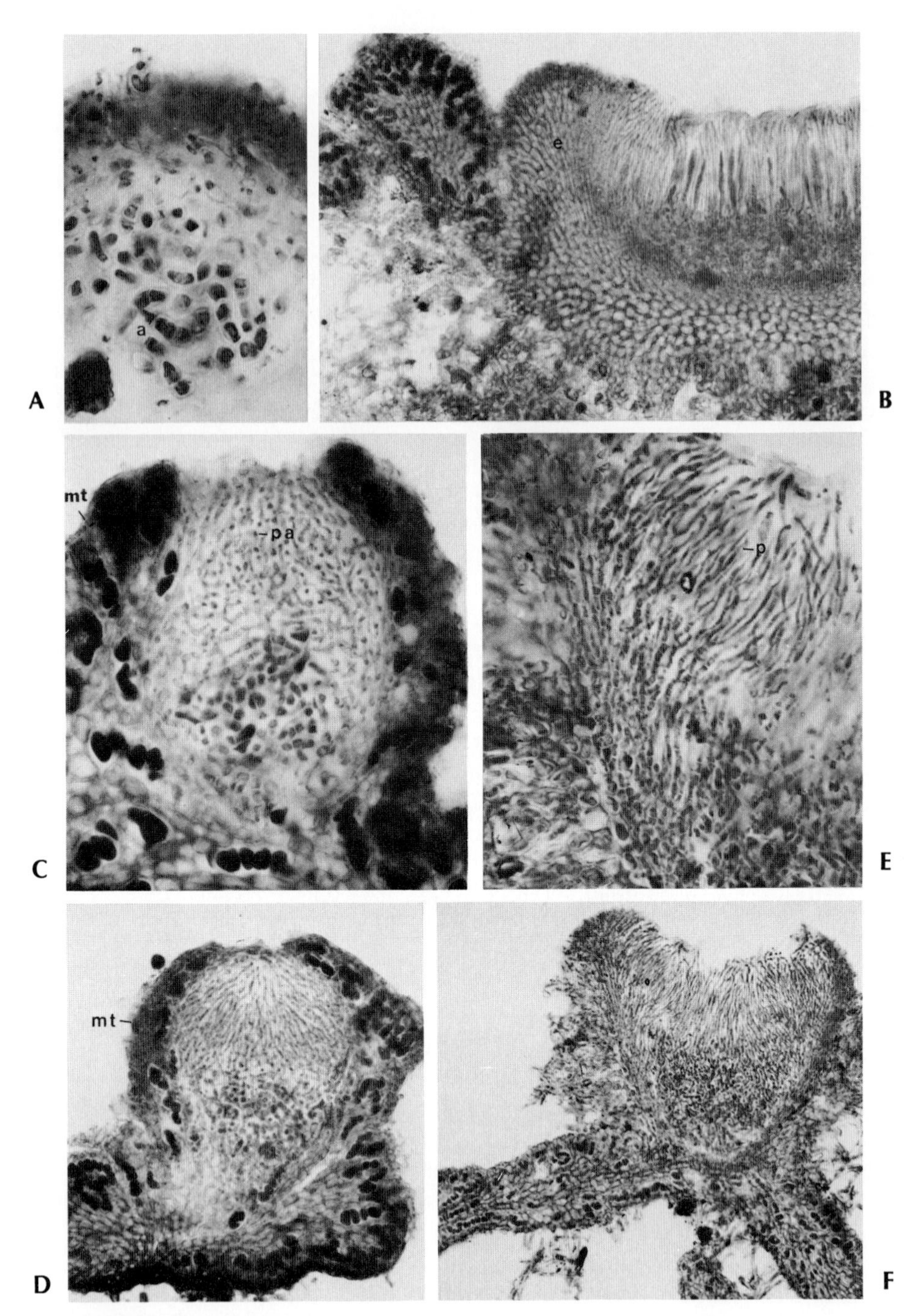

Fig. 10.62 A–F. Development of hemiangiocarpous apothecium in *Placynthium flabellosum*. **A** Slightly bent ascogonia with relatively large cells. X590. **B** Mature apothecium, the excipulum in part pseudoparenchymatous. X200. **C** Primordium with paraphysoids and ascogenous hyphae. X420. **D** Primordium with cavity. X230. **E** Formation of paraphyses X420. **F** Young apothecium with paraphyses, X290. *a* ascogonium, *mt* margo thallinus, *p* paraphyses, *pa* paraphysoids, *e* excipulum. (Microtome sections.)

paraphyses considerably increase in number, and the paraphysoid tissue ruptures to form a covering layer above a cavity filled by mucilage (Fig. 10.62C). This covering layer is itself still covered by the single row of cells of the thallus cortex. The supporting tissue at the base has enlarged primarily with the elongation and division of the cortex cells and is reinforced by the adjacent medullary hyphae. The ascogenous hyphae congregate at the base of the paraphyses. A formative layer is developed below them from the medullary hyphae. Paraphyses are continually being formed from this layer and after the covering layer has burst open

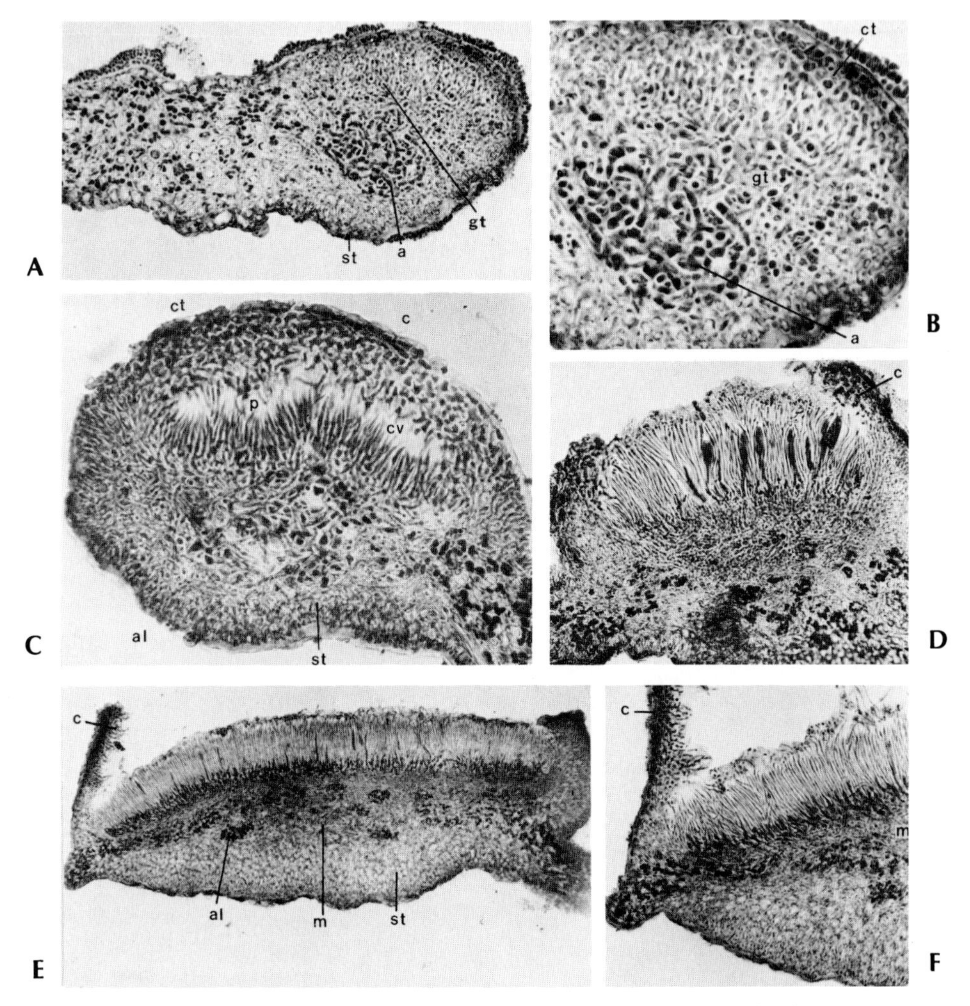

Fig. 10.63 A–F. Development of hemiangiocarpous apothecium in *Hydrothyrea venosa*. **A** and **B** Primordium with ascogonia and generative tissue. X190 and X350. **C** Primordium with cavity. X120. **D** Apothecium after dehiscence of covering layer. X140. **E** Old, centrifugal developed apothecium with supporting tissue. X90. **F** Distal part of the same stage. X140. *a* ascogonium, *al* algal cells, *c* covering layer, *ct* cortex, *cv* cavity, *gt* generative tissue, *m* medulla, *p* paraphyses, *st* supporting tissue. (Microtome sections.)

these hyphae also constitute the excipulum at the proximal end of the fruit body (Fig. 10.63D). The apothecium grows continuously toward the distal end, and no excipulum is formed (Fig. 10.63E, F). The rest of the ruptured covering layer is displaced outward with this elongation of the apothecium in horizontal direction. Growth of the medulla, including groups of algal cells and the basal supporting tissue, accompanies the growth of the developing hymenium and subhymenium. Supporting tissue is formed nearest the border of the apothecium where it joins the excipulum (Fig. 10.63E).

The apothecia of *Peltigera* show the same mode of development, as has been demonstrated by Letrouit-Galinou (1971). The ascogonia consist of very large cells (see Fig.10.65A), which are multinuclear (Moreau and Moreau, 1929). Protruding trichogynes have so far been observed only in *Peltigera rufescens* (Weiss) von Humboldt (Fig. 10.64) (Letrouit-Galinou and Lallement, 1971). In the *Peltigera* species, the paraphysoid tissue is covered by an upper cortex with several layers and no lower cortex of the thallus occurs. The paraphysoid tissue ruptures under the pressure of the invading paraphyses produced by the formative layer (Fig. 10.65D). Below the layer of paraphyses in later stages a darker stained hypothecium may be recognized that is surrounded by a formative layer not

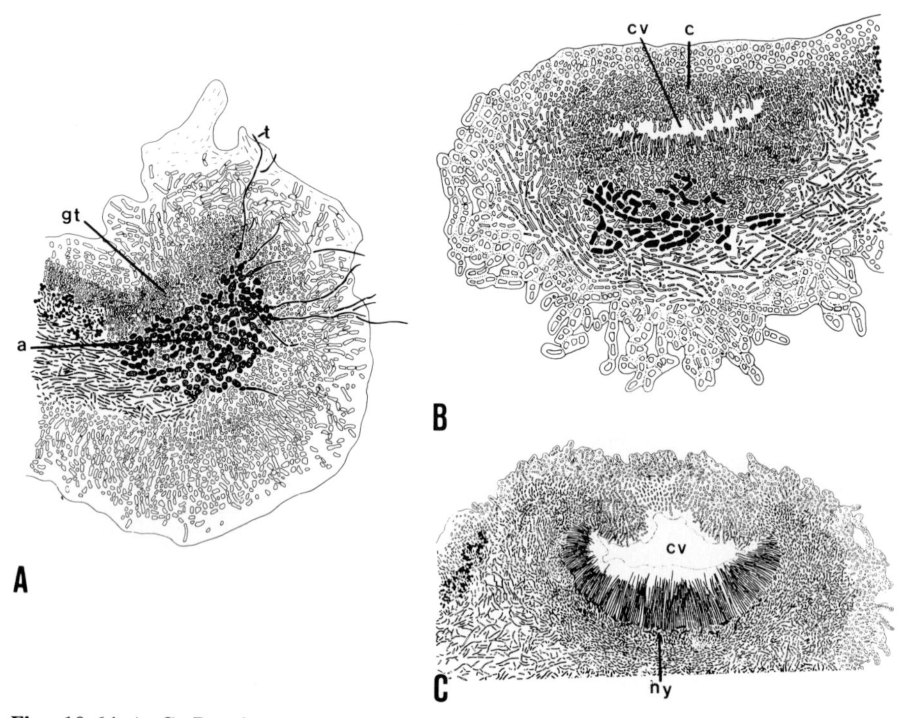

Fig. 10.64 A–C. Development of hemiangiocarpous apothecium in *Peltigera* (schematic). **A** *Peltigera rufescens,* primordium with ascogonia bearing trichogynes. **B** and **C** *Peltigera venosa.* **B** Formation of cavity. **C** Formation of hymenium. *a* ascogonium, *c* covering layer, *cv* cavity, *gt* generative tissue, *hy* hypothecium, *t* trichogyne. (A–D after Henssen and Jahns, 1974.)

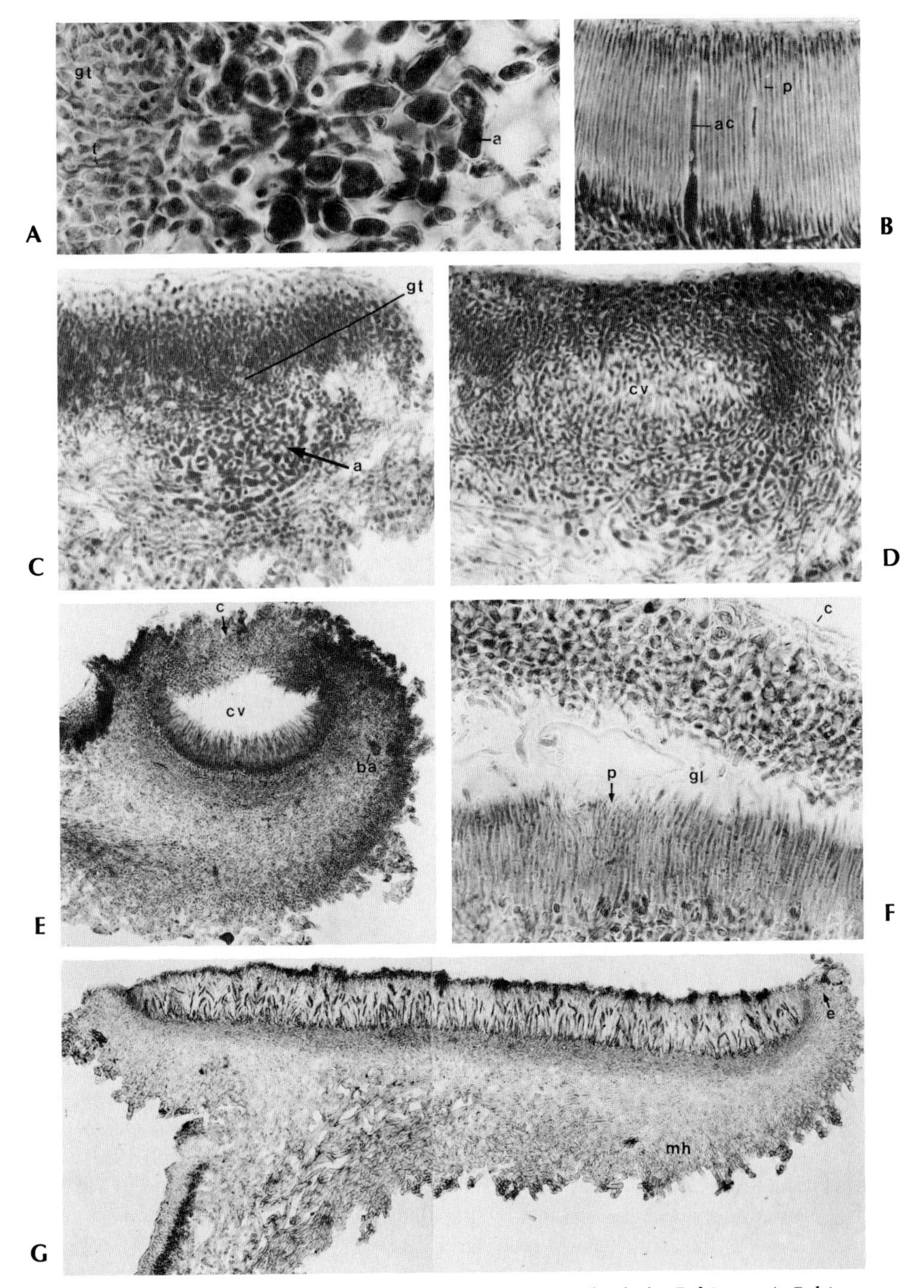

Fig. 10.65 A–G. Development of hemiangiocarpous apothecia in *Peltigera*. **A** *Peltigera rufescens,* portion of ascogonia and covering generative tissue. X520. **B–E** *Peltigera venosa*. **B** Portion of hymenium. X290. **C** Primordium with group of ascogonia and covering layer of generative tissue. X300. **D** Formation of cavity. X300. **E** Later stage with layer of paraphyses. X80. **F** *Peltigera polydactyla,* portion of cavity, covering layer, and young hymenium. X240. **G** *P. venosa,* mature apothecium. X50. *a* ascogonium, *ac* ascus, *c* covering layer, *cv* cavity, *e* excipulum, *gl* gelative mucilage, *gt* generative tissue, *mh* medullary hyphae, *p* paraphyses, *t* trichogyne. (Microtome sections.)

sharply delimited from the medulla (Fig. 10.65E). The formative layer continually gives rise to paraphyses and, after the covering layer has burst open, also forms the hyphae constituting the excipulum. The first asci are only developed at a relatively late stage after the rupture of the covering tissue and push upward between the closely arranged paraphyses in the resultant cavity (see Fig. 10.65B). The end cells of the pointed tips of the young paraphyses may later swell (Fig. 10.65, B, F). The cavity enlarges with the rupture of the gelatinous filling (Fig. 10.65F). The hyphal ends of the covering layer above the hymenial cavity partly grow into this space (Fig. 10.64, B, C). The lower part of the medulla grows simultaneously with the apothecium and in the early stages surrounds the latter in the form of a cup. The apothecium mainly grows in centrifugal direction and, with the accompanying growth of the medulla, finally comes to lie on the upper side of the thallus (Fig. 10.66, A–D). The lower medullary hyphae are oriented anticlinally at both ends of the mature apothecia and develop into agglutinated rows of short cells similar in structure to the excipulum and not sharply delimited by it (Fig. 10.65G). In many species, a stalk is developed simultaneously from the edge of the lobe (Fig. 10.66, D–F). These thalline stalks become erect and raise the apothecia up above the level of the thallus. The excipulum and the covering layer remain visible around the disc of the old apothecium (Fig. 10.66D).

Normally, the medulla on the underside of the apothecium is ecorticate, like the lower side of the thallus, but in some species a cortex is developed beneath the older hymenium. The cortex is continuous in the case of *Peltigera aphthosa* (Linnaeus) Willdenow (see Fig. 10.66E), but in *Peltigera leucophlebia* (Nylander) Gyelnik the cortex consists of isidia-like outgrowths (Fig. 10.66F). This difference is obviously a genetically conditioned specific character (Degelius, 1974). In both species, formation of the cortex results from the stimulus exerted by groups of algal cells lying in the medulla. The primordium of the fruit body arises inside the algal layer, and as a result, a small group of algal cells remains present in the outermost margin of the thallus or below the ascogonial group (Fig. 10.67, A, B). These algal cells come to remain in the medulla underneath the hymenium as growth of the apothecium continues. In the case of *P. aphthosa,* they congregate in a more or less uniform layer in the lower part of the medulla. Beneath the algal layer, a pseudoparenchymatous supporting tissue is formed (Figs. 10.67C and 10.68, A–C). An almost continuous layer of cortex squamules is formed in this species in a later stage. Groups of algal cells lie embedded in medullary hyphae on the pseudoparenchymatous squamules. Fünfstück's (1884) observations on a differentiation of thallus squamules containing an algal layer and a cortex on the upper and lower side have not been confirmed. In *P. leucophlebia,* groups of algal cells immediately become enclosed by medullary hyphae and are pushed down in the vicinity of the thallus lower side. Here thay are transformed into isidia-like structures with a pseudoparenchymatic cortical layer (Figs. 10.67D and 10.68, D–F) (see also Kershaw and Millbank, 1970).

The species of *Solorina* also are distinguished from each other by different types of cortication of the fruit bodies. In these lichens, the apothecia are laminal and immersed in the thallus. In *Solorina saccata* (Linnaeus) Acharius, the disc forms a cup-shaped hollow, and the thick-walled hyphae in the medulla form a

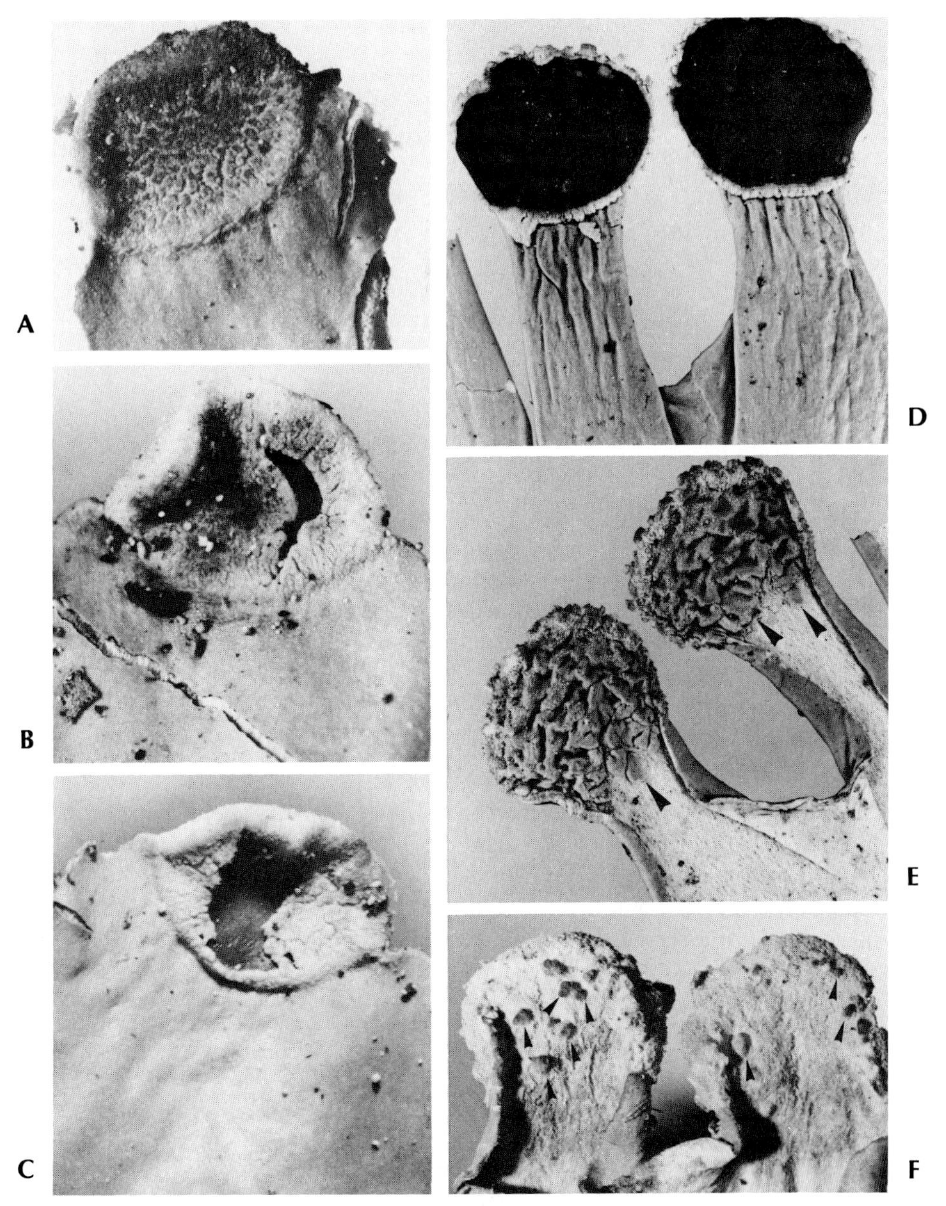

Fig. 10.66 A–F. Habit photographs of *Peltigera* apothecia. **A–C** *Peltigera horizontalis*. **A** Apothecial disc closed by covering layer. X10. **B** and **C** Gradual rupture of the covering layer. X13 and X10. **D** and **E** *Peltigera aphthosa*. **D** Stalked apothecia in surface view. X4. **E** The same apothecia, seen from below, with a continuous cortical layer. X5. **F** *Peltigera leucophlebia,* lower side of apothecia with isidia-like cortical structure. X5. (A–C after Henssen and Jahns, 1974.)

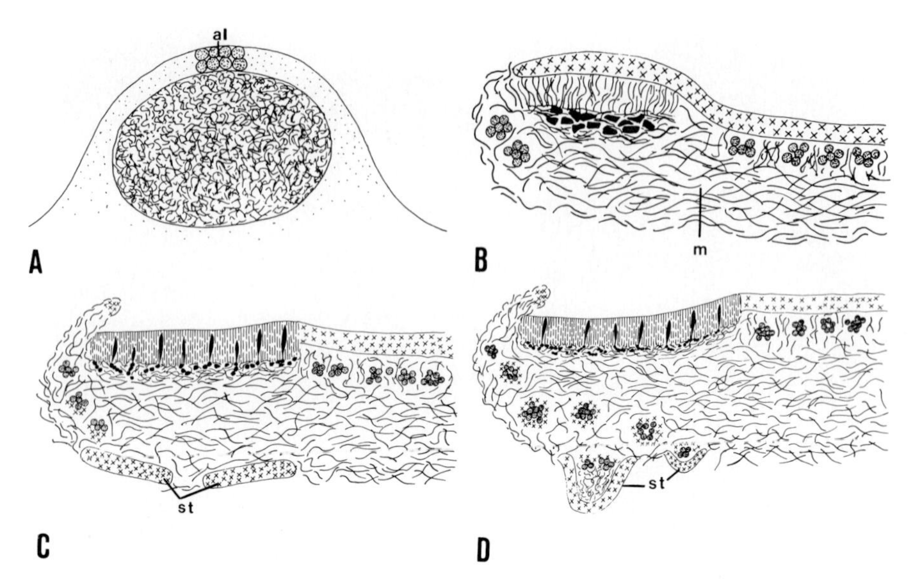

Fig. 10.67 A–D. Development of cortical structures on the lower side of *Peltigera aphthosa* and *Peltigera leucophlebia* (schematic). **A** Primordium seen from above with group of algal cells at the tip of the lobe. **B** Section through young stage. **C** Formation of squamulose cortical structures in *P. aphthosa.* **D** Formation of isidia-like cortical structures in *P. leucophlebia. al* algal cells, *m* medulla, *st* supporting tissue.

pseudoparenchymatous supporting tissue beneath the apothecia (Moser-Rohrhofer, 1969; Keuck, 1977). In *Solorina crocea* (Linnaeus) Acharius, the apothecium with a smooth disc is slightly prominent, and no cortical structure like that found in *S. saccata* is present (Keuck, 1977). A corresponding, more or less pronounced hemiangiocarpous development is known from the marginal apothecia of *Nephroma* (Letrouit-Galinou and Lallement, 1970b; Keuck, 1977). The genus has often been included in the family Peltigeraceae but deviates by the ascus structure (Henssen and Jahns, 1974).

Stictaceae

The Stictaceae form a natural group characterized by the special mode of hemiangiocarpous development arising from a compact primordium and by highly differentiated foliose thalli. Cephalodia are present and sometimes a dimorphic thallus is formed under the stimulus of different phycobionts (James and Henssen, 1976).

Fig. 10.68 A–F. Development of cortical structures in *Peltigera* species. **A–C** *Peltigera aphthosa.* **A** Marginal part of apothecium with supporting tissue formed from medullary hyphae. X85. **B** Cortical structures with underlying medullary hyphae. X160. **C** Squamule of cortical structure. X60. **D–F** *Peltigera leucophlebia.* **D** Portion of apothecium,

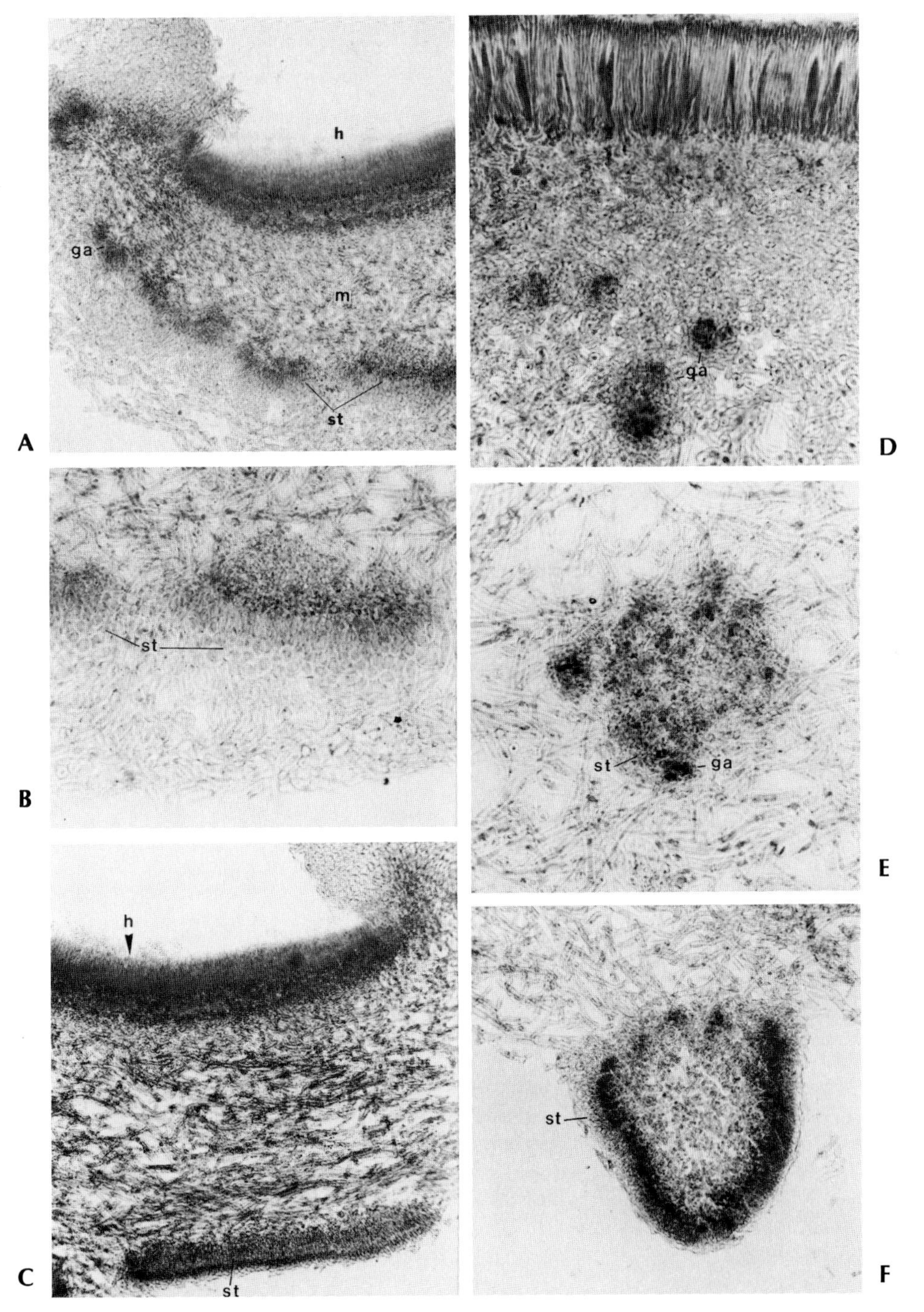

algal cells becoming enveloped by medullary hyphae. X160. **E** Group of algal cells surrounded by densely packed medullary hyphae. X160. **F** Isidia-like cortical structure. X100. *ga* green algae, *h* hymenium, *m* medulla, *st* supporting tissue. (Microtome sections.)

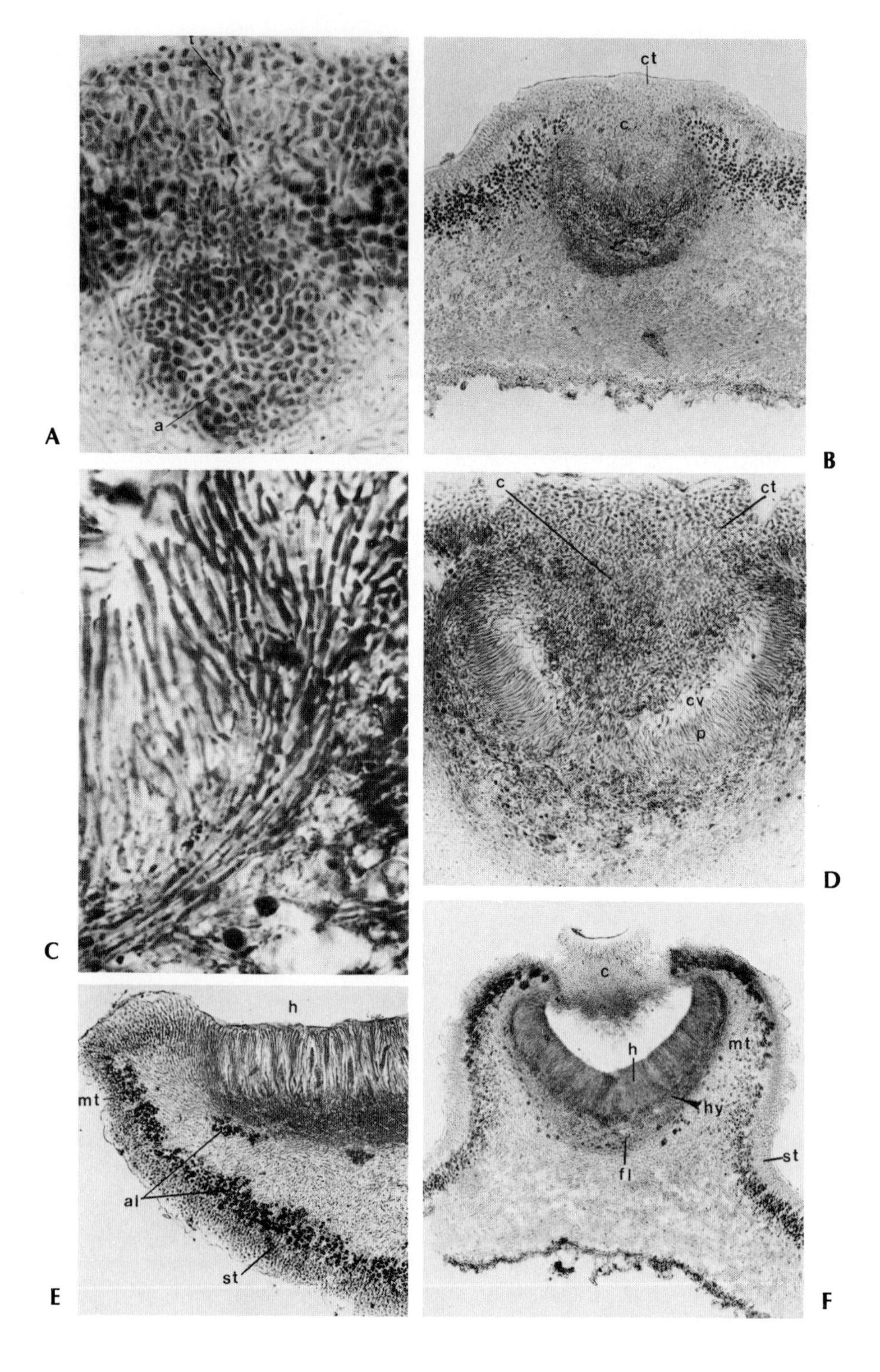
t
a
A
ct
c
B
C
c
ct
cv
p
D
h
mt
al
st
E
c
h
mt
hy
fl
st
F

The developmental morphology in *Lobaria laetevirens* (Lightfoot) Zahlbruckner has been described by Letrouit-Galinou (1971), and by Henssen and Jahns (1974). Additional species of *Lobaria* (Schreber) Hoffmann, as well as species of the genera *Sticta* (Schreber) DeCandolle and *Pseudocyphellaria* Vainio, were studied by Keuck (1977).

In *Lobaria,* the apothecial primordium develops in the lower part of the algal layer. It consists of a nearly pseudoparenchymatous aggregation of cells. Most of the cells are more or less isodiametric and correspond to the generative tissue. The somewhat longer celled ascogonia partly bearing a branched trichogyne are hardly visible between the isodiametric cells of the generative tissue (Fig. 10.69A). The primordium often includes some algal cells that die. The generative tissue and the ascogenous hyphae form a cup-shaped layer in which the paraphysoids lie ball-like and are covered by the thallus cortex. The paraphysoid texture ruptures, and a cavity filled with gelatin is formed gradually (Fig. 10.69, B, D). In comparison with the *Peltigera–Hydrothyria* pattern, the covering layer is very thick in *Lobaria.* During further development, it becomes stretched horizontally and grows out together with the covering cortex of the thallus (Fig. 10.69F). The thalline margin grows simultaneously with the developing primordium. The medullary hyphae are only able to keep pace with the vigorous enlargement; the thallus cortex becomes so intensely stretched that it ruptures into small patches (Fig. 10.69F). A new cortex is formed in the apex by the medullary hyphae (Fig. 10.69E). In *Lobaria amplissima* (Scopoli) Forssell, an excipulum is formed in older apothecia in the usual manner. Figure 10.69C demonstrates the formation of the excipulum hyphae and the paraphyses.

In *Sticta* and *Pseudocyphellaria,* the generative tissue is less compact and the slender, bent or spirally coiled ascogonium can be recognized more easily (Figs. 10.70A and 10.71A). In *Sticta damaecornis* (Swartz) Acharius, the covering layer is only poorly developed and disappears completely when the cavity is formed (Figs. 10.70, B, C and 10.71, B–D). The primordium extends considerably in a vertical direction together with the surrounding thallus tissue and at the same time the ascogenous hyphae grow upward. After dehiscense, the apothecium enlarges laterally with the formation of numerous paraphyses. The excipular hyphae join with the supporting tissue formed by the medullary hyphae within the margo thallinus (Figs. 10.70D and 10.71E). Whereas the apothecium in *Lobaria laetevirens* is lecanorine, in *L. amplissima* and *S. damaecornis* it is composed of an upper excipulum and a lower part, consisting of the thalline support-

Fig. 10.69 A–F. Development of hemiangiocarpous apothecia in *Lobaria.* **A** *Lobaria amplissima,* compact generative tissue with ascogonia. X400. **B** *Lobaria laetevirens,* primordium prior to formation of cavity. X85. **C** *L. amplissima,* formation of paraphyses. X320. **D–F** *L. laetevirens.* **D** Formation of cavity. X160. **E** Marginal portion of old apothecium. X95. **F** Young apothecium prior to discharge of covering layer. X50. *a* ascogonium, *al* algal cells, *c* covering layer, *ct* cortex, *cv* cavity, *fl* formative layer, *h* hymenium, *hy* hypothecium, *mt* margo thallinus, *p* paraphyses, *st* supporting tissue. (Microtome sections. A and C after Keuck, 1977.)

ing tissue. The amount of algal cells below the formative layer may vary considerably in different species. The types of apothecial margin occurring in *Lobaria* have been described by Yoshimura (1971).

The compact primordium resembles that of the Parmeliaceae, but because the further course of development shows no correlation, this similarity has to be considered as convergence. The Stictaceae differ from the Peltigeraceae especially in the structure of the ascogonia and in the shape of the primordium. In the Stic-

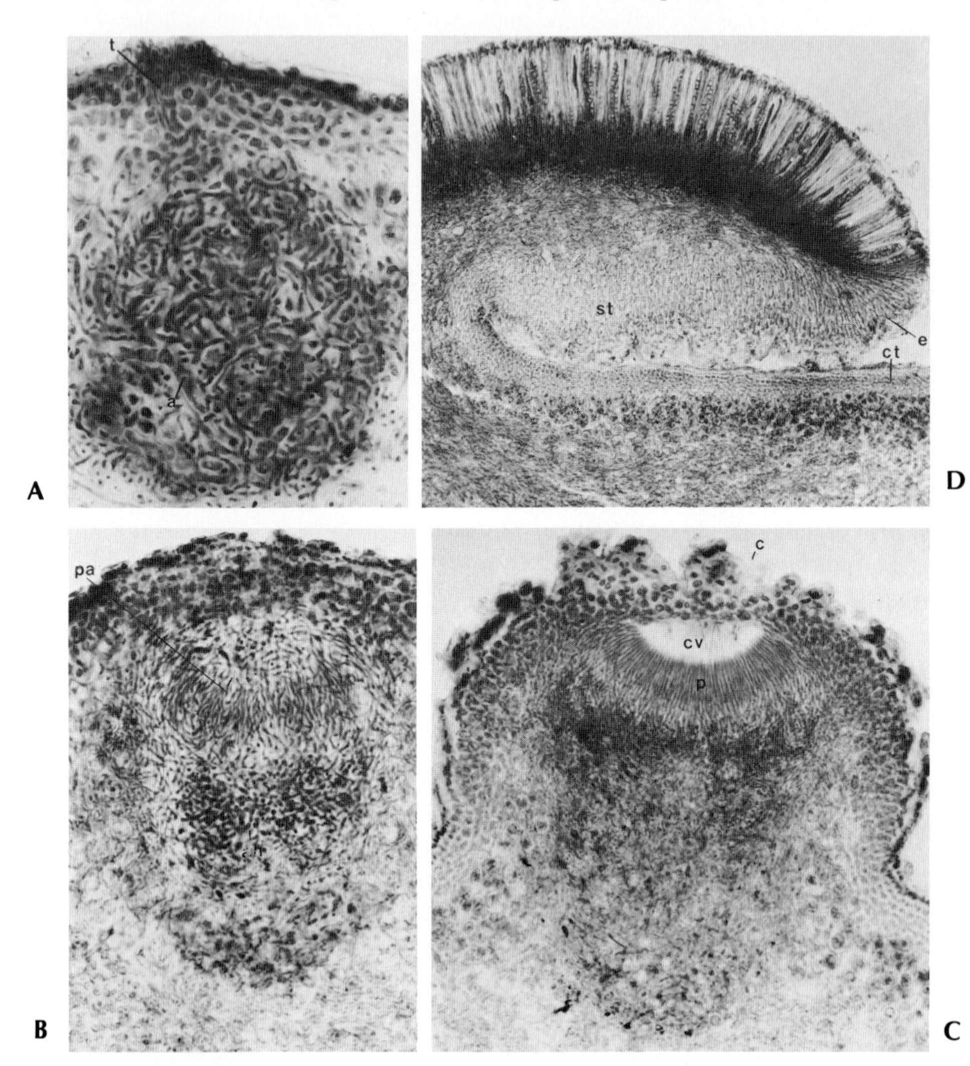

Fig. 10.70 A–D. Development of hemiangiocarpous apothecium in *Sticta damaecornis*. **A** Generative tissue with remains of ascogonia. X380. **B** Primordium with paraphysoid tissue. X90. **C** Young apothecium with cavity X220. **D** Marginal portion of mature apothecium. X45. *a* ascogonium, *ah* ascogenous hyphae, *c* covering layer, *ct* cortex, *cv* cavity, *e* excipulum, *p* paraphyses, *pa* paraphysoids, *st* supporting tissue, *t* trichogyne. (Microtome sections. A–D after Keuck, 1977.)

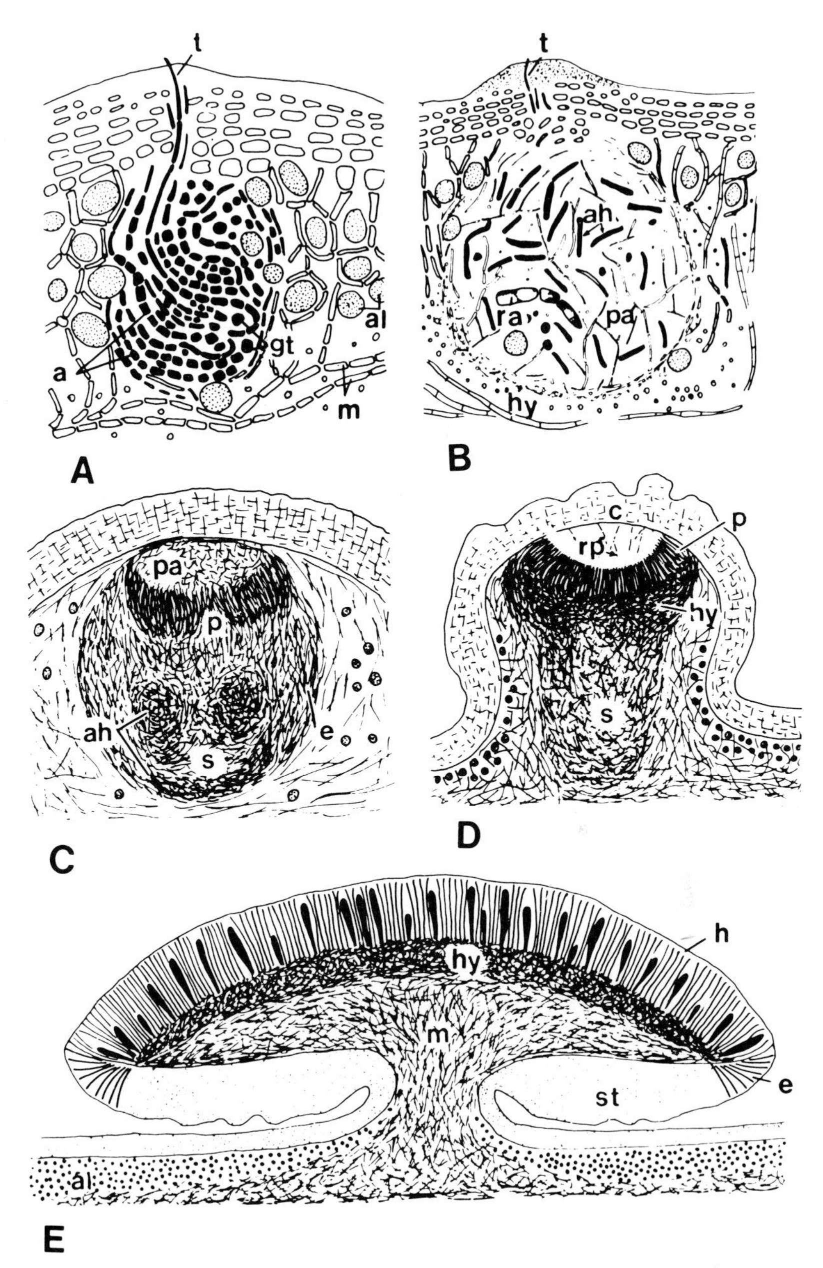

Fig. 10.71 A–E. Development of the hemiangiocarpous apothecium of *Sticta damaecornis* (schematic). **A** Generative tissue with ascogonia. **B** Primordium with ascogenous hyphae. **C** Primordium with remains of paraphysoid layer and paraphyses. **D** Young ascocarp with cavity. **E** Mature apothecium. *a* ascogonium, *ah* ascogenous hyphae, *al* algal cells, *c* covering layer, *e* excipulum, *gt* generative layer, *h* hymenium, *hy* hypothecium, *m* medulla, *p* paraphyses, *pa* paraphysoids, *ra* remains of ascogonia, *rp* remains of paraphysoids, *s* stipe, *st* supporting tissue, *t* trichogyne. (A–E after Keuck, 1977.)

taceae, the hymenial layer is strongly concave in young ascocarps, and the paraphysoids are much better developed and more filamentous than in genera of the Peltigeraceae.

General Conclusions

In the Lecanorales, a number of developmental types of the apothecial centrum and margin may be recognized. The development is gymnocarpous or hemiangiocarpous. In the latter case, the layer covering the young hymenium is composed of a texture originating from the generative tissue and of overlying parts of the vegetative thallus. A limited hemiangiocarpy may be observed in many genera; e.g., *Ramalodium, Parmelia, Usnea,* and *Neuropogon.* Pronounced hemiangocarpous apothecia are characteristic for the genera of the Peltigeraceae and Stictaceae.

In most families of the Lecanorales—as so far known—the primordium consists of a group of ascogonia enclosed by the generative tissue, which originates from the neighboring thallus hypae. The Collemataceae is an exception in which the generative hyphae originate from the ascogonial stalk. The generative tissue differs from hyphae of the vegetative thallus by its stronger stainability and deviating structure. The generative tissue may be a filamentous weft (e.g., Lichinaceae, Pannariaceae, Lecanoraceae) or a compact almost pseudoparenchymatous texture (e.g., Parmeliaceae, Stictaceae, Coccocarpiaceae). The generative hyphae develop to the auxiliary tissue of the fruit body either by division or by a continuous replenishment of cells produced by the neighboring hyphae. The latter case is characteristic for the Coccocarpiaceae but also occurs, to less extent, in genera of the Peltigeraceae. In most types of development, paraphysoids and some rows of the excipulum are the structures first formed by the generative tissue. In the second growing phase of the apothecium, the generative tissue transforms to the formative layer surrounding the hymenium and the hypothecium (the latter in the strict sense). The hyphae of the formative layer give rise to the paraphyses and excipular hyphae. The most active zone is the marginal part, where paraphyses are produced inwards and excipulum hyphae outwards. The formation of true paraphyses and radiating excipulum hyphae develop in this way. This is distinctly seen, for example, in the Collemataceae, Placynthiaceae, and Stictaceae. Such a development of paraphyses and excipulum hyphae corresponds to that in unlichenized discomycetes.

The distinction between paraphysoids and true paraphyses is not always clear. The two types of paraphyses are easily recognized in the Lichinaceae, but this is not the case, for instance, in the Parmeliaceae. Only true paraphyses are formed in the Coccocarpiaceae, and in the genus *Sporopodium,* paraphysoids only. The occurrence of proparaphyses precedes the development of true paraphyses in the Cladoniaceae and Stereocaulaceae as well as in some members of the Lecideaceae. In the strange pycnoascocarps of the Lichinaceae, elongated conidiophores function as paraphyses before the true paraphyses are formed. The function of interascal threads is taken over to a large extent by ordinary thallus hyphae in the thallinocarps of the same family.

The excipulum proprium may be annular or cupulate. The gradual formation of a close cup-shaped excipulum is outstanding in the genus *Ramalodium*, a member of the Collemataceae. The excipulum structure of the Parmeliaceae is highly complex. It consists of a meristematic cupulate layer surrounded by an outer more hyaline texture. The outer layer develops from the meristematic cells by elongation and gelatinization of their walls. The hyphae of the outer layer are reticulately connected, knit to an almost pseudoparenchymatous texture, or stretched and often distinctly contorted. Connecting hyphae arise on the inner side of the meristematic layer, interspersing the hyaline layer between hypothecium and excipulum. The Collemataceae are characterized by the diversity of the excipulum texture and to some extent resemble the differentiation types in the Helotiales. The hyphae may be orientated anticlinally or perclinally and run parallel to each other or are irregularly interwoven. The hyphal cells may be thin-walled, isodiametric, and grouped to form a pseudoparenchyma; the reticulately connected cell lumina may lie embedded in a gelatinous matrix originating from the hyphal wall.

The outer cells of the excipulum often grow out or form hairs or long hyphae. The hyphae may connect the excipulum base with the underlying thallus. Sometimes, a real plectenchyma is formed by such hyphae (e.g., in *Ramalodium*). Hyphae developed from the excipulum cells and called anchoring hyphae grow into the gelatinous thallus in the Collemataceae.

The margo thallinus develops in different ways. The margo thallinus arises by enlargement of the thallus enclosing the growing primordium in the Lichinaceae and Placynthiaceae. In many families, such as the Collemataceae, Parmeliaceae, Lecanoraceae, and Teloschistaceae, the thalline margin develops by a vertical growth of the thallus together with the vertically growing ascocarp. In *Umbilicaria* and genera of the Stereocaulaceae, among others, only the thallus cortex participates in the vertical growth, giving rise to the pseudoexcipulum of the superlecideoid apothecium. The margo thallinus is formed at a later stage of development in the Pannariaceae and Coccocarpiaceae. In this case, medullary hyphae push up between the hymenium and the cortex of the thallus. These hyphae carry with them algal cells during upward growth. The margo thallinus is frequently reinforced by supporting tissues that may develop from medullary hyphae, from cells of the thallus cortex, and/or from the above-mentioned anchoring hyphae. Such supporting tissues may be formed in the outer part of the margo thallinus as well as beneath the subhymenial layers. The supporting tissue often has a structure like that of the cortex of the thallus. The outer cells of the margo thallinus may grow out to form hairs or hyphae. Well-developed supporting tissues or the pseudoexcipulum may simulate the occurrence of a proper margin.

Algal cells more or less covering the hymenium can be observed in *Lichinella* and *Gonohymenia* of the Lichinaceae and in some *Sporopodium* species. In the latter, some algal cells are trapped by the generative hyphae enveloping the ascogonia. These eventual epithecial algae gradually adopt a much smaller shape than the ordinary algal cells of the thallus. The algal layer arises with the particular development of the thallinocarp in the genera of the Lichinaceae and the difference in shape of the algal cells is not so pronounced.

Stipitate fruit bodies are characteristic for the Cladoniaceae and the Stereocaulaceae. The stalks are either podetia developing from the generative tissue, or they are pseudopodetia arising from medullary hyphae or from the whole thallus. Elongated apothecia in which the spurious stalk is a real part of the fruit body are found in *Gomphillus* and in some species of *Umbilicaria*. The latter genus is characterized by the typically umbonate or gyrose discs of the apothecia.

The development of species of Lecanorales on the whole demonstrate strong connection with other orders of ascohymenial lichen families in their development. The primordia of development, including the formation of paraphysoids, correspond especially to the ontogeny in other ascohymenial lichens. A hemiangiocarpous ontogeny also occurs in the Gyalectales and certain members of the Pertusariales (Henssen, 1976). Apart from the formation of a dark-pigmented mantle layer, the development in *Sporopodium* to a certain degree resembles the ontogeny in the genus *Pertusaria* De Candolle.

References

Bachmann, F., 1912. A new type of spermogonium and fertilization in *Collema*. Ann. Bot. 26: 747–767.

Bachmann, F., 1913. The origin and development of the apothecium in *Collema pulposum* (Berh.) Ach. Arch. Zellf. 10: 369–430.

Baur, E., 1898. Zur Frage der Sexualität der Collemaceen. Ber Deutsch. Bot. Ges. 15: 365–367.

Baur, E., 1904. Untersuchungen über die Entwicklungsgeschichte der Flechtenapothecien. I. Bot. Z. 62: 21–44.

Bellemère, A., 1967. Contribution a l'ètude de développement de l'apothécie chez les discomycètes inoperculés. Bull. Soc. Mycol. France 83: I–IV, 395–931.

Corner, E. J. H., 1929a. Studies in the morphology of discomycetes. I. The marginal growth of apothecia. Trans. Br. Mycol. Soc. 14: 261–274.

Corner, E. J. H., 1929b. II. The structure and development of the ascocarp. Trans. Br. Mycol. Soc. 14: 275–291.

Degelius, G., 1931. Zur Flechtenflora von Ångermanland. Arkiv Bot. 24: 1–122.

Degelius, G., 1954. The lichen genus *Collema* in Europe. Symb. Bot. Upsala 13: 1–499.

Degelius, G., 1974. The lichen genus *Collema* with special reference to the extra-European species. Symb. Bot. Upsala 20: 1–215.

Dughi, R., 1954. L'excipulum proprium des apothécies des discolichens. Rev. Bryol. Lichénol. 23: 300–316.

Dughi, R., 1956. Apareils apicaux des asques et taxonomie des *Collema*. C. R. Acad. Sci. Paris 243: 1911–13.

Frey, E., 1936. Vorarbeiten zu einer Monographie der Umbilicariaceen Ber. Schweiz. Bot. Ges. 45: 198–230.

Fünfstück, M., 1884. Thallusbildung an den Apothecien von *Peltigera aphthosa*. Ber. Deutsch. Bot. Ges. 2: 447–452.

Goebel, K., 1926. Morphologische und biologische Bemerkungen. Flora NF 21: 177–188.

Henssen, A., 1963a. Eine Revision der Flechtenfamilien Lichinaceae und Ephebaceae. Symb. Bot. Upsala 18: 1–123.

Henssen, A., 1963b. The North American species of *Massalongia* and generic relationships. Can. J. Bot. 41: 1331–1346.

Henssen, A., 1963c. The North American species of *Placynthium*. Can. J. Bot. 41: 1687–1724.

Henssen, A., 1965. A review of the genera of the Collemataceae with simple spores (excluding *Physma*). Lichenologist 3: 29–41.

Henssen, A., 1968a. A new Lichinodium species from British Columbia. Bryologist 71: 271–274.

Henssen, A., 1968b. Eine neue Lichinella-Art aus Nordamerika (Lichenes). Nova Hedw. 15: 543–550.

Henssen, A., 1969. Die Entstehung des Thallusrandes bei den Pannariaceen (Lichenes). Ber. Deutsch. Bot. Ges. 82: 235–248.

Henssen, A., 1970. Die Apothecienentwicklung bei *Umbilicaria* Hoffm. emend. Frey. Deutsch. Bot. Ges. NF 4: 103–126.

Henssen, A., 1973. New or interesting cyanophilic lichens. I. Lichenologist 5: 444–451.

Henssen, A., 1975. Ontogenesis of the ascocarp in the lichen genus *Steinera*. Abstract of the Seventh International Botanical Congress, Vol. 1, p. 61. Leningrad, Russia.

Henssen, A., 1976. Studies in the developmental morphology of lichenized ascomycetes. *In* D. H. Brown, D. L. Hawksworth, and R. H. Bailey (Eds.), Lichenology: Progress and Problems. New York, Academic Press, pp. 107–138.

Henssen, A., 1977. The genus *Zahlbrucknerella*. Lichenologist 9: 17–46.

Henssen, A., 1979. New species of Homothecium and Ramalodium from S America. Bot. Notiser 132: 257–282.

Henssen, A., 1980. ("1979"). Problematik der Gattungsbegrenzung bei den Lichinaceen. Ber. Deutsch. Bot. Ges. 92: 483–506.

Henssen, A., and P. W. James, 1980. *Parmeliella duplomarginata,* a new lichen from New Zealand. Mycotaxon 11: 217–229.

Henssen, A., and H. M. Jahns, 1974. Lichenes, eine Einfuhrung in die Flechtenkunde. Georg Thieme Verlag, Stuttgart.

Honegger, R., 1978. The ascus apex in lichenized fungi. The *Lecanora, Peltigera,* and *Teloschistes* types. Lichenologist 10: 47–67.

Jahns, H. M., 1970a. Untersuchungen zur Entwicklungsgeschichte der Cladoniaceen unter besonderer Berücksichtigung des Podetien-Problems. Nova Hedw. 20: 1–177.

Jahns, H. M., 1970b. Induktion der Apothecienbildung bei *Cladia aggregata.* (Sw.) Nyl. Ber. Deutsch. Bot. Ges. 83: 33–40.

Jahns, H. M., 1970c. Remarks on the taxonomy of the European and North American species of *Pilophorus*. Th. Fr. Lichenologist 4: 199–213.

Jahns, H. M., 1973. The trichogynes of *Pilophorus strumaticus*. Bryologist 76: 414–418.

James, P. W., and A. Henssen, 1975. A new species of *Psoroma* with sorediate cephalodia. Lichenologist 7: 143–147.

James, P. W., and A. Henssen, 1976. The morphological and taxonomic significance of cephalodia *In* D. H. Brown, D. L. Hawksworth, and R. H. Bailey, (Eds.), Lichenology: Progress and Problems. Academic Press, New York, pp. 27–77.

Kershaw, K. A., and Millbank, J. W., 1970. Isidia as vegetative propagules in *Peltigera aphthosa* var. *variolosa* (Massal.) Thoms. Lichenologist 4: 234–247.

Keuck, G., 1977. Ontogenetisch-systematische Studie über *Erioderma* im Vergleich mit anderen cyanophilen Flechtengattungen. Bibliotheca Lichenologica Bd. 6, J. Cramer, Vaduz.

Lamb, I. M., 1951. On the morphology, physiology and taxonomy of the lichen genus *Stereocaulon.* Can. J. Bot. 29: 522–534.

Lamb, I. M., 1964. Antarctic lichens. I. The genera *Usnea, Ramalina, Himantormia, Alectoria, Cornicularia.* Br. Antarct. Surv. Sci. Rept. 38, pp. 1–34 and Appendix.

Letrouit-Galinou, M. A., 1966. Recherches sur l'ontogénie et l'anatomie comparées des apothécies de quelques discolichens. Rev. Bryol. Lichénol. 34: 413–588.

Letrouit-Galinou, M. A., 1971. Étude sur le "Lobaria laetevirens" (Lght.) Zahlb. (Discolichen, Stictacée). I. Le thalle, les apothécies, les asques. Le Botaniste, Sér. 54: 189–234.

Letrouit-Galinou, M. A., and Lallemant, R., 1970a. Les apothécies et les asques du *Parmelia conspersa* (Discolichen, Parmeliacée). Bryologist 73: 39–58.

Letrouit-Galinou, M. A., and R. Lallemant, 1970b. Le développement des apothécies du *Nephroma resupinatum* (L.) Ach., Lichen, Néphromacée. Rev. Gén. Bot. 77: 331–351.

Letrouit-Galinou, M. A., and R. Lallemant, 1971. Le thalle, les apothécies et les asques du *Peltigera rufescens* (Weiss) Humb. (Discolichen, Peltigeracée). Lichenologist 5: 59–88.

Llano, G. E., 1950. A monograph of the lichen family Umbilicariaceae in the western hemisphere. Navexos P-831. Washington, D.C.

Luttrell, E. S., 1951. Taxonomy of the pyrenomycetes. Univ. Missouri Studies No. 24, pp. 1–120.

Malone, C. P., 1977. Developmental morphology of *Caloplaca ulmorum, C. cerina,* and *Xanthoria elegans.* Mycologia 69: 740–749.

Moreau, F., and M. Moreau, 1929. Les phénomènes cytologiques de la reproduction chez les champignons des lichens. Le Botaniste, Sér. 2: 1–67.

Moser-Rohrhofer, M., 1969. Wachstumszonen des vegetativen Flechtenthallus und des Ascophors einiger Flechtenpilze. Céska Mykol. 23: 15–23.

Poelt, J., 1974. ("1973"). Classification. *In* V. Ahmadjian and M. E. Hale (Eds.), The Lichens. Academic Press, New York, pp. 599–632.

Poelt, J., and H. Wunder, 1967. Über biatorinische und lecanorinische Berandung von Flechtenapothecien, untersucht am Biespiel der *Caloplaca ferruginea*-Gruppe. Bot. Jahrb. 86: 256–265.

Santesson, R., 1944. Contributions to the lichen flora of South America. Ark. Bot: 31A, no. 7: 1–28.

Santesson, R., 1952. Foliicolous lichens. I. A revision of the taxonomy of the obligately foliicolous, lichenized fungi. Symb. Bot. Upsala 12: 1–590.

Scholander, P., 1934. On the apothecia in the lichen family Ümbilicariaceae. Nytt. Mag. Naturo. 75: 1–31.

Stahl, E., 1877a. Beiträge zur Entwicklungsgeschichte der Flechten. 1. Über die geschlechtliche Fortpflanzung der Collemaceen. Felix, Leipzig.

Stahl, E., 1877b. Beiträge zur Engwicklungsgeschichte der Flechten. 2. Über die Bedeutung der Hymenialgonidien. Felix, Leipzig.

Sturgis, W. C., 1890. On the carpologic structure and the development of the Collemaceae and allied groups. Proc. Amer. Acad. Arts Sci. 25: 15–52.

Vergleich de Apothecienentwicklung und Pycnidenstruktur von vier usneoiden Flechten *(U.subfloridana, U.fasciata, U.magellanica, U.sp.)* and *Letharia.* Staatsexamensarbeit, Universität Marburg, Germany (unpublished).

Yoshimura, I., 1971. The genus *Lobaria* of Eastern Asia. J. Hatori Bot. Lab. 34: 231–364.

Zahlbruckner, A., 1907. Lichenes. *In* A. Engler and K. Prantl (Eds.), Die natürlichen Pflanzenfamilien, Vol. A. Spezieller Teil. B. Engelmann, Leipzig, pp. 49–243.

Zahlbruckner, A., 1926. Catalogus lichenum universalis, Vol. 4. Borntraeger, Leipzig.

Taxonomic Index

An asterisk (*) after a page number indicates an illustration.

D

E

R

S

T